TRANSISTOR CIRCUIT DESIGN

PRENTICE-HALL SERIES IN ELECTRONIC TECHNOLOGY

Irving L. Kosow, *editor*

Charles M. Thomson, Joseph J. Gershon, and Joseph A. Labok, *consulting editors*

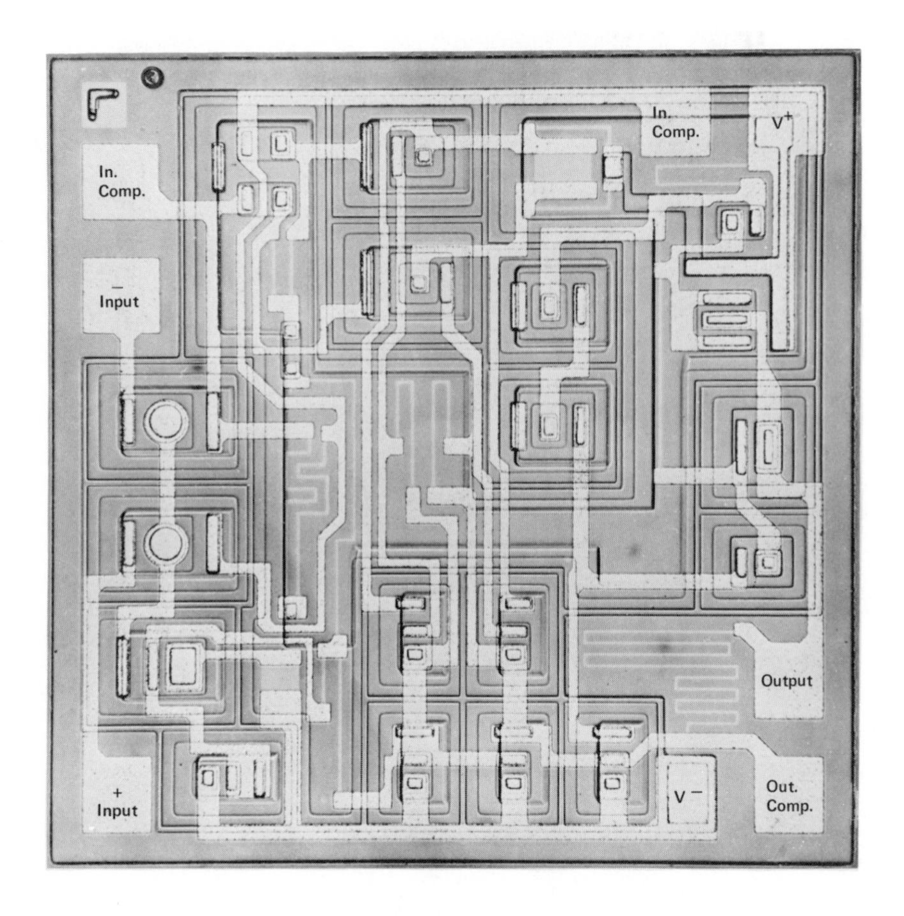

A Photo Micrograph of the
Fairchild μA735 Micropower Operational Amplifier

This integrated circuit amplifier has 15 transistors fabricated on a silicon chip only 1.3 millimeters on a side. The differential input stage on the left has two transistors with their collector load resistors and a third transistor replacing the emitter resistor. The output stage with four transistors operating as complementary emitter followers is on the right side. A 5MΩ feedback resistor is near the lower right corner. When packaged, leads are connected to the eight square terminals along the sides of the chip. The general arrangement of components is similar to the μA735 circuit which is described in Chapter 15 on "Integrated Circuit Amplifiers."

(The photo is supplied through the courtesy of Fairchild Camera and Instrument Corporation.)

TRANSISTOR CIRCUIT DESIGN

LAURENCE G. COWLES

Senior Electronic Engineer
The Superior Oil Company
Houston, Texas

PRENTICE-HALL, INC., ENGLEWOOD CLIFFS, NEW JERSEY

10 9 8 7 6 5 4 3 2 1

ISBN: 0-13-930032-5

Library of Congress Catalog Card Number: 72-3873

Printed in the United States of America

PRENTICE-HALL INTERNATIONAL, INC., *London*
PRENTICE-HALL OF AUSTRALIA, PTY. LTD., *Sydney*
PRENTICE-HALL OF CANADA, LTD., *Toronto*
PRENTICE-HALL OF INDIA PRIVATE LIMITED, *New Delhi*
PRENTICE-HALL OF JAPAN, INC., *Tokyo*

Whatever you can do, or dream you can, begin it.
Boldness has genius, power, and magic in it.

Attributed to Goethe

"There's a way to do it better . . . . find it"

Thomas A. Edison, 1847-1931.

CONTENTS

PREFACE

This book presents a practical outline of transistor circuit design for students, technicians, and engineers who need meaningful design relations with a minimum of mathematical analysis. Because practical transistor circuits must have significant feedback, this work emphasizes the role of feedback in design and includes practical circuit criteria which ensure significant feedback. With significant feedback, the circuit characteristics are fixed by the feedback resistors, and the complicated relations of circuit analysis reduce to simple formulas that can be used as design guides.

The first four chapters are concerned with the transistor as a somewhat variable circuit element and with relations used to evaluate and design circuits. The complex relations usually given in texts are simplified and reduced to more meaningful formulas by representing a transistor by only two parameters. Since a transistor is a highly variable element at best, this simplification focuses attention on the need for feedback and yields a clearer understanding of transistor circuits.

New and useful information is given in Chapters 5 and 6 concerning the design of single-stage and direct-coupled pair amplifiers. The optimum design of an iterated stage is considered more carefully than heretofore because the iterated stage is the basic unit of multi-stage amplifiers. The common emitter stage with both collector and emitter feedback is shown to be simpler to design and better than a stage with emitter feedback only.

The book includes many examples of proven circuit designs with circuit values and shows how these may be scaled or combined to serve many different purposes. The circuits cover the spectrum from dc to microwaves, from small signals to watts, and from discrete to integrated circuits. Although primarily concerned with discrete circuits, which are the building blocks of all circuits, the book includes material on the design of integrated circuits and their application as components of instruments and systems.

In chapters dealing with overall feedback, the first describes the several kinds of feedback and explains the advantages that make feedback an indispensable tool in contemporary design. The second chapter shows how the stability of an amplifier depends on the open-loop gain-frequency characteristic and presents an easy method for calculating the gain and phase margins. The purpose of the latter chapter is to explain techniques for making an amplifier stable with emphasis on the advantages of using uncomplicated circuits to simplify feedback problems.

The chapters on UHF and microwave circuits provide an introduction to the high-frequency applications of semiconductors. Much of this material is concerned with the many uses of diodes as frequency converters, multipliers, tuning capacitors, and signal sources. Sufficient breadth of material is included to prepare the reader for a career in industrial circuit design and applications and to facilitate his study of the current literature.

Because this book is primarily concerned with design as opposed to mathematical circuit analysis, I have found it necessary to lean rather heavily on my former work. Insofar as possible the symbols and terminology in this book are the same as those used in *Transistor Circuits and Applications.* There the reader can find a practically oriented introduction to transistor circuits and a more detailed description of the feedback and circuit relations which are used in this volume. The author's *Analysis and Design of Transistor Circuits* for the engineering designer contains more mathematical material and much more detailed information about transistors and circuits. The advanced book has references, design data, and many circuits that are not covered in the present volume. I have avoided duplications and references by assuming that the reader can refer to the earlier books for additional material.

I wish to express my appreciation to The Superior Oil Company for the freedom to develop and publish a practical understanding of transistor circuits and to thank fellow employees who have helped in preparing the manuscript. My son Christopher has contributed materially to the quality of the book by a critical reading of most of the work, and I am indebted to the reviewers for their suggestions and encouragement. Finally, but not least, I am deeply appreciative to my readers, friends, and family for their interest and approbation.

LAURENCE G. COWLES

LIST OF SYMBOLS

A	current or voltage gain
A'	gain with feedback
A_c	area of core
$A\beta$	loop gain
B_i	current-loss factor
B_v	voltage-loss factor
BV_{CBO}	breakdown voltage, emitter open
BV_{CEO}	breakdown voltage, base open
$BV_{CEO(SUS)}$	breakdown sustaining voltage
BV_{CER}	breakdown voltage with base resistor
C_{be}	base-to-emitter capacitance
C_C	coupling capacitance
C_{gd}	gate-to-drain capacitance
C_I	input capacitance
C_M	Miller effect capacitance
C_N	neutralizing capacitance
C_{ob}	collector-to-base capacitance
C_t	varactor diode capacitance
e_b	ac base voltage
e_G	ac generator or source voltage
e_I	ac input voltage

e_N	rms noise voltage
e_O	ac output voltage
e_p	peak or peak-to-peak voltage
e_S	ac signal voltage
f_β	β-cutoff frequency
F_B	feedback factor
f_c	cutoff frequency (half power, or 3 dB)
f_h	high-frequency cutoff
f_l	low-frequency cutoff
f_M	Miller effect cutoff frequency
f_N	noise factor
f_S	current gain cutoff frequency
f_T	current gain-bandwidth product
G_i	current gain
G_i'	current gain with feedback
G_l	loop gain
g_m	transconductance
g_0	zero gate-voltage transconductance
G_p	power gain
G_v	voltage gain
G_v'	voltage gain with feedback
h	*CB* short-circuit ac input resistance, h_{ib}
H	effective circuit h with series resistors included
h_{ib}	hybrid parameter; see h
I_A	bias current
i_b	ac base current
I_B	dc base current
i_c	ac collector current
I_C	dc collector current
I_{CO}	collector cutoff current, temperature sensitive
I_{DON}	zero bias drain current, I_{DSS}
i_e	ac emitter current
I_E	dc emitter current
i_I	ac input current
I_L	dc load current
I_m	maximum instantaneous current
i_o	ac output current
i_s	ac current in short-circuiting load

L_P	primary inductance
L_S	secondary inductance
n	number of stages, turns ratio, etc.
n_I	input winding turns
n_O	output winding turns
P_{dc}	dc power
P_I	ac input power
P_O	ac output power
R_A	bias resistor (usually adjustable)
R_B	base resistor
r_b	base resistance, internal
r_c	collector resistance, internal
R_C	collector resistor
R_{CC}	collector-to-collector transformer impedance
r_e	emitter resistance, internal
R_E	emitter resistor
R_f	feedback resistor
R_g	generator or source resistor
R_I	input resistance
R_L	load resistance
R_M	Miller equivalent input resistance
R_S	source resistor (of signal source)
R'_S	equivalent ac base resistance
S	S-factor, a substitution for more complicated relations; usually the ratio of two resistors (R_B/R_E or R_f/R_L) that determine the dc current gain of a stage or an amplifier
S'	Shea's stability factor; usually approximately $S + 1$
S_c	corrected S-factor
t	time
T	temperature
T_A	ambient temperature
T_j	junction temperature
T_l	carrier lifetime
t_p	pulse length, seconds
t_r	pulse rise time
t_{rr}	reverse recovery time
t_s	snap time

V_B	dc base voltage to ground
V_{BB}	dc base supply voltage
V_{BE}	dc base-emitter voltage
V_C	dc collector voltage to ground
V_{CC}	dc collector supply voltage
V_D	dc drain voltage
V_{DD}	dc drain supply voltage
V_{DS}	dc drain-to-source voltage
V_E	dc emitter voltage to ground
V_{GS}	dc gate-to-source voltage
V_m	maximum instantaneous voltage
V_P	FET pinchoff voltage
V_R	reverse voltage
V_Z	dc Zener diode voltage
X	reactance
X_E	reactance of emitter capacitor
X_C	reactance of capacitor C
X_L	reactance of inductance L
y_{ie}	*CE* input admittance
y_{oe}	*CE* output admittance
Z_{in}	input impedance
Z_{M}	Miller equivalent input impedance
α	*CB* short-circuit current gain, $-h_{fb}$; approximately 1
β	*CE* short-circuit current gain, h_{fe}; approximately 100
β'	equivalent to $(\beta + 1)$; can be read as β
Δ	a small change of the parameter indicated
Θ	thermal resistance
ω	frequency in radians per second $(2\pi f)$

Note: A primed symbol usually designates a value with feedback.

ABBREVIATIONS

A	ampere	mA	milliampere
BW	bandwidth	mH	millihenry
CB	common base	mV	millivolt
CC	common collector	pF	picofarad
CE	common emitter	P-P	peak to peak
D	drain (FET)	*Q*-point	quiescent point
dBm	decibel referred to 1 mW	S	source (FET)
GHz	gigahertz	SC	short-circuited
Hz	hertz	μA	microampere
IC	integrated circuit	μF	microfarad
kHz	kilohertz	μH	microhenry
kΩ	kilohm		

1

DESIGN

Design is concerned with the synthesis of ideas and structures. A design course gives the student experience in making a choice, developing judgment, and building or inventing a better device. The solution of a design problem requires insight and imagination, and, like a game of chess or bridge, is stimulating to the mind. Design is an exciting and challenging occupation with abundant opportunity for both engineers and electronic technicians.

The art and practice of design are acquired by committing oneself to a design problem and becoming so thoroughly informed about the details and alternatives that the best solution is finally discovered. Each new design problem adds to the designer's fund of experience and store of factual knowledge. Much of the designer's experience, which is obtained by experimental procedures, requires a practical and meaningful understanding of circuits and feedback.

A designer is rewarded by the better understanding that his work gives him of circuits and semiconductors. He is forced to keep abreast of new developments and is given the opportunity to know "the state of the art." The need to keep up-to-date tends to protect a designer from obsolescence, continuously broadens his experience, and ensures that he is too valuable to be given routine assignments. An experienced designer is respected for his skills and is generally given considerable freedom in the pursuit of his occupation.

1.1 THE DESIGN METHOD

The designer is usually given a somewhat vague set of specifications describing how a proposed circuit is to operate, and the process of design begins with his decision to try a circuit he thinks may meet the specifications. He usually selects this circuit from a collection he has acquired, from past experience, or from a search of technical literature. The proposed circuit is studied, changed a little, and constructed on a breadboard for experimental evaluation. From this beginning the improvements and changes in the circuit are usually obtained by experiment.

The ultimate success of the design may come from the designer's long experience or simply from a doggedly determined cut-and-try, supported by careful observation. Each improvement challenges the designer to attempt a simpler and better design. The work on a difficult design can be exceedingly exciting when the designer knows that some minor change, simplification, or original idea will bring about the successful discovery of a new and better circuit.

A trial circuit may be examined by any of several methods. The circuit may be examined mathematically by calculating the gain-frequency characteristics and the effects of supply voltage and temperature. The mathematical method has been taught to the exclusion of other methods because of the need to teach the mathematical analysis of engineering problems. In practice, however, the mathematical method is so complicated that the engineer-designer may overlook and never really understand the practical problems of circuit design. Computer-aided design is a powerful tool for career designers, provided the practical problems and the computer are understood well enough to feed the computer adequate practical data.

Cut-and-try is often the method of design for practical people who do not understand or use the mathematical theory. These persons usually modify someone else's design until a circuit is found suitable for a new application. Circuits are easily modified by changing all component values by a common factor and by adjusting the bias. With concern for the manufacturer's ratings, transistors and supply voltages may be changed. Many times only a minor modification is needed, and with enough experience some cut-and-try designers produce excellent designs with little effort.

The best method of design is any combination of paper work and experiment suited to the designer's skills. In order to calculate the gain and impedance that should be expected of a transistor circuit, a designer needs a small store of approximate formulas for guidance and insight. For the experimental work a designer needs a few instruments and tools, one or two handbooks of transistor data, and a store of circuit designs to provide a starting point. With the examples in this book the reader has a fund of design material that is usually obtained only after considerable practical experience.

1.2 CIRCUIT ANALYSIS FOR DESIGN

The intricate formulas and the 12 hybrid parameters used in circuit analysis are virtually useless in practical design. Moreover, there is no justification for using a complicated format that is impractical for circuit design. Design requires the study of many interrelated factors and is difficult in its own right. These factors include costs, the limitations of transistors, and many practical compromises. Instead of the complicated relations of circuit analysis, the practicing designer always uses approximate relations that focus attention on the parameters that really determine the performance of a circuit. In particular, the transistor parameters are unimportant as long as the transistor has a high current gain, because every practical transistor circuit must use feedback. For these reasons the language and formulas of the designer are, *and must be*, simpler than the language of mathematical analysis.

The designer is concerned with gains and impedances. The gain of a transistor amplifier must always be controlled by providing either significant local feedback or significant overall feedback. When the designer knows there is significant feedback, the gains and impedances are essentially independent of the transistor parameters, and the equations needed for design are greatly simplified.

In a single-stage amplifier with emitter feedback, as shown in Fig. 1.1, the resistor ratio R_B/R_E is called the amplifier *S*-factor. In a stage with collector feedback, as shown in Fig. 1.2, the stage *S*-factor is $S = R_f/R_L$. In either example, if the feedback is adequate, the stage current gain is essentially independent of the transistor current gain and is *S*. Thus, we may use *S* as the amplifier current gain with feedback. If the transistor in either figure is replaced by an amplifier in which the overall gain is controlled by the same resistors, the

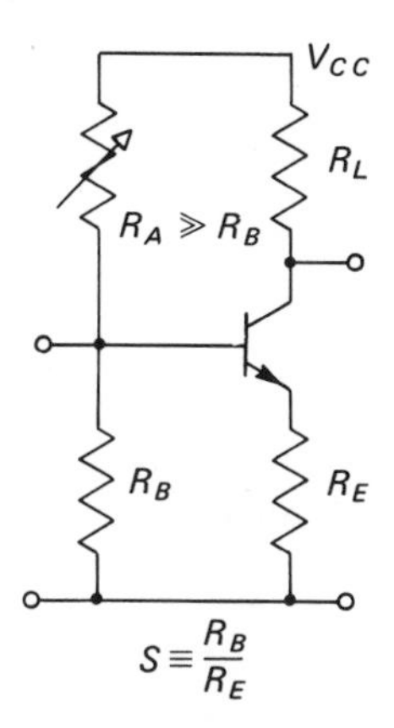

Figure 1.1. Amplifier with emitter feedback.

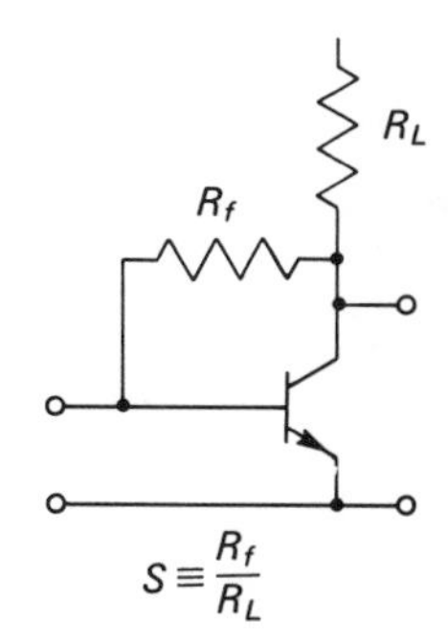

Figure 1.2. Amplifier with collector feedback.

overall current gain of the amplifier is given by the same *S*-factor. In a similar way, the stage input impedance with feedback is essentially independent of the transistor β.

The designer's use of transistor parameters is mainly either for determining when an amplifier has significant feedback or determining the effects of a change of the parameters. For the transistor current gain we use in place of h_{fe} the more easily written and pronounceable β of the *T*-equivalent. Similarly, the input impedance of the common-base (*CB*) transistor is called h instead of h_{ib}, and the input impedance of the common-emitter (*CE*) transistor is written as βh to show its dependence on the current gain instead of concealing this important fact by using the hybrid h_{ie}.

Except to find $\beta = h_{fe}$ in the data sheets, the designer has no need to use the hybrid equivalent circuit or the hybrid parameters. The *CB*-transistor short-circuited input impedance, which occasionally appears in circuit equations, must be calculated by Shockley's relation, which shows that h varies inversely with the emitter current and with the absolute temperature. Thus, by Shockley's relation,

$$h = \frac{26}{I_E} \cdot \frac{T_j}{273} \tag{1.1}$$

where h is in ohms, I_E is in mA, and the junction temperature is in degrees Kelvin. At room temperature Shockley's relation becomes

$$h = \frac{26}{I_E} \tag{1.2}$$

and h is 26 Ω when the emitter current is 1 mA.

1.3 THE TRANSISTOR GAIN-IMPEDANCE RELATION

The designer measures gains with an oscilloscope or a voltmeter and, therefore, he measures voltages and thinks in terms of voltage gains. For the transistor amplifier shown in Fig. 1.3 the voltage gain is the ratio of the output voltage $i_O R_L$ divided by the input voltage $i_I R_I$. Thus,

$$G_v = \frac{i_O R_L}{i_I R_I} \tag{1.3}$$

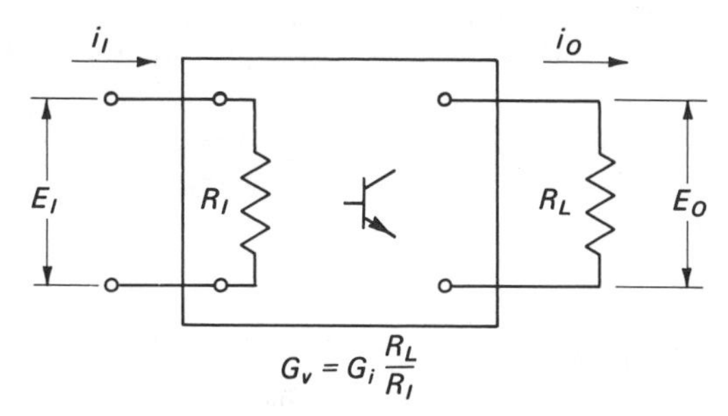

Figure 1.3. Amplifier showing terms of the TG-IR.

Substituting the overall current gain G_i for i_O/i_I in equation (1.3), we find

$$G_v = G_i \frac{R_L}{R_I} \tag{1.4}$$

This, the *Transistor Gain-Impedance Relation*, applies to almost any transistor amplifier, including those with feedback and transformers, so the formula is of great value and utility in design. The word *transistor* is used as a reminder that the relation is not applicable to circuits that include vacuum tubes or FETs. We call equation (1.4) the TG-IR, for short.

This relation shows that if we wish to know the voltage gain of a transistor stage and we know the current gain, we must also know both the load resistance and the input impedance of the stage. Usually we can identify the load impedance R_L. The current gain can be determined by inspecting the circuit. The input impedance of a transistor or of an amplifier can be easily determined by applying a few simple rules. The TG-IR requires nothing new and unifies the understanding of many transistor circuits.

As an example of the use of the TG-IR, consider an amplifier comprised of a series of *iterated R-C* coupled or direct-coupled stages. The input impedance of each stage is the load impedance of the previous stage. Hence, $R_I = R_L$, R_L/R_I in the TG-IR is unity, and the stage voltage gain is numerically equal to the stage current gain S.

In practical amplifiers it is usually necessary to limit the current gain to about 30 per stage by local feedback or by overall feedback. Hence, one does not expect the voltage gain of transistor amplifiers to exceed about 30 per stage. As the TG-IR shows, a voltage gain in excess of the tolerable current gain must be obtained by using an R_L/R_I ratio of more than 1. If the impedance ratio is less than 1, as is often the case, then one must expect voltage gains correspondingly lower than 30.

As a second example, consider a *CC* stage that cannot have a voltage gain in excess of 1. If the current gain is limited to about 50 as the practical upper limit, then it follows that the input impedance R_I cannot be greater than 50 times the load impedance R_L. Where higher ratios of input-to-load impedance are required, one must expect to find additional *CC* stages or step-down transformers or less than unity voltage gain. The reader may obtain additional details concerning Shockley's relation and the TG-IR in the references given at the end of Chapter 3.

1.4 TRANSISTOR CIRCUIT DESIGN

The objective of this book is to give the reader data, circuits, equations, and the understanding needed for transistor circuit design. The book begins with a review of the characteristics of transistors that are important and useful in circuit design, purposely omitting details that only complicate the mathematical relations needed in circuit design. Thus, the opening chapters develop simplified circuit relations and explain their use as design guides.

A third of the book describes the design of single-stage and multistage amplifiers with particular concern for the effects of feedback on the quiescent (*Q*) point stability and on the gains and impedances of an iterated stage. The iterated common-emitter stage is of particular interest because it may be designed to have the maximum power gain obtainable in a direct-coupled series of identical stages.

The central third of the book discusses multistage amplifiers and the use of feedback, which is so important in contemporary circuit design. Chapter 9 has a simplified, practical outline of overall and loop feedback and explains the use of feedback for improving circuit performance. Chapter 10 shows how the loop gain and phase characteristics limit the amount of usable feedback and how a designer may easily shape the phase characteristic to increase the stability of a feedback system.

Material on radio-frequency, video, and microwave circuits brings the designer to a practical understanding of the transistor circuits used in TV, communications, radar, and microwave systems. One chapter outlines the design of linear integrated circuits and gives examples of their use as high-gain packages for both linear and nonlinear applications. Throughout the book the objective is to make contemporary circuit design more understandable and to explain transistor circuits without using complicated mathematical relations.

1.5 DESIGN SKILL AND EXPERIENCE

Skill and efficiency in circuit design are acquired by developing skill in circuit analysis, by studying and experimenting with published circuits, and by acquiring a store of circuits and factual information. Skill in analysis permits evaluation of each step of a design with a minimum of labor. The study of published circuits brings one a store of examples and provides experience in recognizing a good design, as one learns to recognize a fine jewel. The experimental study of interesting circuits often reveals characteristics that have been overlooked and makes the designer more aware of problems that require his attention. Design is a very pleasant combination of analysis, experiment, and invention. There are conflicts and compromises in every piece of technical work, and the manner in which these are resolved may be learned through a study of design.

The practice of circuit design should begin with audio-frequency circuits that do not require either very high or low frequencies. With increasing experience the designer will apply his understanding of temperature problems, feedback, noise, drift, runaway, and other problems. The designer will find these topics covered in this book in sufficient detail for everyday circuit design. When unusual problems are encountered, he should try to find answers to his questions by experimental studies, by consulting library resources, or by obtaining manufacturers' application data.

The designer should make his own collection of circuits, ideas for design, and handbooks of semiconductor data. His search for information and ideas provides day-to-day opportunities for studying new techniques and for self-improvement. The professional designer has abundant opportunity to develop his worth as an employee, provided he takes the initiative in continuing his search for knowledge.

PROBLEMS

1-1. A transistor with a current gain $\beta = 100$ is operated in a resistance coupled stage with a 12-V collector supply voltage and an emitter current of 0.1 mA. Using the TG-IR, show that a single-stage *CE* amplifier has a voltage gain of nearly 400.

1-2. Using the TG-IR, $I_E = 40$ mA, $\beta = 50$, and $R_L = 100\ \Omega$, show what voltage gain, power gain, and input impedance may be expected of a single-stage resistance-coupled transistor amplifier. Consider each type of amplifier: (a) *CB*, (b) *CE*, and (c) *CC*. (d) What emitter current is required to ensure in part (c) that $G_v > 0.67$?

1-3. For the amplifier shown in Fig. P1.3, assume that the transistor has a very high current gain. Neglect R_A, and show that the stage current gain equals the *S*-factor, *S*. (Easy)

1-4. For the amplifier shown in Fig. P1.4, assume that the transistor has a very high current gain. Show that the stage current gain equals the *S*-factor, *S*. (Difficult)

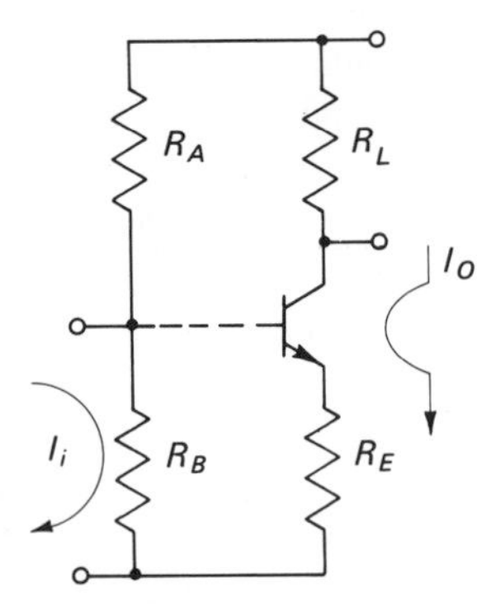

Figure P-1-3.

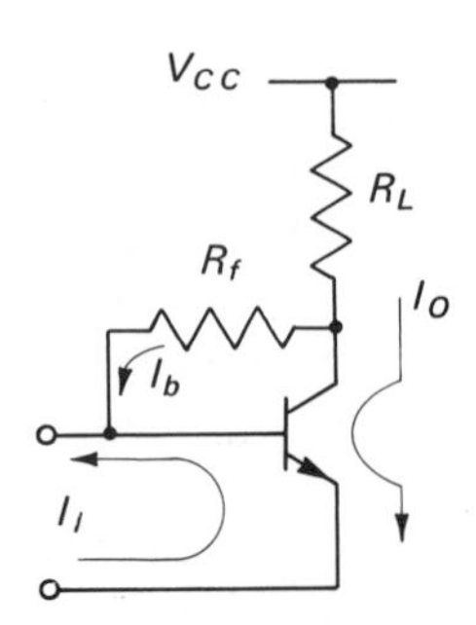

Figure P-1-4.

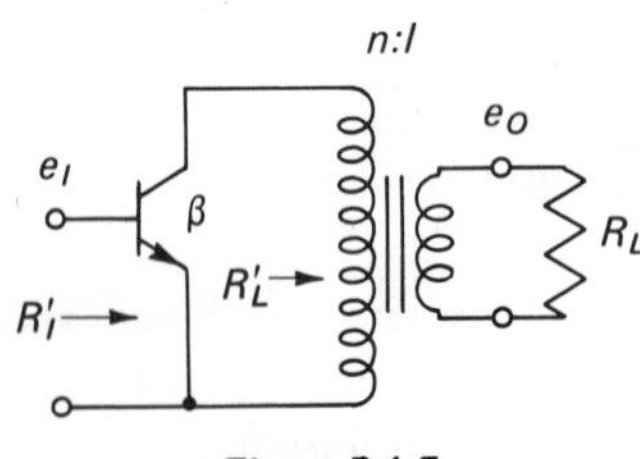

Figure P-1-5.

1.5. (a) Show that the TG-IR correctly represents the input-output current and voltage relations of a step-down transformer that has a 10:1 turns ratio. (b) Write the TG-IR which represents the stage shown in Fig. P1.5.

1-6. Begin your own collection of design ideas, data, and circuits. Find several circuits that you may like to study in the laboratory. Make a list of questions you have concerning one or more of these circuits.

REFERENCES

1-1. *GE Transistor Manual*, 7th ed. Electronics Park, Syracuse, N.Y.: General Electric Co., 1964.

1-2. *RCA Transistor Manual*, SC-13. Harrison, N.J.: Radio Corporation of America, 1967.

1-3. *RCA Silicon Power Circuits Manual*, SP-50. Harrison, N.J.: Radio Corporation of America, 1967.

1-4. *FET Circuit Ideas*. Sunnyvale, Cal.: Siliconix, Inc., 1966, 4 pp.

2

STATIC CHARACTERISTICS AND TEMPERATURE PROBLEMS

The familiar static curves display a transistor's behavior as a pair of coupled diodes and show the relatively large negative temperature coefficient of resistance that distinguishes all semiconductors. The static curves describe the nonlinear characteristics of a transistor and show how the Q-point and the parameters depend on temperature. In this chapter we review briefly the effect of temperature in changing the static curves and examine the temperature-engendered shifts of the dc Q-point of an operating transistor. The main objective of the chapter is to show how the three principal temperature problems confronting the designer can be controlled by circuit design and feedback.

This chapter develops important groundwork for all that follows and should be reviewed, at least for content, so that the reader knows where to find details often needed to understand the more practical and perhaps more interesting material in the rest of the book.

2.1 THE EMITTER DIODE

The static current-voltage characteristics of the emitter diode are usually represented by a curve, with the emitter current as the independent variable. The curve shown in Fig 2.1 represents the fact that an exponential increase of the emitter current produces only a small increase of the emitter voltage drop. In fact, a tenfold increase of the emitter current anywhere on the curve produces only a 60-mV increase of the base-emitter voltage V_{BE}. The V_{BE} curve for a germanium transistor is essentially the same as that for a silicon transistor except that the voltage drop is approximately one half the voltage across a silicon junction, and the voltage change is the same 60 mV for a tenfold increase of emitter current.

The design of a silicon transistor usually results in a maximum β when the V_{BE} voltage is about 0.6 V. Transistors are not often used at emitter currents below $\frac{1}{1000}$ of the value at maximum β. Hence, V_{BE} at low operating currents is 0.6 V − 3(0.06) V = 0.42 V. A transistor turned full ON, as a switch or in a power amplifier, may have a V_{BE} voltage of 0.8 V. These values show that over the entire operating range the V_{BE} voltage is between 0.4 and 0.8 V. Careful measurement of the V_{BE} voltage may give some indication of the transistor Q-point conditions, but the emitter current is usually much more meaningful.

The junction temperature is always specified for a V_{BE} curve because the curve moves markedly downward with an increase of the temperature, as shown in Fig. 2.2. The temperature change of the V_{BE} voltage is important in design and is generally assumed to be about 2 mV/°C for either germanium or silicon devices. If the junction temperature *increases* 100°C, the V_{BE} voltage decreases 0.2 V, and with a fixed bias voltage the emitter current would increase more than 1000 times. Thus, a transistor is generally biased with an approximately constant bias current, or dc feedback must be used to offset the effect of a V_{BE} voltage change.

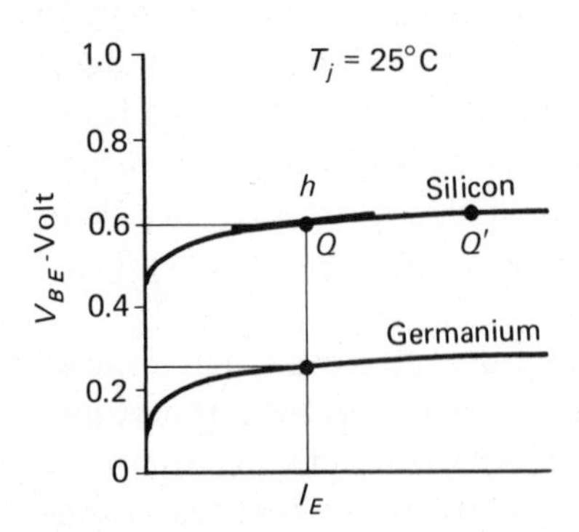

Figure 2.1. Emitter diode dc and ac relations.

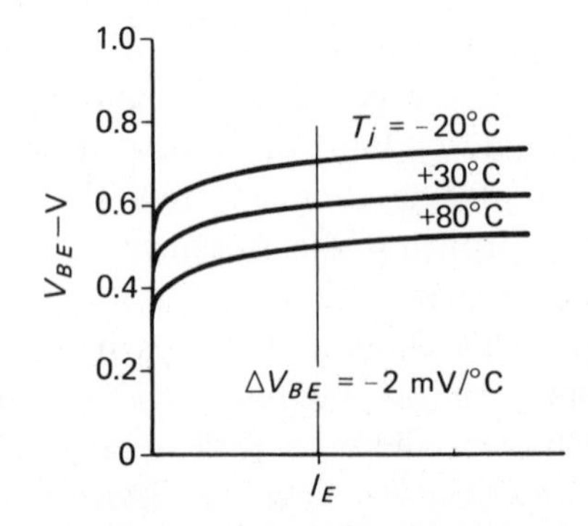

Figure 2.2. Emitter diode base-emitter voltage change with temperature.

The transconductance of a transistor is more than an order of magnitude higher than that of a vacuum tube, but the transistor transconductance is not a useful figure of merit because of the nonlinear relation between the base voltage and the emitter current. Thus, the designer works with the transistor current gain β, which is relatively independent of the collector Q-point current and voltage.

2.2 THE COLLECTOR DIODE AND β

The static collector characteristics of a transistor exhibit the coupled-diode nature of a transistor. As shown in Fig. 2.3, when the base is open, $I_B = 0$, the collector (like a reverse-biased diode) has a small leakage current βI_{CO} that is relatively independent of the collector voltage.

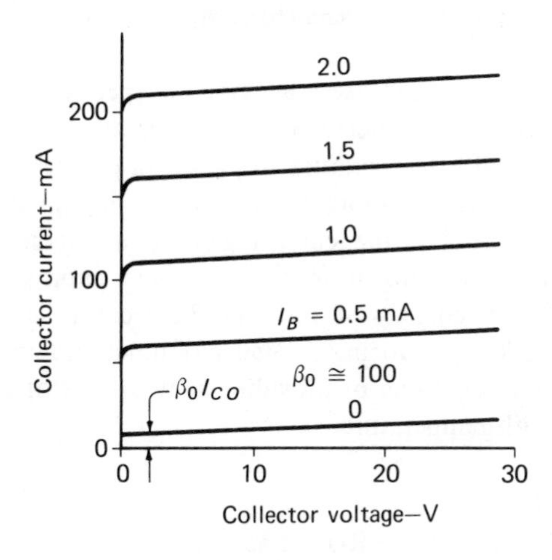

Figure 2.3. Collector current-voltage characteristics at 25°C.

When the base current of a transistor is increased in a series of equal steps, the collector current increases in steps that are approximately β times larger. The collector current depends only on the base current and is independent of the collector voltage. The ratio of the collector-current change to a corresponding base-current change defines the transistor current gain, and, because β is so variable, we need not distinguish between the ac and dc current gains. The transistor current gain β depends on temperature and increases or decreases approximately 6 per cent for each 10°C increase or decrease of the junction temperature.

If the junction temperature of the transistor represented in Fig. 2.3 is increased to 80°C, the collector characteristic curves change approximately as shown in Fig. 2.4. The increased temperature increases the spacing between the

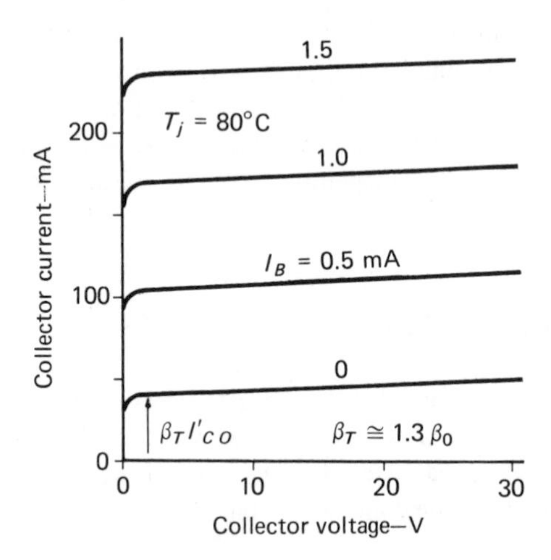

Figure 2.4. Collector characteristics at 80°C.

constant base-current curves because β increased 30 per cent, and all the curves are moved upward because I_{CO} increased with temperature.

Beta varies enough with temperature to present a problem in direct-coupled amplifiers that have two or more stages. When the junction temperature increases from 25 to 150°C, the current gain of a silicon planar transistor approximately doubles, as shown in Fig. 2.5, and approximately halves at −55°C. The temperature changes of V_{BE} and β are controlled in circuit design by the use of feedback, by reducing the range of junction temperature changes, and by avoiding direct coupling or providing some means of compensating for parameter changes with temperature.

2.3 *Q*-POINT TEMPERATURE PROBLEMS

The temperature-engendered shift of the operating *Q*-point is a common limitation in the application of transistor circuits. If an amplifier is poorly designed (e.g., the *S*-factor or the dc voltage gain is too high) or the transistor temperature is high and variable, then the collector current may increase until the amplifier is inoperative. If the operating temperature is too low, the transistor current gain β may decrease until the stage has insufficient gain or is cut off by insufficient bias. The *Q*-point shifts are caused primarily by the temperature-engendered variations of β, V_{BE}, and I_{CO}. Most low-power circuits are designed to use silicon transistors, partly because silicon devices have a negligible I_{CO}. The present discussion of the temperature problems assumes the use of silicon, and we consider first the effects of β changes.

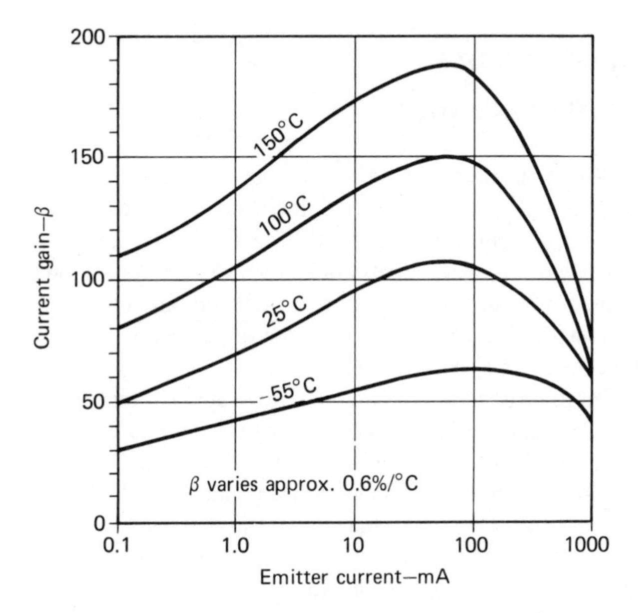

Figure 2.5. Typical transistor current gain curves.

2.4 *Q*-POINT SHIFT WITH β CHANGES

A major source of temperature problems comes from the variation of β. A transistor having a β of 100 at room temperature may have a β of 150 at 125°C junction temperature, and the effective β may be much higher at 150°C. At −55°C the current gain may fall to less than one half the room-temperature value. With feedback, the circuit current gain G_i is approximately

$$G_i = \frac{S}{1 + S/\beta} \tag{2.1}$$

and the stage current gain is made relatively less sensitive to β changes by making S/β less than 1. A low circuit current gain G_i forces the designer to use more stages in an amplifier, so the S/β ratio should be kept as high as is consistent with the expected temperature change. Equation (2.1) can be used to determine how much a change of β changes a given Q-point I_C, but for most design purposes equation (2.1) is sufficient as a reminder that Q-point shifts caused by β changes are controlled by reducing the S-factor.

2.5 *Q*-POINT SHIFT WITH A TEMPERATURE CHANGE OF V_{BE}

As with any germanium or silicon diode, the base-emitter voltage of a transistor decreases 2 mV for each 1°C increase of the junction temperature. This change of the emitter-circuit dc voltage cannot be neglected because it affects a stage having appreciable current or voltage gain. As is shown in Fig. 2.2, the diode curves move down with a junction temperature increase and up with a temperature decrease. In practical circuits the net effect of a temperature rise is to increase the collector *Q*-point current.

A low-power transistor in room-temperature applications may operate over temperatures ranging from 0 to 60°C, and the junction of a power transistor in a commercial application may operate from −30 to 90°C, or higher. These temperature changes produce diode voltage changes of 0.12 and 0.24 V, respectively, and these voltage changes are more than can be tolerated in high-gain CE stages, where the total emitter loop dc voltage may be 1 to 2 V.

As an example of the V_{BE} temperature effect, consider the emitter circuit shown in Fig. 2.6. The emitter is shown with a voltage ΔV_{BE} in series with the base and emitter resistors R_B and R_E. We assume that the transistor is biased by superimposed currents that are not shown, and that ΔV_{BE} represents the increase of the base-emitter circuit voltage caused by a temperature rise after the *Q*-point is established. For S small compared with β, the resulting shift of the collector *Q*-point current ΔI_C may be shown to be given by

$$\Delta I_C \cong S \frac{-\Delta V_{BE}}{R_B} \tag{2.2}$$

Equation (2.2) implies that the equivalent base circuit input-current change equals the current produced in R_B by a voltage change ΔV_{BE}, and that the stage current gain is the stage S-factor. The equation says that an increase of collector current is reduced by using a low S-factor and by designing the emitter circuit with high values of R_B and R_E.

For design purposes the equation can be interpreted as meaning that the change of V_{BE} is equivalent to a change of the stage bias current I_A of Fig. 2.7 by the amount

$$\Delta I_A \cong \frac{-\Delta V_{BE}}{R_B} \tag{2.3}$$

This restatement of equation (2.2) provides a way of experimentally evaluating

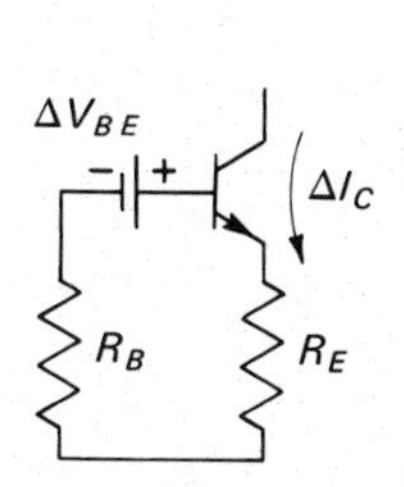

Figure 2.6. Emitter circuit ΔV_{BE} effect.

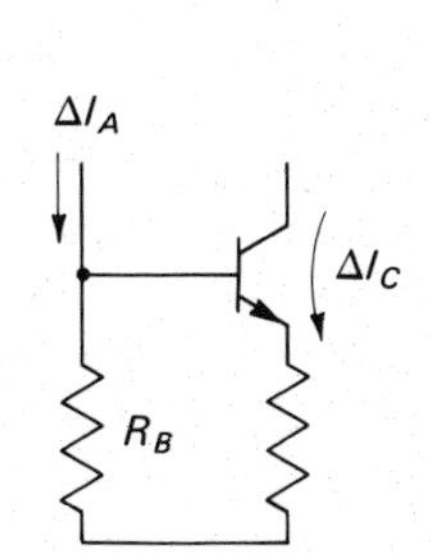

Figure 2.7. Bias equivalent of ΔV_{BE} effect.

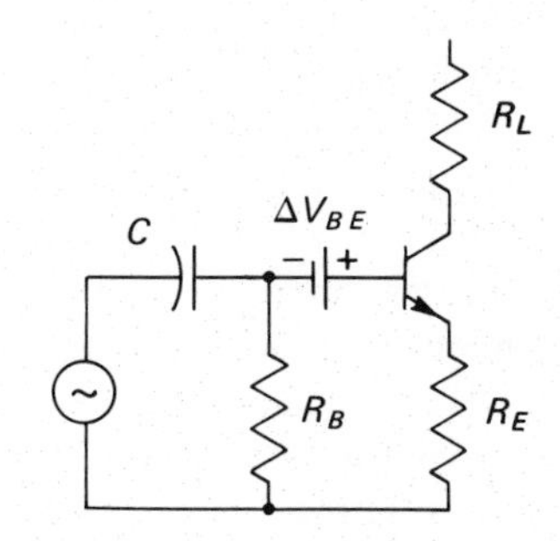

Figure 2.8. ΔV_{BE} effect with emitter feedback.

the effect of an expected V_{BE} voltage change and of comparing this effect with an I_{CBO} leakage current. The effect of a bias increment must be evaluated for its effect on the collectoı Q-point current, and not by comparing the bias increment with the bias current itself. In some circuits a large part of the bias current I_A is used to offset the static V_{BE} voltage, and the collector Q-point current is proportional to a lesser portion of the total bias current. However, for an order-of-magnitude temperature-effect calculation ΔI_A may be compared with the actual bias current I_A.

The reader is cautioned that equation (2.2) does not mean that the ΔV_{BE} voltage changes can be measured across R_B. The voltage change is across the base-emitter junction, and, if β is large compared with S, the emitter voltage changes with the junction voltage. The voltage change across the base resistor is smaller by approximately the ratio of S/β. Equations (2.2) and (2.3) show only how to produce an equivalent voltage change for testing purposes.

2.6 COLLECTOR VOLTAGE SHIFT WITH V_{BE} AND EMITTER FEEDBACK

Transistors and power supplies will usually tolerate the change of collector current that is caused by a V_{BE} voltage temperature change. On the other hand, the constant voltage nature of voltage supplies tends to make a Q-point voltage change a problem for concern in circuit design. An example of this temperature problem is found in the change of the collector Q-point voltage in a resistance-coupled stage that has a resistor as the collector load. Rather than calculating the change of collector current, we find the effect of temperature more easily by a method using the dc voltage gain.

Consider the transistor stage illustrated in Fig. 2.8. The stage is shown with ΔV_{BE} in series with the base and with a zero-impedance ac generator that is capacitor-coupled to the base. With significant feedback, the stage S-factor is small compared with the transistor β; hence, R_B is small compared with the input impedance βR_E that is seen by V_{BE}. Similarly, R_B is large and negligible compared with the ac source impedance. In other words, the dc voltage gain of the stage with V_{BE} as the signal is the same as the ac voltage gain measured with the zero-impedance source. This result means that the desired dc voltage gain may be measured in most amplifiers by removing the emitter capacitor, if used, and introducing a low-impedance ac source. Sometimes, as in this example, the dc voltage gain is easily calculated as R_L/R_E, and the collector voltage shift may be expressed algebraically as

$$\Delta V_C = \Delta V_{BE} \frac{R_L}{R_E} = \Delta V_{BE}(G'_v) \tag{2.4}$$

Equation (2.4) states that with significant emitter feedback the V_{BE} shift of the collector voltage V_C is reduced by using a low dc voltage gain.

A single-stage amplifier with a dc voltage gain of 10 in a low-power room-temperature application produces a $10 \times 0.002 \times 60 = 1.2$ V shift of the collector Q-point. In a commercial power application with a dc gain of 10, the Q-point shift is 2.4 V. Either result may or may not be acceptable, depending on the effect the Q-point shift has on the performance of the stage.

An amplifier with a dc voltage gain of 100 in a commercial application has a Q-point shift of 24 V. If the collector Q-point voltage at the midpoint operating temperature is high relative to the voltage shift, this temperature effect may be acceptable. Generally, however, a Q-point shift exceeding 5 V is more than can be tolerated, so the effect of V_{BE} changes should be examined.

2.7 *Q*-POINT SHIFT WITH COLLECTOR FEEDBACK

For an explanation of the effect of a V_{BE} voltage change in a collector-feedback stage, consider the circuit shown in Fig. 2.9. The voltage change is represented as ΔV_{BE} in series with the feedback resistor R_f. This voltage is shown as a signal on the base side of the junction, for convenience, because the general form of the formula for voltage gain is the same for both CE and CB stages.

By rearrangement, the collector-feedback stage takes the form shown in Fig. 2.10, which shows the similarity between the V_{BE} effect in a collector feedback and a common-collector stage. Observe that R_f replaces R_B, R_L replaces R_E, and the stage voltage gain is unity because the load now falls in the emitter circuit. The effect of a V_{BE} voltage change is to increase the voltage drop in the

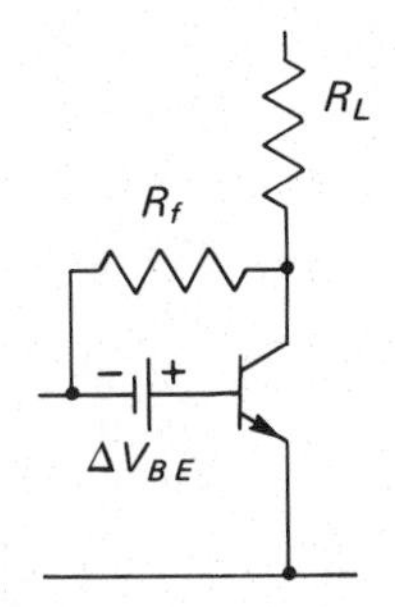

Figure 2.9. ΔV_{BE} effect with collector feedback.

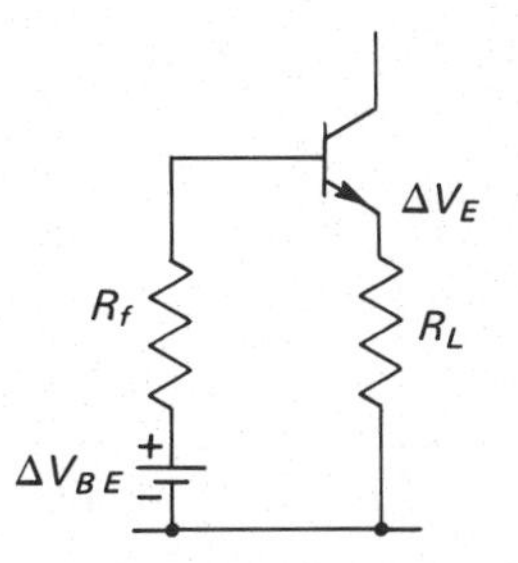

Figure 2.10. CC equivalent ΔV_{BE} with collector feedback.

collector load resistor by an amount equal to ΔV_{BE} itself. In any practical collector-feedback design, $S = R_f/R_L$ is less than β, so that R_f, which is in series with V_{BE}, does not significantly reduce the amount of voltage change transferred to the load resistor. In any event, the collector Q-point shift is 1 or 2 orders of magnitude smaller in a stage having collector feedback than one with emitter feedback. Offsetting this advantage is the fact that the collector Q-point voltage used with collector feedback is sometimes only 1.2 to 1.5 V. In some high-temperature applications the collector-feedback stage may require a more careful analysis than is given here.

2.8 Q-POINT SHIFTS WITH I_{CO}

A third cause of temperature problems is the reverse-bias leakage current in the collector diode. This current is equivalent to a current source connected between the collector and the base and polarized to increase the base and collector currents. This current is known as the collector cutoff current, I_{CO}, and is represented as a constant current source, as shown in Fig. 2.11.

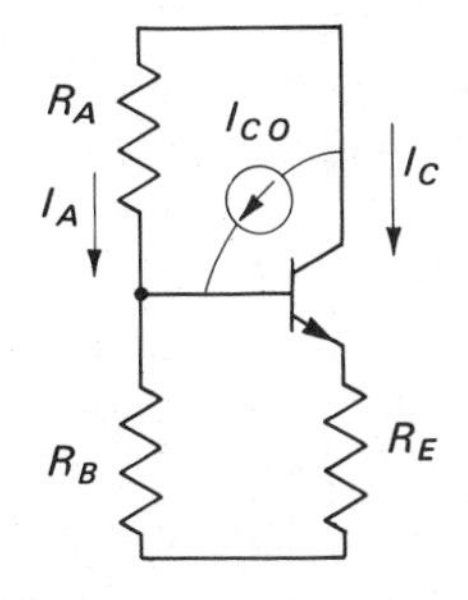

Figure 2.11. Collector cutoff current as an increased bias current.

Germanium and grown junction (an early type) silicon transistors are especially subject to a thermally sensitive collector-to-base leakage current I_{CO} that doubles with every 10 to 15°C increase of the junction temperature. This current, which is not well controlled in manufacture, adds a component to the base bias that increases exponentially with temperature. With some types of transistors the I_{CO} current may be intolerable when equipment is subjected to as little as a 25°C increase of temperature. The I_{CO} drift is kept under control by using low S-factors, by low values of the collector circuit dc resistance, and by designing the amplifier with a relatively high Q-point current. Because the planar passivated silicon transistors have negligibly small I_{CO} leakage currents, they have since 1961 replaced almost all the early types of transistors except for germanium power transistors and some types of high-frequency devices.

The effect of the collector cutoff current I_{CO} on the Q-point current is represented by the equation

$$I_C = S(I_A + I_{CO}) \tag{2.5}$$

The bias current I_A in equation (2.5) enters the circuit at the base and leaves at the emitter, whereas the I_{CO} current enters at the base and leaves at the collector. This fact means that the stage current gains are not the same, but the error from assuming that the gain is S is negligible, particularly when S is small compared with β.

The problem with the I_{CO} current is that a 1-mA leakage current at 25°C, which doubles for each 10°C increase, becomes 200 mA at 100°C, the limiting junction temperature of a germanium power transistor. With a stage current gain of 10, the collector current is increased by 1 A at 90°C. This increase of the collector current may exceed 10 per cent of the collector Q-point current of a germanium power transistor operating at high temperatures. In addition to the temperature-sensitive I_{CO}, there is a voltage-dependent leakage current, which at rated collector voltages may be five times larger, or 50 per cent of the Q-point current. These facts illustrate the need to limit both the S-factor and the collector Q-point voltage of a germanium transistor. Most transistor data sheets supply a value for the temperature-sensitive I_{CO} at 25°C, measured with V_C = 0.5 to 2 V. The voltage-dependent leakage current is given as an I_{CBO} value that is measured at either 50 or 100 per cent of the rated collector breakdown voltage.

An important design problem in germanium-transistor power applications is the control of runaway, which is a sudden steady increase of the collector current and power dissipation. Unless prevented by circuit design or unless the heat dissipation is limited, a power transistor is generally damaged or destroyed by runaway. Runaway is generally the most obvious and most difficult problem in the application of germanium power transistors. Because it is caused by inadequate heat transfer, this problem is discussed in detail in Chapter 13, where we consider the design of heat sinks. Runaway is controlled by using a heat sink that has a low thermal resistance, and by reducing the product $V_C S I_{CO}$. Besides limiting the transistor junction temperature, the designer controls runaway by reducing the collector Q-point voltage V_C, by reducing the S-factor, and by selecting a transistor having a low I_{CO} leakage current.

Present-day silicon transistors have such low, temperature-sensitive components of leakage current that the runaway and I_{CO} temperature problems are almost nonexistent. The manufacturers of silicon power transistors usually give the voltage component of collector-to-base leakage current I_{CBO} and generally omit the I_{CO} value.

2.9 TEMPERATURE-DRIFT SUMMARY

The temperature changes of an amplifier Q-point are caused primarily by three temperature-sensitive parameters of a transistor. The transistor current gain β increases approximately 0.6 per cent/°C, the base-emitter voltage drop V_{BE} decreases approximately 2 mV/°C, and the collector-to-base leakage I_{CO} doubles for each 10 to 15°C increase of the junction temperature. Depending on the circuit design and the temperature range, the Q-point drift may be caused by any one or all three parameters. Silicon transistors at room temperatures may be assumed to have a negligible I_{CO} leakage, and with low S-factors the Q-point drift is mainly caused by the V_{BE} temperature change. Germanium-transistor circuits must be carefully designed to limit the effect of I_{CO} leakage whenever a circuit is used in ambient temperatures exceeding 40°C (104°F), or for power applications in which the junction temperature is increased by collector power dissipation.

The effects of each of the temperature-sensitive parameters in changing the Q-point current are best calculated separately and compared to find the principal cause of a Q-point shift. Generally, the 2-mV/°C base-emitter voltage effect dominates in a low S-factor design unless both the base and the emitter resistors are relatively large. The 0.6 per cent/°C increase of β with temperature tends to dominate the Q-point shift of a stage designed with a high current gain, a high S-factor, and high base circuit resistances. The changes of β and V_{BE} contribute about equally to the temperature drift when the expected change of V_{BE} is approximately 20 per cent of the dc voltage drop in the base resistor R_B.

2.10 Q-POINT SHIFT WITH TEMPERATURE—AN EXAMPLE

An interesting example of the Q-point shift produced by temperature is provided by the circuit shown in Fig. 2.12. This circuit is similar to Fig. 5.2 in

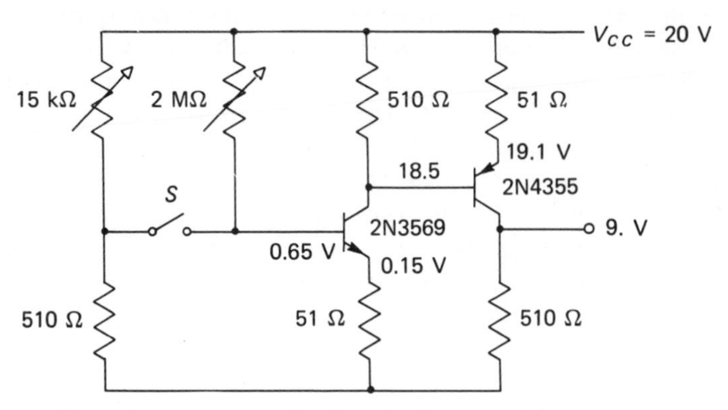

Figure 2.12. A dc amplifier for $\Delta\beta$ and ΔV_{BE} drift calculations.

reference 2-2 except that the method of biasing the input stage may be changed by the switch S. With the switch closed, the first stage has an S-factor of 10, and, when open, the stage gain is β of the transistor. Suppose that we wish to calculate the upper temperature limit at which the amplifier can be operated with the switch either closed or open.

Consider first the case with the switch closed and assume that we are willing to tolerate a 2.5-V shift of the second-stage collector Q-point voltage with temperature. Using equation (2.4), we divide 2.5 V by the overall voltage gain of 70 and get 36 mV as the permissible change of V_{BE} in the first stage. With a V_{BE} change of 2 mV/°C, we find that the junction temperature cannot exceed the bias-point temperature by more than 18°C. If the bias-point temperature is 30°C, then the maximum allowed junction temperature of the first transistor is 48°C, which is a value adequate for room-temperature applications. Because of the lower voltage gain, we assume that the Q-point drift caused by the second transistor can be neglected.

Consider now the case with the switch open, which makes the V_{BE} voltage change a negligibly small change of the bias voltage that determines the base current. With a fixed base current of 20 V/2 MΩ = 10 μA, the collector Q-point drift is caused by the temperature change of β in the first stage. The low S-factor in the second stage makes the current gain change negligible there compared with the β change in the first stage.

By equation (2.1) the Q-point shift with β may be easily calculated, provided we know the value of β for the transistor in the first stage. However, by assuming that the second-stage current gain is 10 and independent of temperature, we may use the overall current gain, which is easily measured. By shunting the 2-MΩ bias resistor with a 22-MΩ ohm resistor, the base current is increased 1 μA and the Q-point collector voltage increases 2 V, which means that the collector current increases by 4 mA. The ratio 4 mA/1 μA indicates an overall incremental current gain of 4000. Observe that this, the incremental current gain, is over twice the ratio of the total collector current, 18 mA, to the bias current, 10 μA. The apparent discrepancy results from the fact that about one half the bias current is required to produce enough voltage drop in the first collector to offset the 0.6-V V_{BE} voltage of the second transistor. When the bias current is increased incrementally, the V_{BE} voltage is not changed, so the entire change is transferred as a change of the second-stage emitter current. The discrepancy* shows the desirability of relying on measured characteristics whenever possible and of using calculated values with considerable distrust.

Assuming once again that a collector Q-point shift of 2.5 V is tolerable, we find that the collector current may increase by 4.9 mA. Inserting the bias current, 10 μA, and the permissible change of the Q-point current into the equation

$$\Delta I_C = \Delta\beta(I_B) \tag{2.6}$$

we find $\Delta\beta = 490$. If the overall current gain increases 0.6 per cent/°C, then a β increase of 490 in a total of 4000 means that the junction temperature may increase 20°C. This calculated result is interesting because it shows that in this design the Q-point drift is a little lower with the switch open than with the switch closed. We may infer, and it is confirmed by experiment, that intermediate values of the first-stage S-factor produce a combined temperature drift that decreases monotonically as S is increased. The result also illustrates the fact that the equivalent drift voltage referred to the input is generally a little less than 2 mV/°C unless the amplifier has some form of drift compensation.

2.11 COLLECTOR POWER AND Q-POINT DRIFT

The collector Q-point shift that may be expected in the amplifier shown in Fig. 2.12 has been calculated on the assumption that the collector temperature varies with the room temperature. With an adequate heat sink the calculations are confirmed by measurements. However, if the transistors are operated without heat sinks, the Q-point is observed to drift approximately 3 V in the first 5-minute period after the amplifier is turned ON. The Q-point drift observed when an amplifier is turned ON generally indicates a junction

*A similar discrepancy exists between the measured voltage gain of the amplifier, 70, and the gain calculated from the R_L/R_E ratios, 100. Here the discrepancy is caused by the fact that h of the first stage is about 10 Ω, so that only the fraction $\frac{51}{61}$ of the input voltage appears as the voltage across the emitter resistor R_E, and the overall voltage gain becomes 83 per cent of the calculated value. The remaining difference between the gain of 83 and the measured 70 is easily explained by the approximations made by assuming transistor βs are so high that the stage current gain is S.

temperature rise that is caused by dc power dissipated in the collector circuit. This initial drift of the Q-point may be found by the methods used earlier in this chapter after calculating the junction temperature rise produced by the power $V_{CE}I_C$ that is dissipated in the collector junction.

The maximum collector input power in the output stage is $V_{CC}^2/4R_L = 0.2$ watt (W), and in the input stage it is $V_{CE}I_C = 0.054$ W. The epoxy-encapsulated transistors have a junction temperature rise of 300°C/W, so the junction temperature of the output stage is 60°C above the ambient, and of the input stage it is 16°C above ambient. For the output stage the V_{BE} voltage changes 0.12 V and produces a 1.2-V increase of the Q-point voltage. The input stage V_{BE} changes 0.032 V and produces a 2.2-V increase. Thus, the calculated values account for the observed drift.

2.12 DRIFT-REDUCTION TECHNIQUES

The Q-point drift produced by temperature changes may be reduced by various techniques. The method chosen usually depends on both the objectives desired and convenience. The available methods include the use of feedback to reduce the dc gain, heat transfer to reduce the temperature, reduced gain, and reduced Q-point power.

The collector power input to the amplifier may be reduced by increasing all resistors in the circuit by the same scale factor. By increasing the resistors shown in Fig. 2.12 by three times, the power input is decreased three times, and the drift is reduced to 1 V when the amplifier is turned ON. Increasing all resistors by ten times reduces the drift to a negligible 0.3 V. However, suppose that the designer is unable to meet impedance or power-output requirements with the increased component values. In this event, the transistors may be changed to metal-encapsulated types and provided with clip-on heat sinks that reduce the temperature rise to one half the rise without heat sinks. Further reduction may be obtained by providing a larger heat sink, but with low-power transistors the lowest obtainable temperature rise is between 80 and 160°C/W. For the amplifier example, the minimum drift obtainable with low-power transistors and a small heat sink is a little less than 1 V.

The transistor V_{BE} shift with temperature is a basic temperature characteristic over which the designer has essentially no control except to reduce the temperature change itself. The V_{BE} change is 2 mV/°C for germanium and silicon transistors alike, but a small reduction may be obtained by selecting transistors. The change of β with temperature may be reduced by choosing the transistor type and selecting a particular device within the type. The manufacturers may offer some assistance in obtaining a suitable transistor having a low β change with temperature.

Q-point drift may be reduced by reducing the dc voltage gain of an amplifier. The best way of reducing the dc gain is usually to add overall feedback.

Feedback may be introduced in the present amplifier example by returning the collector load resistor to the first emitter and lowering the emitter resistor to 5.1 Ω. This change maintains the overall voltage gain of 100 and replaces local feedback in the first stage by overall feedback. Overall feedback reduces the open-loop gain from about 4000 to 100, a factor of 40. This feedback essentially eliminates the second-stage drift but cannot eliminate the input-stage drift. The V_{BE} voltage change in the input stage is an input signal, and, with the same overall dc voltage gain, feedback cannot change the observed Q-point drift. Experiment confirms these statements in that the Q-point drift with feedback is about 2 V when the overall feedback is used. If the overall gain is reduced, as by increasing the first-stage emitter resistor, the drift is reduced in proportion to the gain reduction. The advantage of using feedback to reduce the gain is that the amplifier is improved in other respects.

2.13 SUMMARY

The base-emitter voltage drop of a silicon transistor operated near its intended collector current is about 0.6 V. For germanium transistors the V_{BE} voltage drop is about 0.25 V. For both kinds of transistors the V_{BE} voltage changes approximately 2 mV/°C and decreases with increased junction temperature. Similarly, the V_{BE} voltage increases 60 mV for a factor-of-10 increase of the emitter current. These somewhat conflicting voltage changes explain why the base-emitter voltage is not particularly useful as an indication of transistor or circuit performance. With a fixed emitter current, the V_{BE} voltage may be used to observe or measure the junction temperature. Transistors operated at low current levels may be expected to exhibit a V_{BE} voltage that is about 0.2 V low. At high current levels and high junction temperatures, the V_{BE} voltage may be high by as much as 0.4 V. These variations of the voltage drop with emitter current are useful guides but are rarely an important factor in design.

Amplifier Q-point shift with temperature may be caused by the V_{BE} change or by a β change. The temperature change of β varies with the transistor type and with the operating Q-point, and is generally between 0.5 and 1.5 per cent/°C change of the junction temperature. Together, the β change and the V_{BE} change may be assumed to produce Q-point drifts of less than 2 mV/°C when referred to the input.

Q-point drift is reduced by reducing ambient temperature changes, by reducing or dissipating heat-generated collector power input, and to a limited extent by the application of feedback.

Q-point drifts caused by the I_{CO} leakage may be neglected in planar silicon transistor circuits but must be evaluated in circuits using germanium transistors. Thermal runaway that is found in power amplifiers is caused by I_{CO} and is prevented by reducing the product $V_{C}SI_{CO}$ and the junction temperature.

PROBLEMS

2-1. (a) For the stage shown in Fig. P2.1, what is the stage current gain when the junction temperature is $-55°C$; $25°C$; $100°C$? (b) For a temperature change of β only, what is the dc collector voltage at each temperature? *Note*: Use β as given in Fig. 2.5.

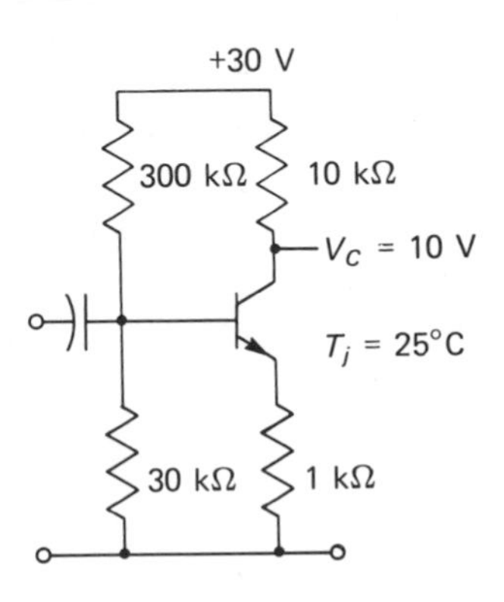

Figure P-2-1.

2-2. (a) For the circuit in Fig. P2.1, neglect all but the ΔV_{BE} temperature effect and find the collector Q-point voltage when the junction temperature is $-55°C$; $25°C$; $150°C$. (b) By what change of bias current can these collector voltage changes be offset?

2-3. Confirm the calculated Q-point ΔV_{BE} voltage shifts given in Section 2.6.

2-4. A germanium power transistor is biased to make the collector current 1 A at $30°C$ when $S = 10$. The junction temperature during operation is $85°C$. The manufacturer gives I_{CO} as 1 mA at $25°C$. What is the collector current at $85°C$ when neglecting the ΔV_{BE} and the β effects?

2-5. A transistor is observed to develop thermal runaway. Is runaway less likely to occur if (a) the S-factor is reduced from 30 to 10, or (b) the heat sink is changed so that the temperature is reduced by $20°C$? Explain.

2-6. (a) Confirm by calculation the Q-point voltage changes given in Section 2.11. (b) What is the Q-point change in the second stage that is produced by increasing the ambient temperature of the second stage alone from 30 to $48°C$? (c) Does the effect of ΔV_{BE} in the first stage offset or add to the second-stage Q-point drift that is observed when the amplifier is turned ON?

EXPERIMENTS

2-1. Construct the amplifier shown in Fig. 2.12 and observe the Q-point drift produced when the amplifier is turned ON. For the output-stage transistor substitute a metal-cased transistor with a small heat sink and observe the improvement. Show that your observations agree reasonably with a calculation. Hint: If a low-power metal-encased *pnp* transistor is not available, consider reversing the power polarity and interchanging the transistors.

2-2. Construct an oven by enclosing a lamp in a wooden box, or immerse the circuit in a hot oil bath, and confirm the temperature effects for the circuit shown in Fig. 2.12, or for a circuit of your own choosing.

DESIGNS

2-1. Design a *CE* temperature alarm that will turn ON a 40-mA 2-V lamp when the transistor reaches approximately 100°C. Use a 12-V supply. Explain the operation, show how to adjust the turn-ON point, and describe the problems that may be expected.

2-2. Design a single-stage amplifier for demonstrating the ΔV_{BE} and the β temperature effects. Predict what should be observed.

REFERENCES

2-1. Cowles, L. G., *Analysis and Design of Transistor Circuits*, Chapter 6.New York: Van Nostrand Reinhold Company, 1966.

2-2. Cowles, L. G., *Transistor Circuits and Applications*. Englewood Cliffs, N.J.: Prentice-Hall, Inc., 1968.

3

EQUIVALENT CIRCUITS AND GAIN-IMPEDANCE RELATIONS

An engineer usually begins the design of a linear circuit by selecting a trial circuit and deciding by suitable tests whether it meets his ac gain and impedance requirements. He may design and evaluate the circuit on paper by calculations, or he may construct the amplifier by trial and error. The latter procedure may seem simpler because the circuit is needed eventually for performance testing. Both methods have advantages, and the best way to design an amplifier is generally a combination of paper work and experiment; but, whatever his method, a designer always saves time by using a few easy calculations. The purpose of this chapter is to review gain and impedance relations that are of practical help as design guides.

3.1 DESIGN GUIDES

The application for which an amplifier is intended generally requires particular values of the input and load impedances. For efficiency and low noise the input signal should connect to the first active element with a minimum of intermediate circuit elements. This requirement means that the amplifier input impedance is usually determined by the emitter current of the first transistor and by feedback. We shall show how the transistor impedance is controlled in design to reflect the required input impedance. The load impedance is generally given, and the power that must be delivered to the load is the concern of Chapter 7.

The current and voltage gains required of an amplifier are determined by comparing the output signal and load impedance with the input signal and input impedance. The Transistor Gain-Impedance Relation (TG-IR) tells the designer how many stages are required in the amplifier, or, conversely, how efficiently a given design meets the specifications. From Shockley's relation we obtain an equation which gives the upper limit of voltage gain that can be produced in a single CE stage. These circuit relations, equivalent circuits, and the feedback relations of later chapters are practical aids for understanding and developing transistor circuits.

3.2 EQUIVALENT CIRCUITS

Transistors are described by a number of equivalent circuits, some of which represent the transistor in such detail as to complicate the design problems. The detail may be provided to represent all that is known of a transistor's characteristics or to show the varied ways in which the frequency response may be influenced by the reactive components. Sometimes the engineer may develop great skill in manipulating a complicated equivalent circuit and predict amplifier performance with a fair degree of accuracy. For exceedingly difficult design problems this may be a valid approach, especially when a computer makes the calculations and types out orderly tables of data. This, however, is not the method by which the great majority of circuit design is actually accomplished.

Generally, a designer has enough leeway within the specifications to permit his using approximate methods to find a good practical design, if not the best design. For such purposes, more is accomplished when the designer uses the simplest possible equivalent circuits and accepts the reduced accuracy in his calculations. Approximate methods show the designer how to implement an experimental design and how to meet the specifications with a minimum of time and effort. For practical applications and most circuit design, the simplest equivalent circuit is the short-circuited (SC) equivalent.

3.3 THE SHORT-CIRCUITED EQUIVALENT CIRCUIT

The SC-equivalent, which represents a transistor by the two most used parameters of the device, is shown in Fig. 3.1 for the *CE* configuration. In this equivalent the transistor is represented by only two parameters, which are its input impedance βh and the active current source βi_b.

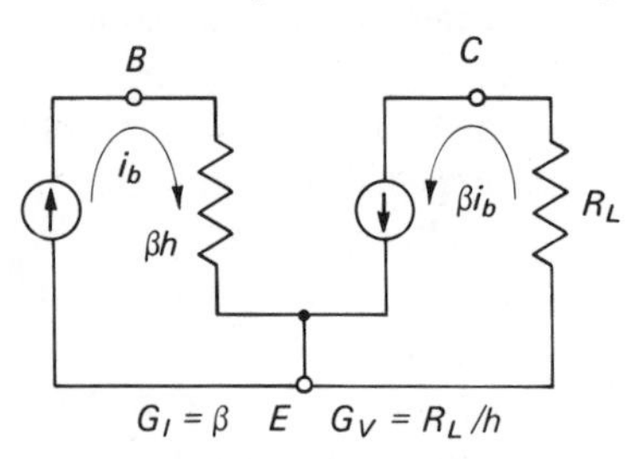

Figure 3.1. *CE* short-circuit (SC) equivalent.

This equivalent is called the *short-circuited* equivalent because the load is assumed to be a short circuit, as compared with the internal output impedance of the transistor. By this assumption two of the usual four parameters are neglected. The internal output impedance of the device is generally 1 to 3 orders of magnitude larger than a practical load and, hence, may be neglected. Furthermore, a short-circuiting load makes the internal feedback and the input impedance independent of the load impedance. This assumption eliminates considerable circuit complication and simplifies the design task at the expense of a small computational error.

Observe that in the SC-equivalent the load current is always β times the input current, the input impedance is β times the transistor h-parameter, and the emitter current i_e is for all practical purposes the same as the load current i_c. Furthermore, we assume without reservation that for low-frequency analysis and design β is always greater than 100. Any recently designed transistor has a current gain that exceeds 50, and both *npn* and *pnp* transistors are available that have minimum βs of 100.

Beta is such a variable quantity at best that a transistor is always used with feedback, often local feedback. We shall show that with significant feedback the designer need only specify a minimum value of β and is then free to assume that the circuit gains and impedances are determined by the feedback resistances.

3.4 THE *CE* STAGE WITH COLLECTOR FEEDBACK

As an elementary feedback example, consider the collector-feedback stage shown in Fig. 3.2. Because we are mainly concerned with practical circuits, we

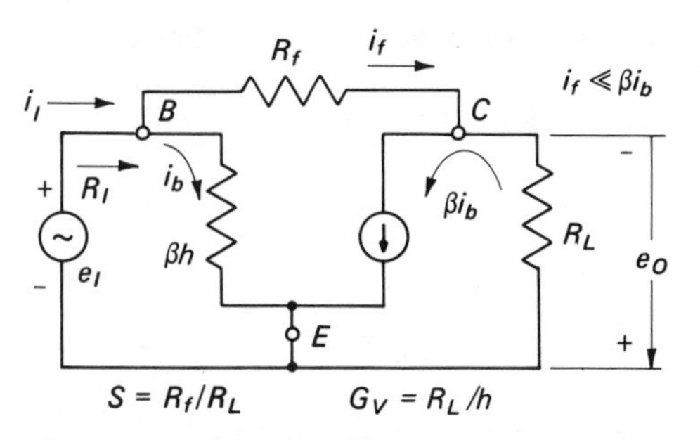

Figure 3.2. SC-equivalent with collector feedback.

assume that R_f is large compared with R_L so that we can neglect the current I_f as compared to I_L. The voltage gain of the stage is

$$G_v = \frac{e_O}{e_I} = \frac{\beta i_b R_L}{i_b \beta h} \tag{3.1}$$

and we find that

$$G_v = \frac{R_L}{h} \tag{3.2}$$

Similarly, by assuming that the output voltage e_O is large compared with the input voltage e_I, the current in the feedback resistor is

$$i_f = \frac{e_O}{R_f} \tag{3.3}$$

The equivalent Miller-effect impedance R_M is

$$R_M = \frac{e_I}{i_f} = \frac{e_I}{e_O} R_f \tag{3.4}$$

and equation (3.4) may be written as

$$R_M = \frac{R_f}{G_v} \tag{3.5}$$

Substituting G_v from equation (3.2) into equation (3.5) gives

$$R_M = \frac{R_f}{R_L} h \tag{3.6}$$

The S-factor of the collector feedback stage is defined as

$$S \equiv \frac{R_f}{R_L} \tag{3.7}$$

and equation (3.6) reduces to

$$R_M = Sh \tag{3.8}$$

As shown by equation (3.8), collector feedback reduces the stage input impedance by shunting the transistor input impedance βh with the Miller equivalent Sh, so the input impedance with feedback is the parallel equivalent resistance

$$R_I' = \frac{S\beta}{S+\beta} h = \frac{\beta}{S+\beta} Sh \tag{3.9}$$

and, because the stage current gain is reduced by the same factor,

$$G_i = \frac{S\beta}{S+\beta} = \frac{\beta}{S+\beta} S \tag{3.10}$$

With adequate feedback, S is small compared with β, so the input impedance and the current gain reduce to Sh and S, respectively, and both are independent of β. However, in practical designs we must accept a lesser amount of feedback, and we say that the feedback is significant when

$$S = \frac{\beta}{3} \tag{3.11}$$

With significant feedback the impedance and gain variations caused by β are reduced by a factor of 4, and to a close approximation we may say that the stage current gain is S and the input impedance is Sh.

Thus, with collector feedback the designer may produce a desired input impedance Sh by his choice of S and h and a desired voltage gain by his choice of R_L. Significant feedback is assured by specifying a minimum value for β that is at least three times the S-factor. The voltage gain, which is not affected by feedback, is precisely determined by h and the emitter current. However, we show in Chapter 5 that if the stage is driven by a signal source that has an impedance R_S large compared with Sh, the voltage gain is approximately

$$G_v = \frac{R_f}{R_S} \tag{3.12}$$

and both the current and voltage gains are controlled by the feedback resistors.

3.5 SHOCKLEY'S RELATION IN DESIGN

Shockley's relation is often used in design because it gives the value of h that controls the input impedance and voltage gain of a transistor used without feedback. Shockley's relation in the form

$$h = \frac{26}{I_E} \tag{3.13}$$

usually gives h within a 10 per cent error for emitter currents within a factor of 10 of the design-center current at which a *low-power* transistor is intended to be used. However, Fig. 3.3 for a low-power transistor shows that h tends to be higher than predicted by Shockley's relation when at high emitter currents the dc resistance becomes comparable to h.

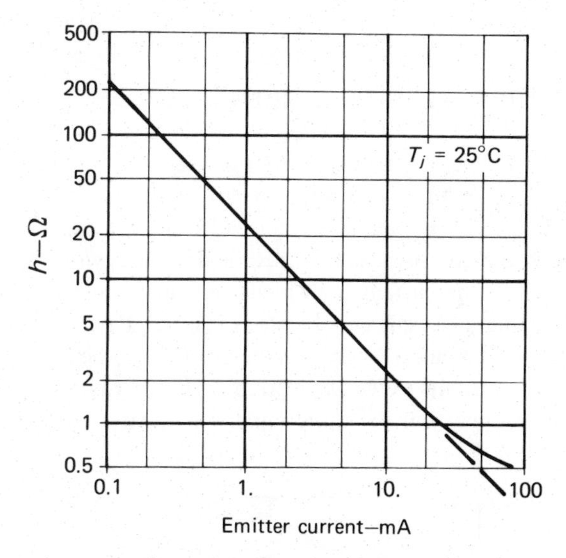

Figure 3.3. Shockley's relation.

Equation (3.13) gives h for a junction at room temperature, whereas equation (1.1) shows that h is proportional to the absolute temperature of the junction. When a power transistor is operated near its maximum current and temperature ratings, h may generally be assumed to be 50 to 100 per cent higher than the value given by equation (3.13). In power applications a precise value of h is not usually needed, and a considered estimate of the combined temperature and resistance increase is generally close enough for design purposes.

Equations (3.2) and (3.9) illustrate an application for Shockley's relation in calculating the gain and the input impedance of a *CE* stage. Equation (3.9) may be generalized for any *CE* stage by saying a high input impedance usually requires a high value of h and, therefore, a low emitter current. Once the input impedance is selected, the stage voltage gain is determined by the current gain and the load impedance.

Later we shall show that the voltage gain of a series of iterated stages is determined only by the stage current gain and is independent of h. For this reason the emitter current of the input stage is usually determined by the required input impedance, and the emitter currents of succeeding stages may be chosen somewhat arbitrarily. With any reasonable collector supply voltage and with significant feedback, the gain of iterated stages is determined only by the S-factors. Thus, with iterated stages the designer should select transistors that have reasonably high values of β at the chosen emitter currents.

An exception to the previous statements is found in amplifiers operated on a low collector supply voltage when an exceptionally high voltage gain is desired in a one- or two-stage amplifier that does not have local feedback. In this exceptional case the stage voltage gain depends on h and the supply voltage, as shown in Section 3.7.

3.6 GAIN AND THE COLLECTOR CURRENT

In choosing the collector current of the transistors for a practical circuit, the designer should consider the way in which β varies with the collector current. As shown in Fig. 3.4, β has a maximum value at a current that depends on the purpose for which the transistor is intended, and β varies with the temperature, as shown in Fig. 2.5. The current for which β is a maximum is usually given in the manufacturer's data. At currents between 0.1 and 1 per cent of the current given for a stated β, the current gain may be assumed to be at least 30 per cent of the stated value. Currents exceeding a given "typical" value should be avoided generally, because β may fall off quite rapidly at emitter currents above the β peak.

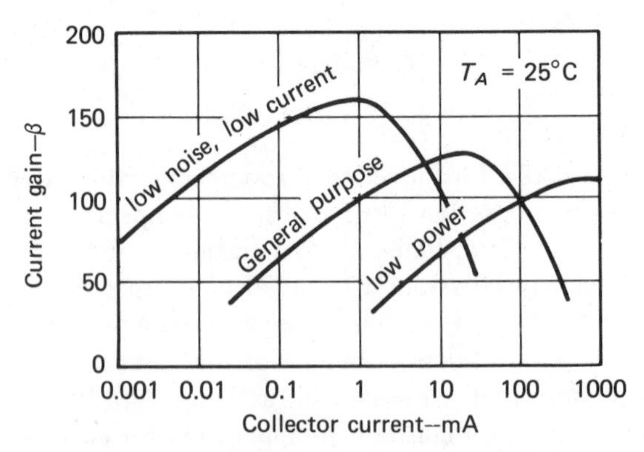

Figure 3.4. Transistor current gain curves.

The variations of β with the current and temperature are usually the main reason for selecting a particular collector current for a low-power iterated stage. Because the gain of a stage varies with the S-factor of each stage, a circuit design should use high S-factors and transistors that have β values high enough to give the desired amount of local feedback throughout the operating temperature range.

3.7 GAIN WITH A LOW-VOLTAGE SUPPLY

By replacing h in equation (3.2) by h as given by Shockley's relation, we obtain

$$G_v = \frac{[I_E R_L]}{0.026} \tag{3.14}$$

Equation (3.14) gives an upper limit to the voltage gain of a *CE* or *CB* stage. Notice that the voltage $[I_E R_L]$ is placed in brackets as a reminder that the numerical value of the voltage is to be used so that the equation will be dimensionally correct. Because the theoretical limit is never quite attained, the reciprocal of 0.026 from Shockley's relation may be replaced by 25, which reduces equation (3.14) to a practical working guide:

$$G_v = 25[I_E R_L] \tag{3.15}$$

Equation (3.15) may be used to calculate the voltage gain of a resistance-coupled *CB* or *CE* stage without feedback when the dc voltage across the load resistor is known. The equation states that the voltage gain of a *CB* or *CE* stage cannot exceed 25 times the product of the emitter current and the load resistance. In other words, the equation shows that the voltage gain cannot exceed 25 times the equivalent dc voltage drop in the load resistor. We say *equivalent voltage drop* because the relation holds even in those cases in which the dc collector current bypasses the load resistor.

Equation (3.15) is particularly useful in the design of micropower integrated circuits when high voltage gain is required of an amplifier using a 3-V or lower collector supply. Suppose that a high-gain amplifier is to be operated with two mercury cells as the supply voltage. Assuming an end point voltage of 2.7 V and allowing for a collector Q-point voltage that is one third the supply voltage, we are left with about 2 V as the equivalent dc voltage in equation (3.15). We conclude that the voltage gain cannot exceed about 50 for each *CE* stage. If the transistors are to be operated at as low emitter currents as practical, say about 10 μA, then $h = 2600\ \Omega$, and $R_L = 50h = 130{,}000\ \Omega$.

An experimental confirmation of these calculations is made by using the circuit shown in Fig. 3.5 and trying different values of R_A and R_L. The experiment shows that the estimated voltage gain is attainable, but that without feedback control the circuit is impractically sensitive to the supply voltage. The example is interesting, however, because it shows that an integrated circuit like the μA735, which is designed to operate on a 3-V supply, must have at least three *CE* stages to develop an open-loop voltage gain of 20,000.

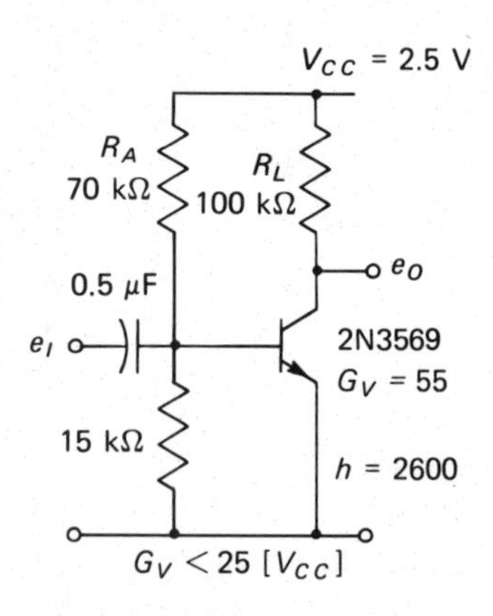

Figure 3.5. Amplifier showing V_{CC} limitation.

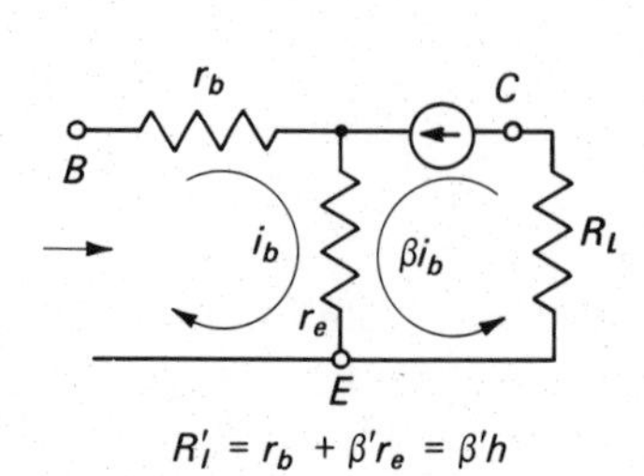

Figure 3.6. SC common-emitter *T*-equivalent.

3.8 THE CIRCUIT *H*-PARAMETER

Whenever a transistor is represented by a *T*-equivalent circuit, the *h*-parameter may be separated into two components, r_b and r_e, as shown in Fig. 3.6. The input impedance of the *CE* transistor is the sum of the voltage drops in the input divided by the base current; that is

$$R_I' = \frac{i_b r_b + (\beta + 1) i_b r_e}{i_b} \tag{3.16}$$

and

$$R_I' = r_b + \beta' r_e \tag{3.17}$$

Because $R_I' = \beta h$, equation (3.17) may be solved for $h = R'/\beta$, whence

$$h = r_e + \frac{r_b}{\beta'} \tag{3.18}$$

The usefulness of this form of the h-parameter is that whenever resistors are connected in series with the base and emitter, the effect of these resistors on the gain and input impedance may be simply calculated by adding the values of the external resistors to r_e and r_b of equation (3.18). In other words, with external resistors h becomes H, where

$$H = r_e + R_E + \frac{r_b + R_B}{\beta'} \tag{3.19}$$

and the input impedance is βH, where

$$\beta H = \beta R_E + R_B + \beta h \tag{3.20}$$

The stage current gain remains β, but the voltage gain G_v' is

$$G_v' = \frac{R_L}{H} \tag{3.21}$$

The form of H is shown by equation (3.19) to make an external emitter resistor β times more effective than a base resistor in increasing the input impedance or in decreasing the voltage gain. The H-parameter is occasionally used in such ways to substitute for a more lengthy calculation. In applying equation (3.20) the last term βh is generally negligible, but the form of the h-parameter, $r_e + r_b/\beta'$, shows that the impedance βH with external resistors has two components, R_B and βR_E. Similar relations, which apply when a transistor is operated as a *CB* or *CC* stage, are explained in both *Applications* and *Analysis*, to which the reader is referred for more information.

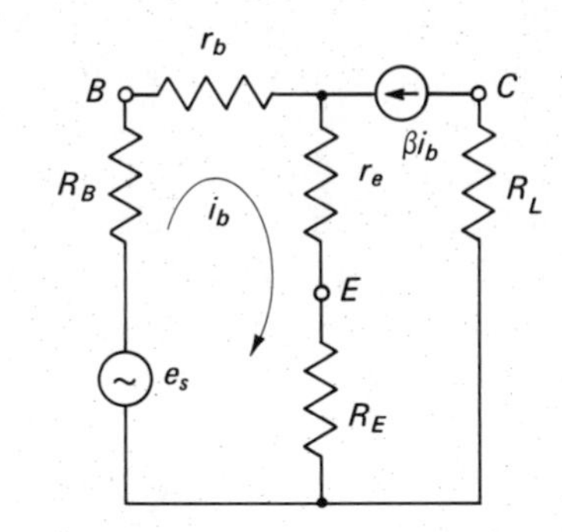

Figure 3.7. H-parameter equivalent circuit.

As an example of the use of the H-relation, equation (3.19), consider the circuit shown in Fig. 3.7. If we wish to calculate the impedance seen by the source e_S, we note that both R_B and R_E are in series with the base and the emitter, respectively, and write the input impedance as

$$R_I = \beta H \tag{3.22}$$

or

$$R_I = \beta\left(R_E + \frac{R_B}{\beta'}\right) + \beta h \tag{3.23}$$

If the stage has significant feedback, then $R_E = 3h$, and with adequate feedback h is negligible compared with R_E. In other words, if R_E is to serve any useful purpose, we may neglect the term βh in equation (3.23). Also, by rearranging equation (3.23) and dropping the βh term, we have

$$R_I = \beta R_E + R_B \tag{3.24}$$

and it follows from equation (3.24) that if R_B is SR_E while β is greater than $5S$, the first term is at least five times larger than the second term. In other words, with S-factors small compared with β, the transistor input impedance is an open circuit compared with R_B. We conclude that when β is greater than $5S$, the Thévenin equivalent impedance from the base to the common ground is R_B to within a 20 per cent error.

Consider now the amplifier shown in Fig. 3.8 and assume that the transistor β is significantly larger than the ratio of R_B to R_E. When we calculate the voltage gain of the stage, the source shunts R_B so that the voltage gain R_L/H reduces to R_L/R_E. The simplicity of these calculations of the input impedance and gain comes from the use of the SC-equivalent circuit and the circuit H-parameter.

3.9 THE TRANSISTOR GAIN-IMPEDANCE RELATION

Transistor circuits are characterized by current gains that can be evaluated by observing the ratios of resistors used in the circuit. For this reason, the voltage gain of a complicated circuit may be found by the TG-IR, which is derived and explained in Chapter 1. The purpose of the present discussion is to show how the TG-IR is of special help in circuit design.

Consider a two-stage transistor amplifier in which the collector of the first stage is connected to the base of the second, as shown in Fig. 3.9. The

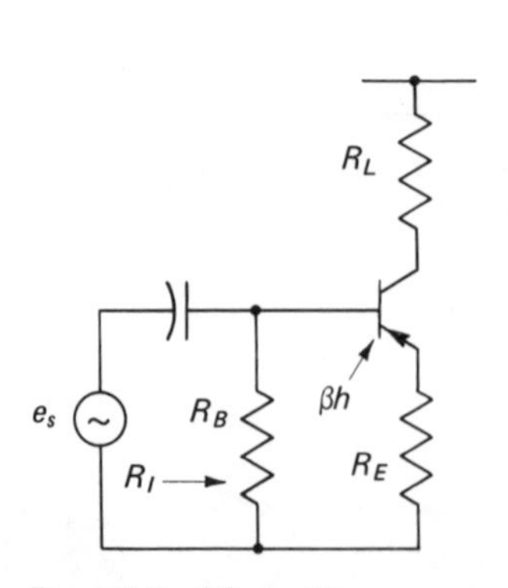

Figure 3.8. *CE*-amplifier example.

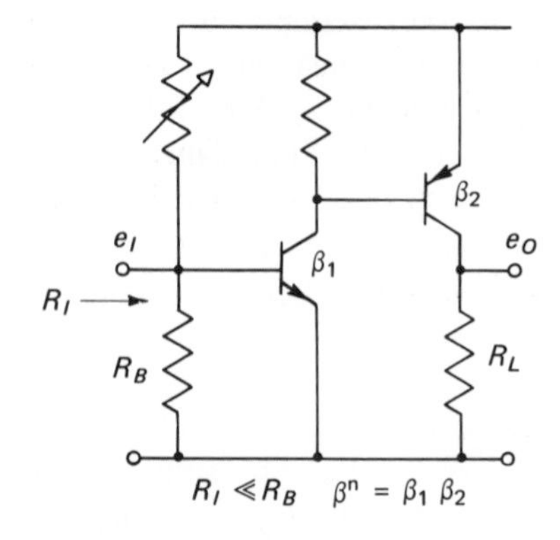

Figure 3.9. *CE-CE* pair amplifier.

base-emitter junction of the second stage has an input impedance that is several orders of magnitude lower than the collector impedance of the first stage. Hence, the two-stage current gain is the short-circuit current gain of the first stage multiplied by the current gain of the second stage. This fact may be represented mathematically by writing the TG-IR with the overall current gain shown as a product; thus,

$$G_v = \beta_1\beta_2\frac{R_L}{R_I} \tag{3.25}$$

Because the transistor current gains are variable and indeterminate, there is no practical reason for showing the betas separately. Hence, for an n-stage amplifier the TG-IR may be written as

$$G_v = \beta^n\frac{R_L}{R_I} \tag{3.26}$$

In a practical amplifier it is generally necessary to use an interstage resistor R_B and, perhaps, an emitter resistor R_E, as shown in Fig. 3.10. Either or both of these resistors reduce the stage current gain. Therefore, the TG-IR representing a practical situation is better written to express the overall current gain as the product of the S-factors,

$$G_v = S^n\frac{R_L}{R_I} \tag{3.27}$$

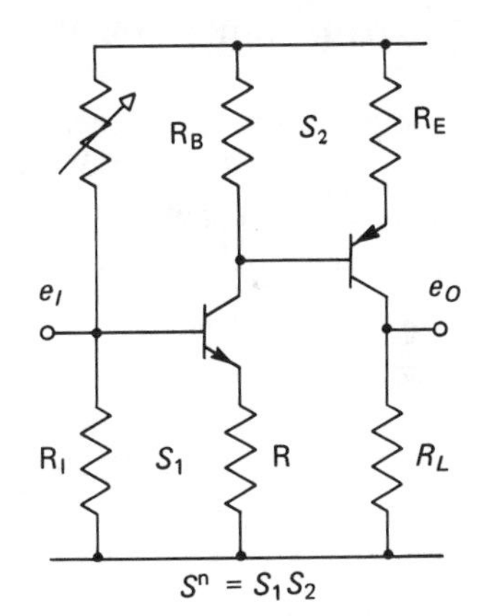

Figure 3.10. *CE-CE* pair with local feedback.

When examining a given circuit or planning a design, it is generally easy to estimate the S-factors of the individual stages and, thereby, to write the overall current gain S^n. With the output and input impedances given, equation (3.27) gives the overall voltage gain closely enough for almost any practical purpose. With this equation the designer knows at least the minimum number of stages required to satisfy a specified voltage gain. If transformers are used, the product of the inverse turns ratios should be included as a factor in the overall current gain.

In this book the gain of an amplifier is usually given as the iterated voltage gain. Because the input and output impedances of an iterated stage are equal, the voltage and current gains are also equal. Therefore, when iterated stages are capacitor-coupled in a series, the current gain of the series is numerically the product of the iterated voltage gains of each stage. However, when stages are direct-coupled, we usually remove one or more resistors and reduce the interstage loss. As a result, direct-coupled stages may be expected to give as much as two times the gain per stage of capacitor-coupled stages.

The TG-IR is not applicable when FETs are used either alone or with transistor stages. The voltage gain of the FET stages must be calculated separately and included in the total overall gain. The voltage gain of an FET stage is the product of g_m and the load resistance. The g_m of the FET varies with the Q-point current and may be corrected by the method given in Chapter 16.

The transconductance of a transistor is about 40 times that of a high figure-of-merit FET, so there is little reason to use FETs except to take advantage of their high input impedance. Even when the transistor voltage gain is reduced by feedback, there is still more gain attainable in a transistor stage than in an FET stage. Moreover, for the same voltage gain the FET stage requires about 10 times the collector power required in a transistor stage. An FET-transistor pair has the advantages of high power gain and significant feedback.

3.10 THE TG-IR IN DESIGN—AN EXAMPLE

Suppose that a designer wishes to build a phonograph amplifier, as shown in Fig. 3.11, that delivers 8 W to an 8-Ω load when the input signal is 2 mV and the amplifier input impedance is 15 kΩ. The specifications show that the output signal must be 8 V root mean square (rms), so the voltage gain is 4000.

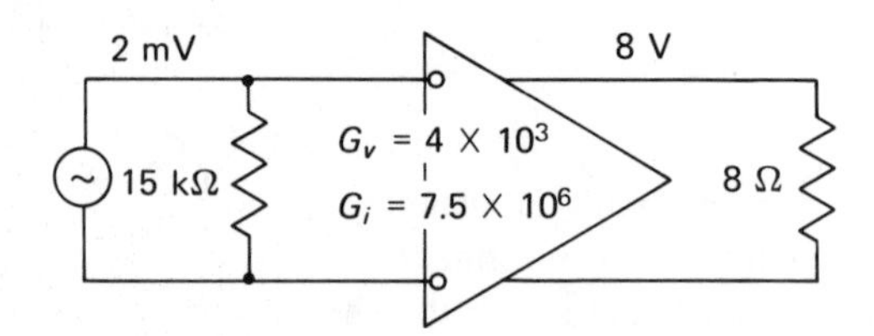

Figure 3.11. TG-IR for phono-amplifier.

Substituting in the TG-IR gives

$$4000 = S^n \frac{8}{15{,}000} \tag{3.28}$$

and solving for S^n gives

$$S^n = 7.5 \times 10^6 \tag{3.29}$$

The best way to decide on the number of stages required in a trial design is to assume several values for the S-factor and find values of n that satisfy equation (3.29). For example, we try $S = 20$ and $n = 5$ and find that

$$S^n = 2^5 \times 10^5 = 3.2 \times 10^6 \tag{3.30}$$

We conclude that five stages will not allow enough excess gain to offset the circuit losses. We try $S = 20$ and $n = 6$ and find that

$$S^n = 2^6 \times 10^6 = 64 \times 10^6 \tag{3.31}$$

Six stages provide nearly a factor-of-10 excess gain, which we estimate is required for a high- and low-frequency boost and for feedback in the power stages. The designer now knows that the amplifier needs at least six stages. The calculations suggest also that an eight-stage design may not be making full use of the transistors and should be redesigned.

As a second example of the TG-IR in design, consider the two-stage amplifier shown in Fig. 3.12. Each stage has an S-factor of 20. Hence, the expected voltage gain is

$$G_v = 20^2 \frac{10{,}000}{10{,}000} \tag{3.32}$$

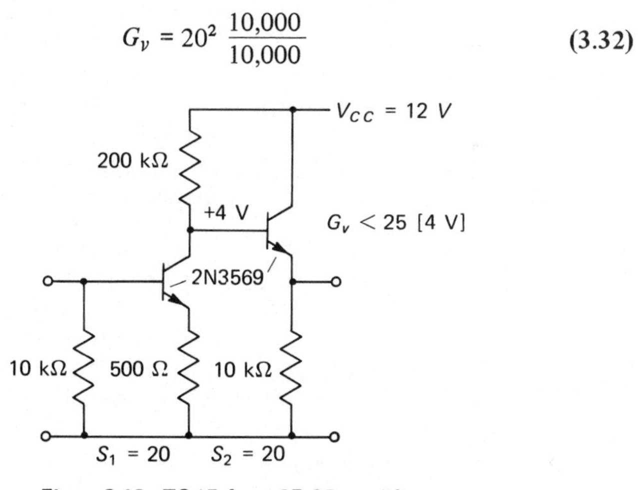

Figure 3.12. TG-IR for a *CE-CC* amplifier.

When the amplifier is constructed, the designer finds that the overall voltage gain is considerably lower than expected. If the *CC* stage is operated with the base and emitter Q-point voltages at about one half the 8-V supply voltage, only 4 V is available as the voltage drop in the 200-kΩ collector resistor. From equation (3.15) we find that the maximum voltage gain obtainable in the first stage is

$$G_v = [25V_L] = 100 \tag{3.33}$$

Suppose that the designer wishes to keep the *CC* output stage and chooses to add another *CE* input stage. From the TG-IR we now have

$$400 = S^3 \frac{10{,}000}{10{,}000} \tag{3.34}$$

and

$$S^3 = 400, \qquad S = 7.4 \tag{3.35}$$

On finding that S is so much lower than the value of 20, which was first proposed, the designer may decide to increase the input impedance, decrease the output impedance, or increase the voltage gain. By knowing the rules that guide circuit design, a designer avoids many blind alleys and wasted effort.

3.11 DESIGN-GUIDE SUMMARY

The mathematical relations of this chapter are used to evaluate circuits and to save time in circuit design.

The simplest equivalent circuit for circuit design is the SC-equivalent. Because the load is short circuiting, the transistor is represented by two parameters. For a *CE* transistor the input impedance is βh and the current gain is β.

The gain and impedance calculation of circuits having a resistor in series with the base and the emitter may be simplified by using the circuit H-parameter,

$$H = h + R_E + \frac{R_B}{\beta} \tag{3.36}$$

For a *CE* or *CC* stage the voltage gain is $G_v = R_L/H$, and the input impedance is $R_I = \beta H$. For a *CB* stage the voltage gain is the same R_L/H, and the input impedance is H.

Shockley's relation shows how the h-parameter is determined by the emitter current and explains why the stage gain of micropower circuits is sometimes limited according to the relation $G_v = [25\ V_L]$. As a general rule, transistor amplifiers (without feedback) develop higher gains with a high value of the supply voltage, and, for a maximum gain, each stage should use an emitter current at which the transistor operates near the β maximum.

The TG-IR is perhaps the most important and useful design guide for linear-circuit design. The TG-IR shows the designer the current gain that must be provided to develop a required voltage gain between given input and output impedances. The TG-IR cannot be used with FET and vacuum-tube stages, but transformers may be included if they are represented as a current gain or loss. With local feedback the current gains are the ac S-factors of each stage.

Collector feedback makes the CE-stage current gain and input impedance independent of β. With adequate feedback the current gain is S, the S-factor, and the input impedance is Sh. Significant local feedback is obtained by selecting a transistor which has a current gain that is at least three times the stage S-factor.

PROBLEMS

3-1. (a) Does the amplifier shown in Fig. P3.1 have more or less than significant feedback? (b) What is the input impedance? (c) For what value of R_f does the amplifier have significant feedback?

3-2. (a) With the aid of the SC-equivalent circuit and the TG-IR, describe the characteristics of the amplifier shown in Fig. P3.2. Assume that equation (3.15) does not apply. (b) How are the characteristics changed when equation (3.15) is taken into consideration?

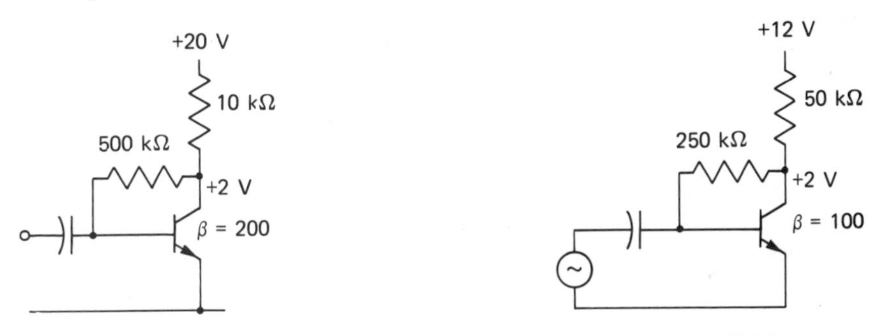

Figure P 3.1

Figure P 3.2

3-3. An amplifier shown in Fig. P3.3 has $R_I = 10$ kΩ and a voltage gain of 46 dB. (a) What feedback resistor should be used to obtain significant feedback? (b) What is the input resistance with feedback?

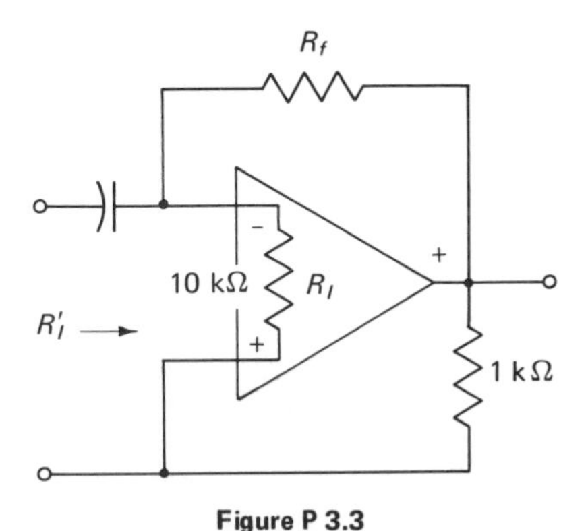

Figure P 3.3

3-4. (a) What is the input impedance of a transistor operated with $I_C = 50$ mA and a junction temperature of 100°C? (b) Repeat for $I_E = 0.1$ mA and a junction temperature of −55°C. Use Fig. 2.5.

3-5. (a) A three-stage amplifier must have $R_I = 10$ kΩ and $R_L = 10$ Ω with $V_{CC} = 30$ V. What are the possibilities of using three *CC* stages, or three *CE* stages, or two *CC* and one *CE* stage? Repeat, assuming that $V_{CC} = 3$ V.

3-6. A hearing aid operating on 6-V direct current is to have $G_v = 40{,}000$ when $R_I = 100$ Ω and $R_L = 10$ *k*Ω. How many *CE* stages are required? Explain.

3-7. An amplifier with 42-dB voltage gain is to have $R_I = 100$ kΩ and $R_L = 8$ Ω. If $S = 20$ in all stages, how many *CE* stages are required? How many *CC* stages would you expect to use in the amplifier?

DESIGNS

3-1. Design and construct a single-stage amplifier that operates on two 2.7-V mercury cells in series and gives a maximum voltage gain. The input impedance must exceed 10 kΩ and the collector current must not exceed 10 μA. Assume that the end point voltage is 2.5 V.

3-2. Design an experimental circuit with which you can demonstrate the application of the circuit *H*-parameter, which is discussed in Section 3.8.

REFERENCES

3-1. Cowles, L. G., *Transistor Circuits and Applications*, Chapters 2 and 4. Englewood Cliffs, N.J.: Prentice-Hall, Inc., 1968.

3-2. Cowles, L. G., *Analysis and Design of Transistor Circuits*, Chapter 3. New York: Van Nostrand Reinhold Company, 1966.

4

AMPLIFIER DESIGN PRINCIPLES

This chapter is concerned mainly with the *CB* and *CC* amplifiers, which develop power gain only when associated with an impedance change, as at the input and the output of an amplifier. The *CB* and *CC* stages are both shown to have advantages as broad-band input and output stages. Both are easily designed and are relatively free from *Q*-point drift with temperature.

The *CE* stage is the only bipolar connection that develops power gain in an iterated series of identical stages. We examine the problems of coupling a series of *CE* stages so that the iterated gain is maximized, and we consider the frequency effects caused by the coupling and transistor input capacitance. The chapter concludes with the design of a two-stage video amplifier that has power gain with equal source and load impedances.

4.1 AMPLIFIER PRINCIPLES

An amplifier is a series of stages comprised of active elements appropriately coupled and supplied with dc power and bias. The mechanism by which an active device amplifies a signal is represented by one term of an equivalent circuit. That term, the active element, is a current or voltage in the output that is proportional to a current or voltage in the input. Because active devices are nonlinear, an equivalent circuit generally represents a device only over the limited range in which the amplifier response is approximately linear.

Field-effect devices and vacuum tubes can be represented by the equivalent circuit shown in Fig. 4.1, in which the drain current i_D on the load side of the amplifier is proportional to the gate voltage e_G on the input side. In these devices the input resistance, though high, varies with the temperature and other factors that make the input current quite unpredictable. For this reason the designer shunts a fixed resistor R_G across the input side to make the input load fixed and predictable. Generally, both R_G and the input current may be neglected, so for field-effect devices the transconductance g_m is the main term in the equivalent circuit that is used for design.

The equivalent circuit of a bipolar transistor, shown in Fig. 4.2., differs in two respects from that of the field-effect devices. In a bipolar transistor the input impedance is low and the active element is a current on the load side that is proportional to the input current. With local feedback the designer is able to make the current gain S and the input impedance R_I relatively independent of the device, the temperature, etc. Because amplifiers may include field-effect devices and because amplification is customarily expressed as a voltage gain, the current gain of a transistor amplifier is usually converted to a voltage gain by means of the TG-IR.

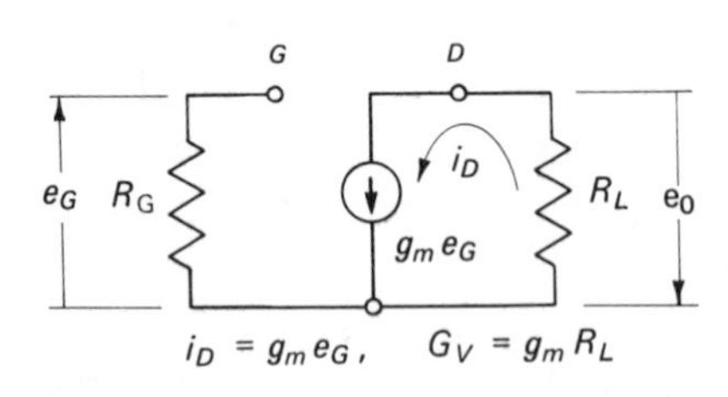

Figure 4.1. Equivalent circuit for FETs and vacuum tubes.

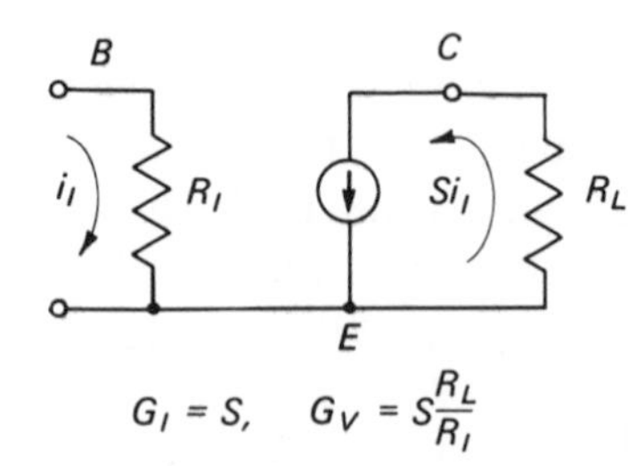

Figure 4.2. SC-equivalent of *CE* transistor with local feedback.

The TG-IR seems to imply that the characteristics of a complicated multistage amplifier depend only on the terminating impedance and the overall current gain developed by the transistors. This implication, which is valid for simple low-frequency amplifiers, neglects the frequency response, temperature rise, and other effects that must be considered in design. Thus, the designer needs to understand the choices that can be made by operating transistors *CB*, *CC*, or *CE*.

4.2 COMMON-BASE AMPLIFIERS

When a transistor is operated with the base as the common terminal, the amplifier has a unit current gain $\alpha = 1$, and an input impedance h that is determined by the emitter current according to Shockley's relation. Because the input characteristics are those of a forward-biased diode, h varies with the signal current. Hence, for a good wave form the *CB* transistor is generally operated with a fixed resistor in series with the emitter to provide feedback. With a resistor R_E in series with the source, the TG-IR says that the voltage gain is

$$G_v = \frac{R_L}{H + h} \tag{4.1}$$

By making R_E large compared with h, the voltage gain

$$G_v = \frac{R_L}{R_E} \tag{4.2}$$

depends only on the ratio of the load resistor to the series emitter resistor.

The principal advantages of the *CB* amplifier are a very linear response from a current source, an ability to withstand high peak collector voltages, and a frequency response that is constant almost to the f_T frequency of the transistor. Because the *CB*-stage Q-point and gain depend on α instead of β, the amplifier offers a high stability of the Q-point, power gain, and input impedance.

A *CB* stage has power gain only when the output impedance is higher than the input impedance, but the ease with which a high-frequency low-impedance source may be coupled into a high-impedance load makes the *CB* stage attractive for coupling cables into amplifiers or low impedances to high impedances. The collector signal is in phase with the input signal. Hence, there is no Miller cutoff; but with a high load impedance there may be a high-frequency cutoff caused by the collector capacitance C_{ob} across the load.

The *CB* amplifier shown in Fig. 4.3 is used to couple a video cable into a circuit impedance of 1 to 3 kΩ. The base is grounded so that both positive and negative supply voltages are required. By adjusting the emitter resistor, we bias the collector Q-point at 4 V, a little less than one half the supply voltage, because a

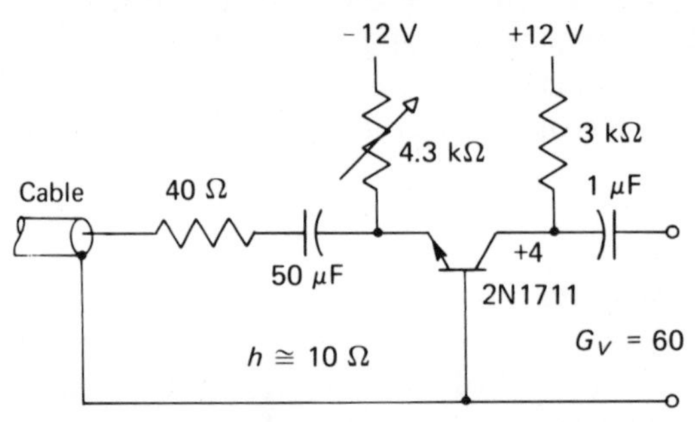

Figure 4.3. Video *CB* amplifier with base grounded.

low voltage allows slightly higher peak output signals when the amplifier is coupled to an external load. The emitter current is about 2.7 mA, which makes $h = 10\ \Omega$. The 10-Ω dynamic resistance with the 40-Ω series resistor terminates the cable in 50 Ω without reflecting a variable impedance, as when the input is tuned.

Signal linearity is ensured by making the emitter current at least 2 mA so that h is less than 13 Ω. This choice of the emitter current makes h less than 13 per cent of the total loop impedance, and a higher emitter current may be used if still better signal linearity is required. The no-load voltage gain of the amplifier is $\frac{3000}{50} = 60$, or nearly 36 dB. The 2N1711 transistor has an output capacitance of about 20 pF, which places the high-frequency cutoff at approximately 3 MHz. By reducing the collector load impedance to 300 Ω and accepting reduced gain, the cutoff can be moved to about 30 MHz. The 2N1711 has a minimum gain-bandwidth product of 70 MHz and a relatively high output capacitance.

A disadvantage of operating a *CB* transistor with the base grounded is that high-current positive and negative power supplies are needed to bias the transistor. By adding more components, the base may be operated off ground, as shown in Fig. 4.4, so that only a single polarity power supply is required.

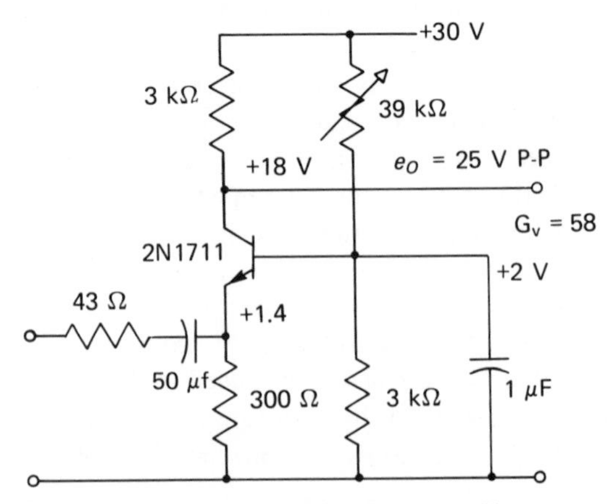

Figure 4.4. Video *CB* amplifier with base off ground.

The amplifiers in Figs. 4.3 and 4.4 should be compared for their similarities and for their differences. Both amplifiers use 3-kΩ collector resistors, but the second amplifier has about twice the emitter current. The higher emitter current decreases the dynamic input impedance slightly, but the main difference is that the second amplifier is linear to about three times the peak signal level of the first amplifier.

The base capacitor shown in Fig. 4.4 is not required in low-frequency applications. Removing the capacitor lowers the gain less than 3 dB, depending on the transistor β, but the high cutoff frequency is considerably reduced. At low frequencies the voltage gain is reduced only to the extent that H in equation (4.1) is increased by the external base resistor R_B. If $\beta = 200, R_B/\beta$ is 15 Ω, so removing the capacitor changes H from 43 + 6 + 0 to 43 + 6 + 15. Hence, the stage gain is reduced by the ratio 64/49 = 1.3, 2 dB. At frequencies above the β cutoff, the current gain decreases linearly with frequency, and at f_T/n the current gain is n. In the example, if $n = 60$, then $H = 43 + 6 + 50$, and the gain is down 6 dB at about 1 MHz.

4.3 A LINEAR SUMMING AMPLIFIER

The amplifier shown in Fig. 4.5 is designed to add signals from a series of signal sources without intröducing cross-feed from one source to another. This circuit takes advantage of the low input impedance and linear response of a *CB*-stage amplifier. This summing amplifier has a wide-band frequency response, excellent wave form up to 10-V peak-to-peak output signals, and a voltage gain of 1 from each source to the output. In low-frequency applications additional gain may be obtained by increasing the supply voltage and the load resistor.

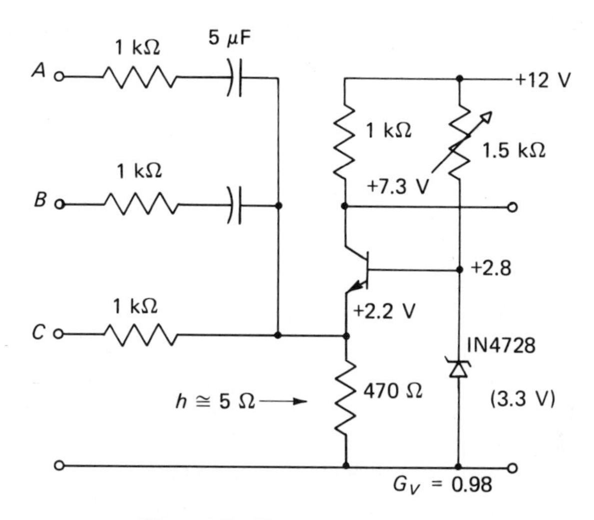

Figure 4.5. *CB* summing amplifier.

The emitter current is fixed by the emitter resistor and by the Zener diode. Because the 5-mA emitter current makes h equal to 5 Ω, the signal at the emitter summing point is the fraction 5 Ω/R_S of the signal voltage. The fraction of the emitter signal that appears at one of the other inputs is attenuated further by R_S and depends on the impedance presented by a signal source. Even with high source impedances, the crossover signal from one channel to another is less than 1 per cent.

The input connections for signal sources A and B are shown as individual 5-microfarad (μF) capacitors. To prevent signal loss, the impedance of these capacitors should be about 1000 Ω, or less, at the lowest frequency of interest. If a capacitor is used in common with all inputs, the capacitor must have an impedance that is small compared with the emitter input impedance at the lowest frequency of interest. The input impedance is 5 Ω. Hence, the separate capacitors are a factor of 200, smaller than a common capacitor.

Any number of inputs may be added so long as the signal sum does not overload the collector circuit, which tolerates up to nearly 10 V peak-to-peak. The Zener diode is used to make the emitter circuit H-factor a low impedance. For β greater than 100, the impedance of the Zener diode need only be less than 100 Ω, and, therefore, the diode is operated on the knee of the curve where the dc current is low and the dc voltage is about 15 per cent below the diode voltage rating. An adjustment of R_A is needed only for large-signal operation.

The input C is shown without a capacitor, as this input may be used to change the collector dc voltage level in response to a level-changing command. The inherent stability of the *CB* stage makes this circuit a practical way of combining static (dc) controls or ac signals.

4.4 *CB*-AMPLIFIER SUMMARY

The *CB* amplifier is characterized by a low input impedance, a linear response, and excellent Q-point stability. The amplifier has a unit current gain, so there is power gain only when the load impedance is greater than the input impedance. Practical load-to-input impedance ratios generally result in voltage gains of 3 to 30 and power gains of 10 to 30 dB. A series of iterated *CB* stages cannot give any additional gain with the same source and load impedances unless interstage transformers are used to couple stages.

The *CB* amplifier is easily designed and is relatively independent of temperature. It offers a wide-band frequency response and freedom from the Miller effect, and should always be considered when coupling low-impedance sources to high-impedance circuits. The high-frequency limit of the amplifier is fixed either by the gain-bandwidth product f_T of the transistor or by the collector-to-base capacitance C_{ob} and stray capacitance across the load.

4.5 COMMON-COLLECTOR AMPLIFIERS

The *CC* amplifier, or emitter follower, is a broad-band amplifier used to couple high-impedance circuits into low-impedance loads. The amplifier transfers the input voltage to the output load without requiring voltage step-down, as in a transformer. Thus, the *CC* stage provides an impedance step-down and a current gain that makes the ac emitter voltage follow the base signal with a loss of only 1 or 2 per cent.

The current gain of a *CC* transistor alone is β, which means that a *CC* stage should have significant feedback to limit the dc current gain. Without *S*-factor control, the *Q*-point of a *CC* stage varies with β and temperature, and both the input impedance and the ac load current vary similarly with β and temperature.

A *CC* amplifier is made to be independent of β by using a low *S*-factor circuit design similar to the example shown in Fig. 4.6. For operation within a limited temperature range, the *S*-factor of the *CC* stage may be designed to be as high as one half the minimum β that is expected over the temperature range. Hence, with transistors that have current gains exceeding 100, a *CC* stage may have an *S*-factor as high as 50. Because the load is in the emitter circuit, the dc emitter voltage is usually high enough to make the V_{BE} voltage change with temperature relatively unimportant.

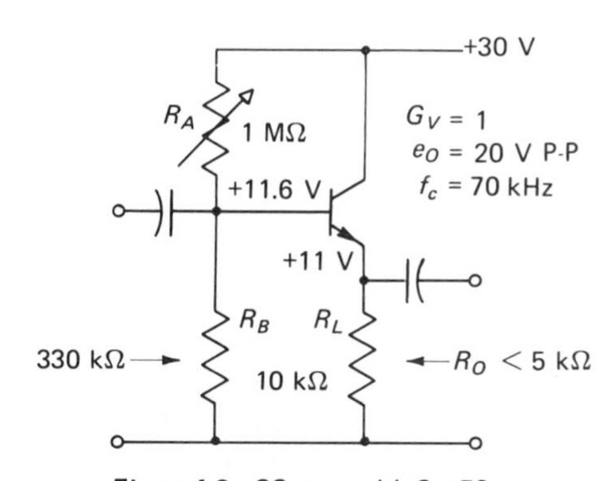

Figure 4.6. *CC* stage with $S = 50$.

For large-signal outputs we may make the bias resistor R_A about two thirds the base resistor R_B so that the emitter *Q*-point voltage is a little less than one half the collector supply voltage. This *Q*-point allows a maximum signal output. With the dc emitter voltage at the midpoint, the power dissipation is the same in the transistor as in the resistor and is easily calculated. The *S*-factor of the *CC* stage is the parallel value of R_A and R_B divided by R_L, and the input impedance is then SR_L.

The base-to-emitter voltage gain of the CC stage at low frequencies is

$$G_v' = \frac{R_L}{R_L + h} \tag{4.3}$$

and a simple calculation shows that h may be neglected any time the dc voltage drop across the emitter resistor is large compared with 26 mV. This calculation shows that the base-to-emitter voltage loss is made negligibly small by making the dc emitter voltage large. Generally, the voltage loss from the base to the emitter can be neglected, and the loss from a high impedance source to the base may be calculated by neglecting the transistor as a load on the bias resistors. Obviously, the voltage loss exceeds 0.5 if the source resistance exceeds the parallel equivalent of the two bias resistors.

The design of an emitter follower is very simple. The designer need only select the S-factor and calculate the transistor power dissipation. The S-factor is selected in the same manner as for a CE stage, except that higher values are generally acceptable. If the transistor is direct-coupled to a prior stage, the equivalent base resistor of the CC stage is the load resistor of the prior stage. If the stage is capacitor-coupled, the base and bias resistors are determined in the same way as for a CE stage, except that the base voltage should be one half the supply voltage whenever large output signals are required.

The output impedance of an emitter follower is the low impedance we see as we look back into the emitter. Neglecting the emitter resistor R_E. the impedance may be calculated as H, where

$$H = h + \frac{R_S}{\beta} \tag{4.4}$$

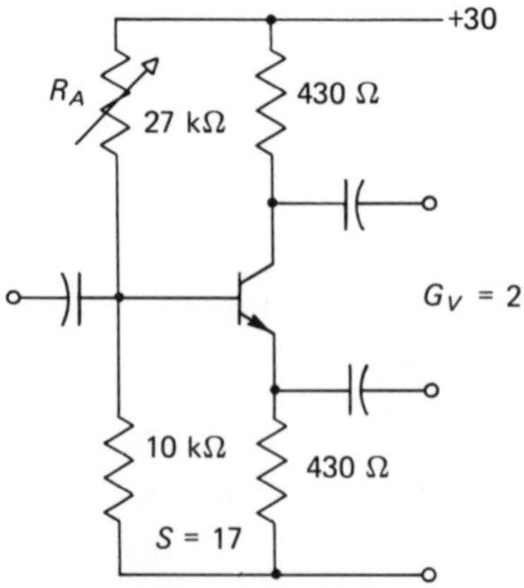

Usually h may be neglected, so that the internal output impedance is the source impedance reduced by β of the transistor. This result means that with S-factor control the internal output impedance is at least an order of magnitude lower than the value of R_E.

A CC amplifier may be used as a power amplifier when the load impedance is low and a low driving impedance is desired. In a similar application, a CC stage may be used as a phase inverter to drive a low-power push-pull amplifier, or simply to offer a choice of signal polarity. The phase inverter illustrated in Fig. 4-7 is designed as am emitter follower, except

Figure 4.7. Phase inverter.

that equal-value resistors are used in the collector and emitter circuits and the Q-point voltage is reduced. For a maximum signal on both phases, the dc emitter voltage should be about one fourth the supply voltage. Hence, R_A is approximately two and one half to three times R_B. The signals on the collector and the emitter both equal the base input signal, so a phase inverter has a net gain of 2. An emitter follower is sometimes observed to have a resistor in the collector lead, even though no use is made of the collector signal. The purpose of the resistor is either to limit the collector dc power dissipation by lowering the collector voltage, or to prevent second breakdown by limiting the collector voltage when the collector current is high.

4.6 *CC*-STAGE FREQUENCY LIMITATIONS

When operated from a relatively low-impedance source, the *CC* stage has a high-frequency cutoff at a frequency that equals the transistor current gain-bandwidth product divided by the S-factor. This cutoff frequency is

$$f_c \cong \frac{f_T}{S} \tag{4.5}$$

When operated from a high source impedance, the transistor output capacitance C_{ob} shunts the input and may lower the cutoff frequency below that given by equation (4.5).

4.7 *CC*-AMPLIFIER SUMMARY

A *CC* stage is useful for coupling impedances with step-down ratios of 10 to 50. The input impedance is S times the load impedance, and the input voltage is transferred to the load with a negligible base-to-emitter voltage loss. The power gain of the *CC* stage is about 10 to 15 dB. Additional gain and higher impedance ratios are obtained by cascading as many as four and five *CC* stages.

CC stages are either used as high-impedance input stages or as low-impedance output stages. Because the emitter follows the base and the dc Q-point voltage is unimportant in small-signal applications, an emitter follower may often be direct-coupled to a *CE* stage without concern for biasing. The emitter signal voltage is the same as the prior-stage signal. Hence, the same dc Q-point voltage is usually satisfactory.

The upper frequency limit of a *CC* stage is determined either by the gain-bandwidth product f_T of the transistor or by the loading effect of C_{ob} at the input. A *CC* stage reflects impedances both ways and may cause unexpected resonance effects. Above the β cutoff frequency, a *CC* stage may exhibit negative resistance effects and break into oscillation.

The outstanding advantages of the *CC* stage are its simplicity, wide frequency response, and the ease with which it may be direct-coupled to almost any point in a circuit, regardless of the dc voltage.

Additional advantages of the *CC* stage are that the ac ground on the collector eliminates the Miller effect, and the emitter feedback reduces the effect of the transistor input capacitance. These effects increase the upper cutoff frequency about 2 orders of magnitude above the cutoff frequency of a *CE* stage with an equal input impedance. Whenever an amplifier has a limited frequency capability, a *CC* stage should be considered, either to lower the circuit impedance for succeeding stages or to drive the load directly as a replacement for a *CE* stage.

4.8 ITERATED *CE* STAGE DESIGNS

An important unit for amplifier design is the iterated stage, which is one of a series of identical stages. Once an iterated stage is designed, we know the stage gain and how the stage is affected by the related stages. With this information a designer can adjust his design for different source or load impedances, knowing how the adjusted circuit will perform. In this and succeeding chapters we consider mainly the design of iterated stages.

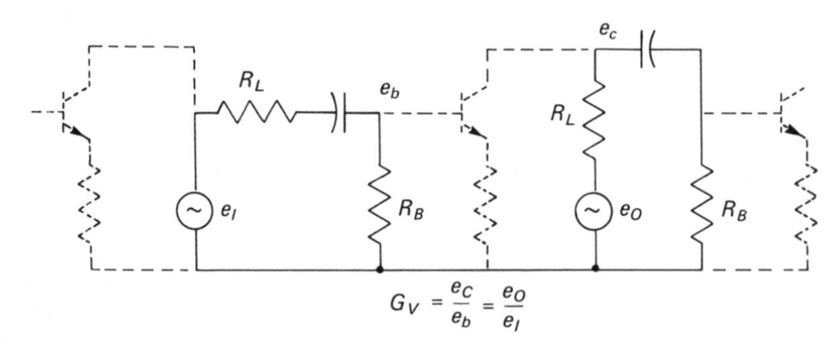

Figure 4.8. Iterated stage as constructed.

A single intermediate stage, one of a series of iterated stages, is represented in Fig. 4.8. The stage is driven by a prior stage that has the internal impedance R_L, the same as the load impedance of the intermediate stage, and is loaded by the input impedance R_B of a following stage. The voltage gain of the stage may be measured by comparing the collector voltage e_c with the base voltage e_b when R_B is in place, just as the gain is measured in a series of such stages. However, if R_B is removed, the output voltage e_c increases to the open-circuit ac voltage e_O by exactly the same factor by which the source voltage e_I is larger than the base voltage e_b. In other words, the iterated gain may be calculated or measured as the ratio e_O/e_I, and the circuit is simplified as in Fig. 4.9 by the removal of the following stage.

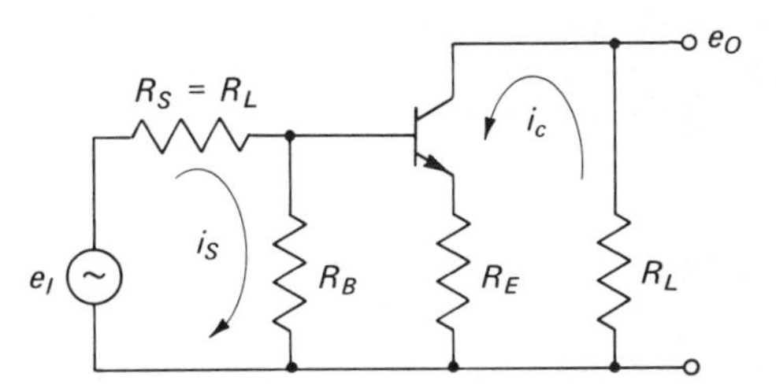

Figure 4.9. Iterated stage for gain calculation.

Consider the single stage shown in Fig. 4.9 and assume that the input impedance may be represented by the resistor R_B. The input impedance seen by the source e_I is $R_S + R_B$. If the current gain i_c/i_S of the stage is S then the TG-IR states that the voltage gain e_O/e_I is given by

$$G_v = S\frac{R_L}{R_S + R_B} \tag{4.6}$$

For an iterated stage the source and load impedances have the same value, $R_S = R_L$, and equation (4.6) reduces to the form

$$G_v \cong S\frac{R_L}{R_L + R_B} \tag{4.7}$$

Equation (4.7) shows that when R_B and R_L are equal, the stage voltage gain is $S/2$. In Chapter 5, when we consider the practical problems of *CE*-stage design, we find that in a well-designed amplifier R_B is approximately equal to R_L. Equation (4.7) shows that design economy is achieved by making the S-factor as high as possible while keeping R_B less than R_L, but fails to show that low values of R_B lead to biasing problems. Moreover, there is no advantage in making a stage difficult to operate in order to get a small gain increase, because a single emitter capacitor increases the gain more than can be obtained by the optimum design of even two or three iterated stages. In Chapter 5 we show that the iterated gain of a collector-feedback stage may approximately equal the equivalent ac S-factor. The TG-IR shows that the iterated voltage gain of a stage cannot exceed the equivalent current gain, no matter how obtained.

Summarizing iterated-stage design, we can assert that practical circuits are obtained by making $R_B = R_L$ and making the S-factor as high as the required Q-point and bias stability permit. The voltage gain of an iterated stage is approximately $S/2$. Direct coupling of iterated stages allows removal of R_B, thus increasing the iterated-stage gain to approximately S, the S-factor. An emitter-bypass capacitor increases the ac gain to β and thus increases the iterated voltage gain to approximately β.

Useful power gain cannot be obtained with iterated *CB* or *CC* stages because these stages develop power gain only when the input and load impedances are unequal. Because almost any amplifier stage is one of a series, the *CE* stage is of more general use than either the *CB* or *CC* stages.

4.9 INTERSTAGE COUPLING METHODS

The principal means of interconnecting transistor stages are by the use of transformers, resistance-capacitance coupling, direct coupling, and double-tuned (resonant) transformers. Transformer coupling permits power to be transferred at a high efficiency from one impedance level to another but has the disadvantage that good low-frequency and wide-band transformers are bulky and expensive. Because the signal power is conserved, a transformer may be considered as a device for increasing the current gain of an amplifier. When coupled between a current source and a load, the transformer current gain is proportional to the turns ratio and is in addition to the individual-stage current gains. Transformers are sometimes used for voice-frequency applications, as in hearing aids, to obtain increased power gain by impedance matching when there is a need to conserve dc collector power. Transformers are used also in an amplifier that is capable of delivering the required power only at an impedance level that is different from the given load impedance.

Resistance-capacitor coupling is used at audio frequencies when the desired low-frequency response can be obtained without requiring bulky capacitors. Because of the low dc voltages in transistor amplifiers, electrolytic capacitors exceeding 100 μF are often used for interstage coupling. With interstage impedance levels of 10 kΩ the designer has considerable freedom of choice in selecting the capacitor. At the low audio frequencies, say 30 Hz, the required capacitor is about 1 μF, which offers a choice between a small sized electrolytic or a large paper capacitor. On the other hand, a smaller paper capacitor may be used by raising the circuit impedance level or by raising the low-frequency cutoff. Field-effect transistors permit coupling impedances that are higher by approximately a factor of 10 and allow the use of relatively small 0.1-μF mylar capacitors.

Direct coupling of transistor stages has several important advantages. Some form of direct coupling is, of course, the only way that permits the amplification of dc and subaudio frequencies. Direct-coupled amplifiers are readily available as integrated circuits, and for many applications these circuits simplify design and are priced far below the cost of a custom-designed amplifier. On the other hand, special-purpose-instrument amplifiers and many audio amplifiers are better and simpler when custom designed with only one or two stages.

An amplifier that is direct-coupled does not necessarily have the same high gain at dc as at the midband frequencies. Two- and three-stage amplifiers are often direct-coupled to eliminate components and to obtain higher gain than is otherwise possible. The designer then stabilizes and reduces the Q-point drift by dc feedback but uses less feedback at the midband frequencies. This type of direct-coupled amplifier is recognized by finding a capacitor that is used to reduce feedback at the signal frequencies.

At radio frequencies amplifier stages are often coupled by bringing a tuned *L-C* circuit as the collector load into the proximity of a tuned *L-C* circuit in the following stage base circuit. High-*Q* tuned circuits transfer power to each other at high efficiency in a relatively narrow frequency band. The design of a loosely coupled tuned circuit is very different from the design of a closely coupled transformer, and correct tuning is important for proper operation of coupled circuits. However, tuned coupled circuits provide voltage and current changes, as with a transformer but the purpose of a tuned circuit is generally to provide a narrow frequency band, dc isolation, and a power loss to reduce interaction between alternate stages. The design of tuned coupling circuits is outlined in Appendix A-4.

4.10 FREQUENCY RESPONSE WITH *R-C* COUPLING

One reason for using resistance-capacitor coupling is that the capacitors may be sized to eliminate unwanted low- and high-frequency signals or noise. The low cutoff frequency of an *R-C* coupled amplifier is determined by the series-coupling capacitor. The high cutoff frequency is determined by the shunt capacitors and by the input capacitance of the transistor itself.

The elements of a typical *R-C* coupled wide-band amplifier are illustrated in Fig 4.10. At midband frequencies the reactance of the coupling capacitor C_C is negligibly small compared with the circuit impedances, so the midband signal loss of the coupling network may be calculated by replacing C_C by a short circuit. However, because the midfrequency loss is taken into account whenever the iterated gain is known, the designer rarely, if ever, actually calculates the loss factor.

The effect of the coupling capacitor at low frequencies is calculated by replacing the transistor *TR*-1 and the load resistor R_L by the Thévenin equivalent voltage source, as shown in Fig. 4.11. The low frequency at which the coupling capacitor has sufficient reactance to produce a 3-dB loss is the frequency at which the ac current through C_C is 71 per cent of the midfrequency value. As is well known, the loop current shown in Fig. 4.9 is reduced to 71 per cent of the maximum value when the reactance of the coupling capacitor equals the total loop resistance $R_L + R_B$. When expressed by an equation, the low cutoff frequency is

$$f_l = \frac{1}{2\pi C_C(R_B + R_L)} \tag{4.8}$$

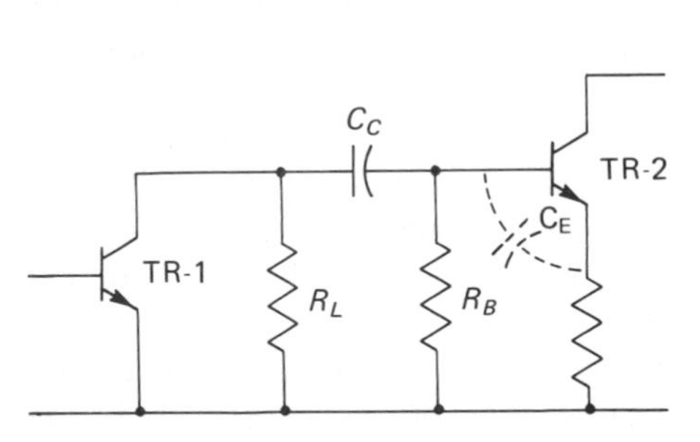

Figure 4.10. *R-C* coupling circuit.

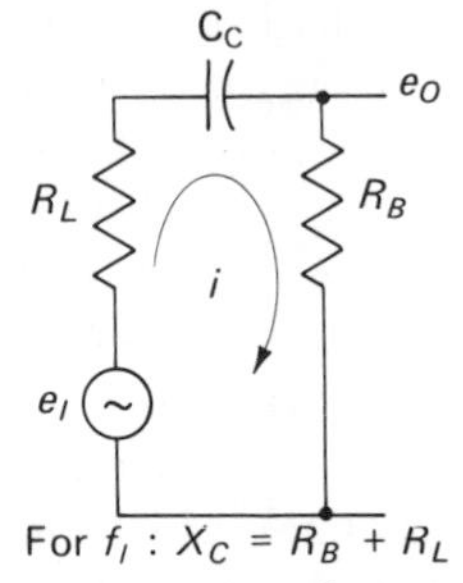

Figure 4.11. Low-frequency equivalent circuit.

Generally, the simplest way to find f_l is to find the frequency on a reactance chart.

A capacitor is sometimes connected to ground from the collector or the base in order to eliminate frequencies above an upper frequency limit f_h. Usually, the shunt capacitor is smaller than one fifth of the coupling capacitor, in which case the frequency band is so wide that the low and high cutoff frequencies may be calculated separately. For this wide-band case, the base and load resistors are effectively in parallel, as in Fig. 4.12 and the shunt capacitor reduces the signal only by reducing the circuit impedance below the parallel equivalent of R_B and R_L. The 3-db loss falls at the frequency f_h, for which the reactance of the capacitor equals the equivalent parallel resistance. Algebraically, we have

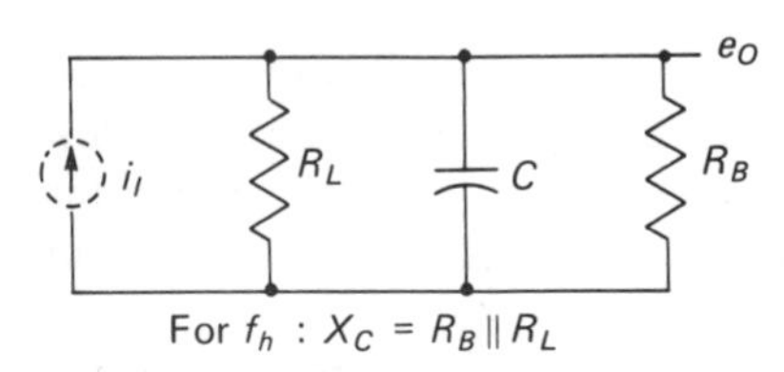

Figure 4.12. High-frequency equivalent circuit.

$$f_h = \frac{R_L + R_B}{2\pi C(R_L R_B)} \tag{4.9}$$

Again, however, f_h is more easily found by referring to a reactance chart.

In many instances the high-frequency cutoff of an *R-C* coupled amplifier is surprisingly low, even when the shunt capacitor is not in evidence. In these cases the high-frequency cutoff is caused by the input capacitance of transistor *TR*-2 (Fig. 4.10) and the Miller effect. The high-frequency response of transistors is discussed in more detail later. For the present we assert that the high-frequency cutoff of a high-frequency *CE* stage is no higher than the gain-bandwidth product of the transistor f_T divided by the stage *S*-factor. The Miller effect, which increases with the stage voltage gain, may lower the high-frequency cutoff as much as an order of magnitude below f_T/S.

By suitably proportioning the shunt and series capacitors in an *R-C* coupled amplifier, we may reduce the bandwidth to approximately twice the midband frequency, an equivalent Q of 0.5. If R_L and R_B are equal, the maximum Q is 0.36 when the total shunt capacitance is twice the coupling capacitance. Either increasing or decreasing the capacitor ratio only increases the bandwidth. Narrower frequency bands are best obtained by using tuned L-C collector loads.

4.11 THE DESIGN OF A VIDEO AMPLIFIER

As an example of practical circuit design consider the problem of building an amplifier to amplify a video signal in a long cable. Assume that the cable impedance is given as 50Ω in both directions. Assume also that the amplifier is required to handle small signals and to have a simple design.

We know that a *CB* stage would be satisfactory as an input stage, but a load impedance that is needed for voltage gain is too high to be coupled into the outgoing cable. We also know that a *CC* stage can be used to couple the *CB* stage into the cable. As a first trial we decide to direct-couple the *CC* stage shown in Fig. 4.6 to the *CB* stage shown in Fig. 4.4. We soon discover that the *CC* stage should have all impedances lowered by a factor of 10 or more in order to couple into the outgoing cable.

As a second trial design we decide to use a 100-Ω emitter resistor in the *CC* stage in order to make the internal output impedance an order of magnitude lower than the cable impedance. The amplifier output impedance is easily adjusted by inserting a series resistor. If the *CC* stage is direct-coupled to the *CB* stage shown in Fig. 4.4, the S-factor is 30, a reasonable value. However, in the final circuit, Fig. 4.13, the collector load resistor is lowered to 1000Ω in order to raise the amplifier cutoff frequency, and the *CC* S-factor is reduced to 10. The final amplifier has a power gain of approximately S^2, which is 20 dB.

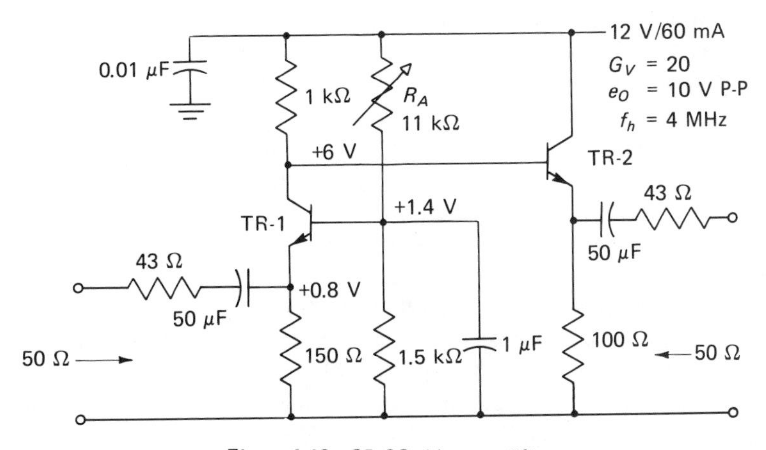

Figure 4.13. *CB-CC* video amplifier.

The maximum transistor power dissipation occurs when the power loss in the transistor equals the loss in the emitter resistor. A 12-V supply voltage was chosen because six volts across 100 Ω is 0.36 W, a reasonable value for TR-2 and for a 1-W resistor. A performance test shows that the amplifier has a voltage gain of nearly 20 and a high-frequency cutoff at 4 MHz. The design seems like a satisfactory circuit for trial as a cable amplifier.

4.12 SUMMARY

A multistage amplifier is usually a series of separately designed stages, each with local feedback. The required voltage gain of an amplifier may be obtained by using iterated *CE* stages or by using *CB* stages in which the load impedance is greater than the input impedance. Impedance step-down may be obtained at the input, the output, or between stages by using *CC* stages. The *CB* and *CC* stages have the advantage of a wide-band frequency response but the disadvantage of relatively low power gain. The *CE* stage has the advantage of providing voltage gain in iterated stages or in stages having a moderate input-output impedance step-down. The *CE* stage offers greater power gain but a much more restricted frequency response than when the same transistor is used in a *CB* or *CC* stage.

The *CB* stage has a voltage gain that is approximately the ratio of the load impedance to the circuit H-factor. Because the current gain of the *CB* stage is 1, there is voltage and power gain only when the load impedance exceeds the impedance H of the emitter loop. The *CB* stage has a very wide-band response and is often used to couple low-impedance cables into high-impedance amplifiers.

The *CC* stage is used to couple high-impedance circuits into low-impedance loads. A *CC* stage transforms the base voltage to the load with a negligible loss, provided the S-factor of the stage, generally 10 to 50, is significantly smaller than the transistor β. The main signal loss in a *CC* stage is caused by the bias network as a load on the signal source. A *CC* stage has a wide-band frequency response, is free of the Miller effect, and is easily direct-coupled to adjacent stages.

Gain in a single-stage transistor amplifier is obtained between equal source and load impedances only by using a *CE* stage. Practical iterated circuit designs produce stage gains between $S/2$ and S, depending on the interstage coupling loss. When ac feedback is not required for ac gain stability, the stage gain may be increased to β by bypassing the feedback.

The simplicity, stability, and broad-band advantages of the *CC* stage should be exploited by using as many *CC* stages in an amplifier as practical. The voltage-gain requirements of an amplifier (without transformers or *CB* stages) must be supplied by *CE* stages, and the impedance step-down from input to output usually can be supplied by *CC* stages. The best division of an amplifier into *CE* and *CC* stages may be determined by separating the voltage-gain and current-gain requirements and using only enough iterated *CE* stages to provide the required voltage gain. The impedance change from the input to the output should be supplied by *CC* stages at the input for high input impedance or at the output for low output impedance.

PROBLEMS

4-1. Compare the amplifiers shown in Figs. 4.3 and 4.4, as suggested in Section 4.2. What changes should be made in order to operate the amplifier shown in Fig. 4.4 on a +15-V supply?

4-2. Show by the TG-IR that neither a *CB* nor a *CC* stage produces a voltage gain in a series of iterated stages. Show that a *CE* stage may produce an iterated voltage gain.

4-3. Explain what happens if the point *C* in Fig. 4.5 is (a) connected to ground; (b) connected to +12 V.

4-4. What are the low-frequency and high-frequency cutoffs, f_l and f_h, for the amplifier shown in Fig. P4.4? If the second stage has a collector-to-base capacitance of 30 pF, what is the frequency f_h?

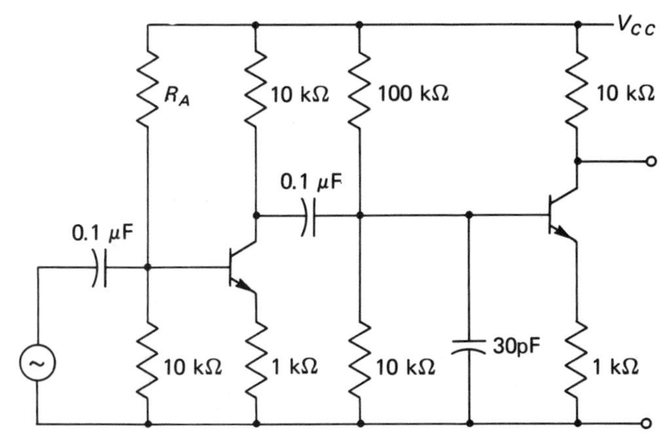

Figure P 4.4

4-5. When the *CC* stage shown in Fig. 4.6 is driven by a current source, the cutoff frequency is at 35 kHz. (a) What value of C_{ob} will account for the observed cutoff frequency? (b) How should the design be changed in order to increase this cutoff frequency by a factor of 2?

DESIGNS

4-1. Design and evaluate a 500-Ω line amplifier that operates on a +20-V supply.

4-2. Construct a two-stage *CE-CE* amplifier and show that you can account for the observed cutoff frequencies f_l and f_h.

4-3. Design a preamplifier for your oscilloscope that gives a voltage gain of 10 without degrading the frequency response. Assume that the amplifier is to be used with source impedances that are less than 10 kΩ.

REFERENCES

4-1. Related materials may be found in textbooks on transistor and vacuum-tube electronics.

4-2. Terman, F. E., *Electronic and Radio Engineering*, 4th ed., Chapter 8, "Voltage Amplifiers for Audio-Frequencies." New York: McGraw-Hill Book Company, 1955.

5

COMMON-EMITTER AMPLIFIERS

The *CE* stage is generally preferred for amplifier applications because this circuit has a high input impedance and offers power gain when used in a series of identical stages. This chapter begins with a review of the single-stage *CE* amplifier without feedback to illustrate the problems that make feedback necessary in practical circuit design. Single-stage designs using collector feedback are discussed and are shown to be improved by the addition of emitter feedback. A combination of collector and emitter feedback has several advantages over the emitter feedback that has been so popular.

5.1 *CE* STAGE WITHOUT FEEDBACK

The simplest form of the *CE* amplifier, which is shown in Fig. 5.1, is impractical for many applications. However, this amplifier is sometimes used because it provides high gain and requires only two resistors and a small coupling capacitor. Moreover, a study of the gain and impedance characteristics of the amplifier reveals useful information about single-stage amplifiers and the problems of their design.

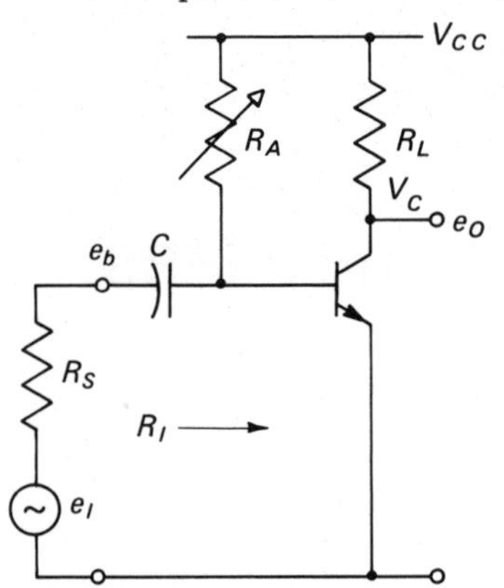

Figure 5.1. Common-emitter amplifier.

This *CE* amplifier has only a load resistor R_L and a single bias resistor R_A. The bias current is proportional to the supply voltage. Therefore, the collector current is proportional to β and to the supply voltage. Since a *CE* stage without feedback is not generally satisfactory for large-signal operation, the Q-point may be set to accommodate the expected temperature change of β and its effect on the Q-point. Because the bias current varies with the supply voltage, the collector voltage tends to remain a fixed fraction of the supply voltage as long as β is constant. Generally, for small-signal operation the collector Q-point voltage V_C should be between $V_{CC}/4$ and $V_{CC}/2$.

The collector Q-point voltage V_C is the supply voltage V_{CC} less the dc voltage loss in R_L. Written as an equation,

$$V_C = V_{CC} - \beta \frac{V_{CC}}{R_A} R_L \tag{5.1}$$

and

$$V_C = V_{CC}\left(1 - \beta \frac{R_L}{R_A}\right) \tag{5.2}$$

Equation (5.2) shows that V_C is zero if $R_A/R_L = \beta$, which suggests a way of measuring β. If the collector Q-point voltage V_C is one half the supply voltage, then $V_C = V_{CC}/2$ and $R_A = 2\beta R_L$. The last result shows that, for a fixed collector voltage, R_A depends on β. Therefore, the bias resistor must be selected after the transistor is placed in the circuit. Now, if V_C is made $V_{CC}/2$ and β subsequently doubles with increased temperature, the Q-point moves to $V_C = 0$.

Since β may easily vary over a 2:1 range, these relations show that the *Q*-point voltage may vary more than can be tolerated in many applications. In room-temperature conditions, 0 to 60°C, the *Q*-point can be set to accommodate the usual temperature changes, but a *CE* stage is much more satisfactory when stabilized by collector feedback.

The advantage of this *CE* stage, if any, is that the iterated voltage gain is relatively high. The main disadvantage is that the gain varies with β, and the input impedance varies with both β and h. However, we can obtain an estimate of the input impedance that is useful for design purposes. By the TG-IR, the base-to-collector gain of a *CE* stage is

$$G_v = \beta \frac{R_L}{R_I} \tag{5.3}$$

The desired relation is obtained by solving equation (5.3) for the input impedance and substituting for G_v the value given by equation (3.15). These changes give

$$R_I \cong \frac{\beta R_L}{[25 I_E R_L]} \tag{5.4}$$

Simplifying further, we can assume that β equals 50 and reduce equation (5.4) to the approximation

$$R_I \cong \frac{2R_L}{[V_{CC}]} \tag{5.5}$$

The usefulness of equation (5.5) stems from the fact that it gives an approximate value for the intrinsic input impedance of any *CE* stage biased as a linear amplifier. The equation shows that for ordinary values of the supply voltage the input impedance is an order of magnitude smaller than the load impedance. This means that the input impedance of an iterated stage is small compared to the collector load resistance and explains the small signals often observed at the input of a *CE* stage.

By equation (5.3), a *CE* stage designed to operate between an equal source and load impedance cannot have a voltage gain in excess of the transistor current gain. Therefore, any measurement indicating a voltage gain exceeding β implies that the amplifier input impedance is less than the load impedance. Such an amplifier will show a gain of less than β when operated in a series of stages between equal load and source impedances. The excess gain indicated by equation (3.15) is better used as local feedback than lost in the impedance mismatch implied by equation (5.5).

5.2 *CE*-STAGE SUMMARY

The characteristics of a single-stage amplifier may be summarized as follows:

1. The operating *Q*-point of a *CE* stage depends on the transistor selected, so the bias resistor must be individually selected with concern for the expected junction temperature changes. As a rule, a *CE* stage is unsatisfactory without feedback if the expected temperature range causes β to change more than 2:1.
2. The iterative voltage gain of a *CE* stage is approximately β, and a voltage gain exceeding β implies that the source impedance is less than the load impedance.
3. Both the input impedance and the voltage gain of a *CE* stage tend to vary with the collector supply voltage. The intrinsic input impedance is usually less than one tenth the load impedance and may be calculated by equation (5.5). A higher input impedance may be obtained by inserting a series base resistor and accepting decreased voltage gain or by using emitter feedback.
4. Because of the disadvantages, a *CE* stage should not be used without feedback.

5.3 THE *CE* STAGE WITH COLLECTOR FEEDBACK

Feedback brings the output signal of an amplifier back to the input through a resistor network and forces the gain of the amplifier to depend, partly at least, on the feedback resistors. The extent to which the gain becomes less dependent on the transistors is nearly proportional to the gain reduction produced by the feedback.

The simplest way of applying local feedback in a *CE* stage is to connect the bias resistor to the collector, as shown in Fig. 5.2. This method of biasing a stage makes the bias current dependent on the collector current. If the transistor β increases after the bias is set, the increased collector current reduces both the collector voltage and the bias current. Because the bias change tends to reduce the collector current change, the stage has current feedback, and the current gain is less than β, the current gain without feedback.

The current gain with collector feedback is determined by the transistor β and by the *S*-factor, which is defined as

$$S \equiv \frac{R_f}{R_L} \tag{5.6}$$

In Section 5.6 we show that the current gain is given by the corrected S-factor S_c, where

$$S_c = \frac{S\beta}{S + \beta} \tag{5.7}$$

When S is small compared with β, the current gain S_c is approximately S. When S is not small compared with β, the corrected S-factor is calculated by equation (5.7), as if calculating parallel resistances. When S and β are approximately equal, the current gain is approximately $S/2 = \beta/2$. Generally, with collector feedback the current gain can be estimated from the S-factor, since β is too variable to make a carefully calculated S_c really meaningful. Most collector feedback designs make S small compared with β, and for these designs the current gain is a little less than the S-factor.

5.4 THE COLLECTOR Q-POINT VOLTAGE AND THE S-FACTOR

The collector Q-point voltage of a CE stage without feedback can be adjusted without changing the amplifier current gain. With feedback, the Q-point voltage is adjusted by changing the feedback resistor R_f, and this changes S_c, the current gain. A relation showing how V_C varies with S_c is easily derived, and a curve representing the relation is shown in Fig. 5.3.

The base current in a collector feedback stage is I_C/β, which is only about one per cent of the collector current. Thus, a calculation of the collector voltage for the circuit in Fig. 5.2 is greatly simplified by assuming that the current in the

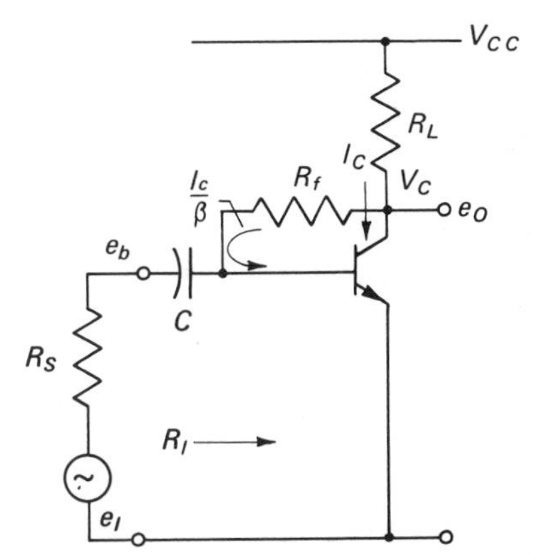

Figure 5.2. *CE* amplifier with collector feedback.

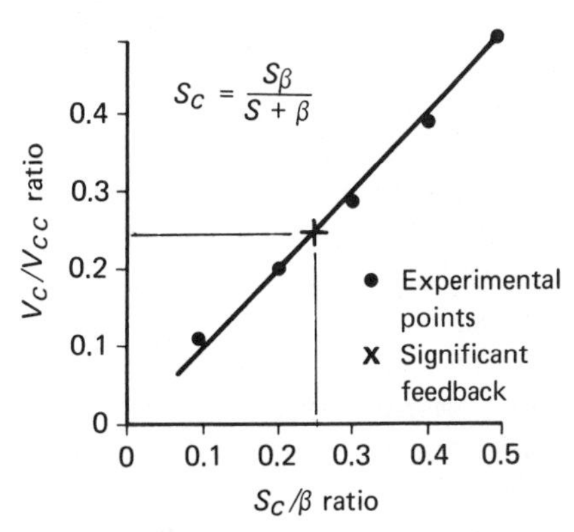

Figure 5.3. Collector voltage ratio with collector feedback.

load resistor is I_C instead of $I_C + I_B$. With this approximation the collector voltage with collector feedback is given by the equation

$$V_C = V_{CC} - I_C R_L \tag{5.8}$$

Replacing I_C by βI_B, we obtain

$$V_C = V_{CC} - \beta I_B R_L \tag{5.9}$$

and

$$V_C = V_{CC} - \beta \frac{V_C}{R_f} R_L \tag{5.10}$$

Solving equation (5.10) for V_C, we find

$$V_C = \frac{S}{S + \beta} V_{CC} \tag{5.11}$$

or

$$V_C = \frac{S_c}{\beta} V_{CC} \tag{5.12}$$

Equation (5.12) shows that the collector Q-point voltage is adjusted during design by selecting a suitable value for S_c. If a Q-point suitable for a large signal application is desired, S must equal β; and for room temperature conditions and with care in adjusting the bias, the S-factor may be as high as $\beta/2$ or even β. However, since V_C depends on β, the Q-point generally varies with temperature too much for large signal applications, and the most satisfactory and all-purpose Q-point is usually obtained with significant feedback.

Significant feedback requires $S = \beta/3$, which makes the current gain about 3/4 S and the collector voltage about $V_{CC}/4$. Thus, significant feedback is ensured by selecting a value of R_f, which makes V_C equal to one-fourth the supply voltage.

With a relatively fixed value of S_c, which is obtained by making $S < \beta$, V_C varies inversely with β, and a temperature change that doubles β only halves V_C. Thus, feedback produces a smaller and more desirable Q-point temperature change than is produced in a *CE* stage without feedback. With the Q-point adjusted for significant feedback the transistor β can increase or decrease a factor of 2 or 3 without producing faulty small signal performance.

A basic difficulty with collector feedback is that a low S-factor produces a low Q-point voltage, as shown by equation (5.12). When the collector voltage is less than 10 V, the base-emitter voltage drop cannot be neglected, and the observed V_C is about .5 V higher than the value given by equation (5.12). For collector voltages below 10 V, the collector voltage is given more accurately by the equation

$$V_C = \frac{S_c}{\beta} V_{CC} + 0.5 \text{ (volts)} \tag{5.13}$$

The improvements brought about by collector feedback are approximately in proportion to the S/β ratio. For example, if S is $\beta/10$, the stage gain is only about one-tenth as sensitive to a given change of β*, of temperature, or of the supply voltage. On the other hand, as the preceding analysis has shown, the designer's choice of the S-factor is somewhat limited, and the S-factor of practical circuit designs is usually between $\beta/3$ and $\beta/10$.

The collector voltage change with temperature is mainly caused by the change of β with temperature† and is usually 0.6%/°C, or less. Thus, the temperature change is proportional to the Q-point voltage and, when $V_C = 10$ V, is about 60 mV/°C. In a direct-coupled amplifier the voltage change with temperature may be reduced by reducing V_C, but this reduces the dc gain also. However, there may be an advantage in using a very low collector voltage, at which β decreases when V_C is reduced by temperature. Thus, the decreasing value of β tends to offset the Q-point temperature shift.

5.5 COLLECTOR FEEDBACK EXAMPLES

An example of the iterated collector feedback stage is shown in Fig. 5.4. The supply voltage is 32 V, and the feedback resistor was adjusted to make the collector voltage 8 V. Because the S-factor is $\beta/3$, we know that $\beta = 3(62) \cong 180$. The corrected S-factor is $\beta/4$, so we expect a voltage gain of $\beta/4 = 45$. The measured voltage gain is 43.

If the ambient temperature changes from 25 to 50°C, the collector Q-point of the amplifier drops about 1 V. Hence, the lower Q-point permits a peak signal level of about 5 V. At low temperatures the Q-point may increase to about 9 V.

*This can be shown by differentiating equations (5.1) and (5.12). By differentiating equation (5.12), the effect of a small change of β with collector feedback is given by

$$\frac{dS_c}{S_c} = \frac{S_c}{\beta}\frac{d\beta}{\beta}, \text{ and } \frac{dV_C}{V_C} = -\frac{d\beta}{\beta}$$

†For a typical planar silicon transistor β increases about 0.6%/°C. For a few, the temperature change may be as low as 0.2%/°C.

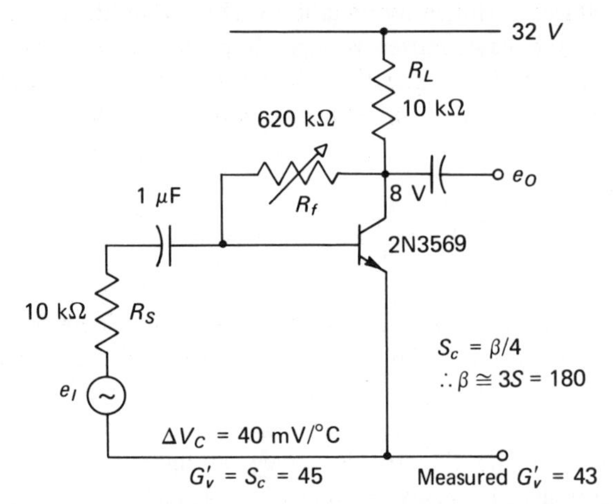

Figure 5.4. Collector-feedback stage with significant feedback.

For large temperature changes the measured Q-point temperature change is about 40 mV/°C, which is negligible in single-stage applications.

The circuit in Fig. 5.5 shows an example of a low S-factor design. The amplifier is operated on 90-V dc, obtained from 120-V ac by use of a half-wave rectifier and a capacitor input filter. The bias-circuit S-factor is 2.5, and equation (5.13) with $\beta = 100$ gives 2.7 V as the expected Q-point voltage. The stage is obviously intended for low-signal room-temperature applications. With the load switch open, the expected voltage gain is shown in Section 5.8 to be R_f/R_S = 100. With the load switch closed, the ac S-factor becomes R_f/R_L = 100, so the corrected ac current gain is about 50. This stage has the interesting

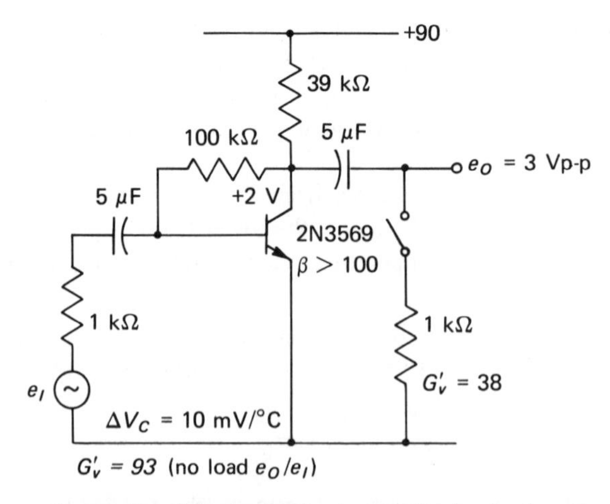

Figure 5.5. Collector-feedback stage (high gain, low S).

characteristic that the voltage gain is nearly independent of the load impedance, because the ac output impedance is reduced to about 1 kΩ by feedback. We note that the intrinsic no-load voltage gain is $[25V_{CC}] = 2500$, so the voltage gain and the output impedance are reduced approximately 30 times by the feedback.

5.6 *CE*-STAGE CURRENT GAIN WITH COLLECTOR FEEDBACK

The ac current gain of the stage represented in Fig. 5.6 may be calculated by dividing the change of the collector current by a corresponding change of the input current ΔI. The calculation is simplified by finding the collector current that exists when the input current is $\Delta I = 0$ and calculating the collector current that exists when ΔI is increased to make $V_C = 0$. The current and voltage values that exist with each input current are shown in Table I and are explained below.

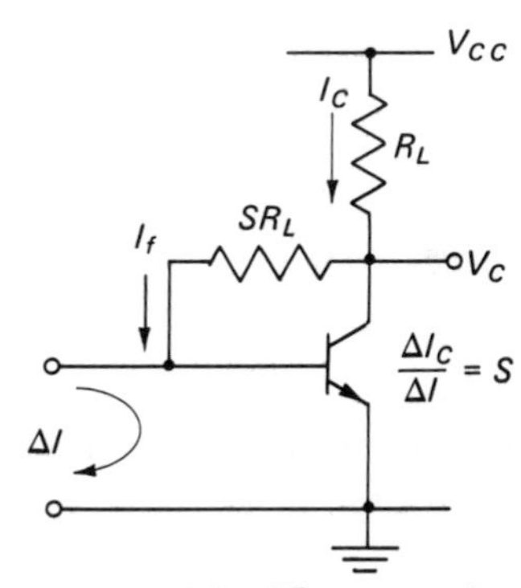

Figure 5.6. *CE* current gain.

TABLE I

Voltage and Current Relations In a Collector-Feedback Stage

Input Current ΔI	Collector Voltage V_C	Collector Current I_C	Collector-Current Change ΔI_C
0	$\frac{S_C}{\beta} V_{CC}$	$\frac{V_{CC} - V_C}{R_L}$	0
$\frac{I_C}{\beta}$	0	$\frac{V_{CC}}{R_L}$	$\frac{V_C}{R_L}$

When $\Delta I = 0$, the collector voltage is V_C, as given by equation (5.12). If I_C is increased to make $V_C = 0$, then the increase of collector current ΔI_C multiplied by the load resistance R_L is equal to the voltage change V_C. This statement written as an equation is

$$\Delta I_C R_L = V_C = \frac{S_c}{\beta} V_{CC} \tag{5.14}$$

From equation (5.14) we obtain

$$\Delta I_C = \frac{S_c V_{CC}}{\beta R_L} \tag{5.15}$$

When $V_C = 0$, the feedback current I_f is 0 and the input current is

$$\Delta I = \frac{I_C}{\beta} = \frac{V_{CC}}{\beta R_L} \tag{5.16}$$

The current gain of the stage G_i is the ratio $\Delta I_C / \Delta I$ which by equations (5.15) and (5.16) is

$$G_i = S_c \tag{5.17}$$

Equation (5.17) shows that the signal current gain is reduced by feedback to S_c, as given by equation (5.7).

5.7 INPUT IMPEDANCE WITH COLLECTOR FEEDBACK

The effect of feedback in lowering the input impedance is easily overlooked. This input impedance calculated in equation (3.8) as an example of the Miller effect, is

$$R_I = S_c h \tag{5.18}$$

Without feedback the input impedance is βh, so the feedback lowers the impedance from βh to $S_c h$. This input impedance is small compared with the iterated impedance, as shown in the next paragraph.

Consider the amplifier shown in Fig. 5.2, which has collector-to-base feedback. The signal voltage e_I is connected to the base through the series resistor $R_S = R_L$. The ac base voltage is e_b, and the output voltage is e_O. The voltage gain e_O/e_b is the intrinsic voltage gain of the CE transistor, which is given by equation (3.15) as

$$\frac{e_O}{e_b} = [25 I_E R_L] \tag{5.19}$$

The current gain of a collector feedback stage is given by equation (5.17) as S_c, which makes the overall iterated voltage gain $e_O/e_I = S_c$. Because the intrinsic voltage gain is much higher than the overall voltage gain, the base signal is much smaller than the input signal. And for a given output the base signal e_b is smaller than the input signal e_I by the ratio of the overall gain divided by the intrinsic gain. Thus,

$$e_b = \frac{S_c}{[25 I_E R_L]} e_I \tag{5.20}$$

Practical values of S_c seldom exceed 50; hence, e_b is small compared with e_I if $[I_E \ R_L]$ is 10 or larger. Therefore, for all practical purposes the input impedance with collector feedback is small compared with the iterated impedance $R_L = R_S$.

5.8 A COLLECTOR FEEDBACK GAIN RULE

If a collector feedback stage is replaced by a high-gain operational amplifier using the same feedback resistors, the voltage gain is given by the resistor ratio R_f/R_S. Similarly, under conditions specified by the collector feedback gain rule, the gain of a collector feedback stage may be evaluated by the same resistor ratio.

When the S-factor of a collector feedback stage is small compared with β, the collector Q-point voltage is a small fraction of the supply voltage, and equation (3.15) for the intrinsic voltage gain may then be written as

$$\frac{R_L}{h} = [25 \ V_{CC}] \tag{5.21}$$

Solving equation (5.21) for h and multiplying both sides by the S-factor, we find

$$Sh = \frac{R_f}{[25 \ V_{CC}]} \tag{5.22}$$

Assume also that the stage has an intrinsic voltage gain which is much larger than the resistor ratio R_f/R_S. From this statement we have

$$[25 \ V_{CC}] \gg \frac{R_f}{R_S} \tag{5.23}$$

Combining equations (5.22) and (5.23) we obtain the inequality

$$R_S \gg Sh \tag{5.24}$$

which shows that the amplifier input impedance is small compared with R_S. Therefore, the input signal must be small compared with the source signal, and the overall voltage gain is given by the resistance ratio R_f/R_S.

The *collector feedback gain rule* states that if S is small compared with the transistor current gain β and if R_f/R_S is small compared with the intrinsic voltage gain $[25\ V_{CC}]$, the voltage gain G_v' of a collector feedback stage is

$$G_v' = \frac{R_f}{R_S} \tag{5.25}$$

and the current gain is

$$G_i' = S \tag{5.26}$$

In a practical example it is easy to determine whether the gain rule is applicable by verifying the inequality (5.23) and by comparing the S-factor with β. When the gain rule applies, equation (5.25) shows that a collector feedback stage can be used as a low-gain operational amplifier. The gain rule is useful for evaluating amplifier gains and as a test for adequate feedback. With a 200 V collector supply the intrinsic gain is about 2500, and with a Darlington pair the transistor current gain can be at least 2500.

We observe that the gain rule does not apply when S is comparable with β, even though the base signal is small compared with the input signal. For such cases the gain should be calculated by using the corrected S-factor in the *TG-IR*; thus,

$$G_v' = S_c \frac{R_L}{R_S} \tag{5.27}$$

The stage shown in Fig. 5.5 presents an example for applying the gain rule. With the load switch open the intrinsic voltage gain is approximately 2500, whereas R_f/R_S is 100, and the S-factor is small compared with β. The gain rule indicates a voltage gain of 100, and we measure 93. With the switch closed, the corrected ac S-factor, approximately 50, is not small compared with β, so the gain rule does not apply. Equation (5.27) gives the expected voltage gain as 50 when the gain rule gives 100. The measured gain is 38, and the agreement with equation (5.27) improves as the supply voltage is increased to 200 V.

5.9 COLLECTOR FEEDBACK CIRCUIT DESIGN

When a collector feedback stage is operated with the collector resistor R_L as the load impedance and $R_S = R_L$, the iterated voltage gain is approximately the same as the dc S-factor. If high gain is obtained by increasing the S-factor to approximately equal β, the stage lacks significant dc feedback and is generally impractical.

On the other hand, if a collector feedback stage is capacitor coupled to an external load R_L' with $R_S = R_L'$, the iterated gain can be as much as twenty times the dc S-factor. As shown in Fig. 5.5, the higher gain is obtained by constructing the stage with the collector resistor R_L approximately ten times the external load resistor. The low-impedance load effectively bypasses feedback at signal frequencies while the dc stability is retained. With the ac feedback removed, the gain depends on β, and the ac input impedance seen by the signal e_I is approximately the same as the iterated impedance R_S.

With collector feedback the coupling capacitors may need to be surprisingly large. The input impedance is so low that the input capacitor should be calculated by assuming that the series impedance of the base loop is no larger than the impedance of the source. Similarly, the internal output impedance of an amplifier may be small compared to the collector load resistor. The low output impedance may cause problems in circuit design, especially if the following stage is nonlinear when driven by a low-impedance source. The interstage impedances make both low-and high-frequency filtering difficult because the impedances are low and variable. When the amplifier frequency response is to be precisely controlled or a high interstage impedance is desired, a combination of collector and emitter feedback is preferable.

If the base signal is observed when the source impedance is large compared with the input impedance Sh, the voltage wave form may indicate considerable, even harmonic distortion. This distortion is to be expected in amplifiers that have feedback. Nevertheless, the output signal is linearly related to the input current, which is the signal.

5.10 DIRECT-COUPLED *CE* STAGES

The amplifier shown in Fig. 5.7 is an interesting example of a collector-feedback stage that is direct-coupled to a high-gain CE stage. The collector-feedback stage TR-1 offers a low-drift Q-point and is direct-coupled to the base of the output stage TR-2. The amplifier may be used as a high-gain low-drift dc

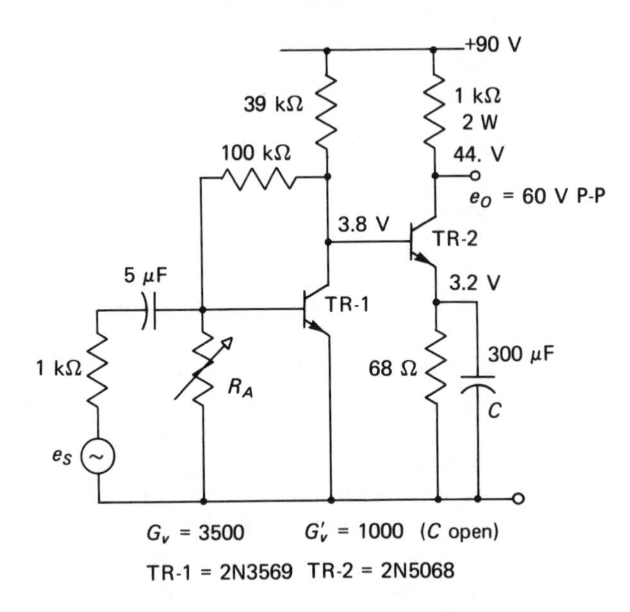

Figure 5.7. Direct-coupled two-stage *CE* amplifier.

amplifier with the emitter capacitor removed or as a large-signal ac amplifier with C in place. The output transistor does not require a heat sink, but the turn-on drift is reduced if a heat sink is used.

The first stage of the amplifier is identical with the stage shown in Fig. 5.5. With the emitter capacitor removed, the input impedance of TR-2 is about 68β, or approximately 4000 Ω. From experience with the amplifier shown in Fig. 5.5, we should expect a first-stage voltage gain between 40 and 90. The second-stage voltage gain is $1000/68 = 15$. The overall voltage gain should be about $60(15) = 900$. With the emitter capacitor in place, the expected voltage gain is the current gain product $S_c\beta_2 \cong 50(50) \cong 2500$. The measured gains are 1000 and 3500, respectively. With the capacitor removed, the feedback in the second stage is almost significant.

As a dc amplifier, the two-stage pair offers a low-drift second-stage collector Q-point. If the junction temperature of TR-1 increases, the input current to TR-2 decreases and offsets the temperature increase of the collector current in TR-2. If TR-1 is mounted in thermal contact with TR-2, the Q-point drift may be varied or reversed in direction by adjusting the amount of thermal feedback. With minor adjustments the Q-point drift may be reduced to less than 1 V for an ambient temperature range of at least 40°C. A similar low-power direct-coupled amplifier that has a Darlington pair with collector feedback is described elsewhere. (See reference 5-1, Fig. 5.7.)

5.11 COLLECTOR-FEEDBACK SUMMARY

1. Collector feedback reduces the *CE*-stage gain from β to S_c, where S_c is the corrected *S*-factor given by equation (5.7). Significant feedback is obtained by making $S_c = \beta/4$, which means that the *S*-factor is $R_f/R_L = \beta/3$.
2. The gain and impedance relations describing a collector-feedback stage are summarized by writing the TG-IR as follows:

$$G_v' = \frac{S_c R_L}{R_S + S_c h} \qquad (5.24)$$

   For an iterated-stage design, $R_L = R_S$, the input impedance $S_c h$ is small compared with R_S, and the voltage gain S_c is relatively independent of β and temperature.
3. Because the collector *Q*-point voltage is proportional to the S/β ratio, a collector-feedback stage is used mainly in small-signal applications. With significant feedback the *Q*-point voltage is $V_{CC}/4$, and lower values of S may be used if the resulting *Q*-point voltage $(S/\beta)V_{CC}$ is at least 2 V. A collector-feedback stage operates over a wide range of supply voltage, and the bias circuit does not require an additional power-hum filter.
4. The temperature shift of the collector voltage is mainly produced by the temperature change of β. With silicon transistors, the collector V_C decreases approximately 0.6 percent for each centigrade degree increase of the junction temperature. While feedback does not reduce the percent change of V_C with temperature, a low *S*-factor may be required to limit the overall dc gain and *Q*-point drift in a direct-coupled amplifier. The optimum *Q*-point for a single-stage amplifier is usually obtained with significant feedback which usually permits a 2 or 3 to 1 change of β after adjusting the *Q*-point bias.
5. The voltage gain of a stage with significant feedback is approximately three-fourths the *S*-factor, or $\beta/4$. When designed with a low *S*-factor and a capacitor-coupled load, an iterated stage may have a voltage gain of nearly $\beta/2$.
6. The collector feedback gain rule states that if S is small compared with β and R_f/R_S is small compared with $[25\ V_{CC}]$, the stage voltage gain is approximately R_f/R_S.

   The intrinsic voltage gain of a collector feedback stage is approximately $[25\ V_{CC}]$, which is high enough to make a single-stage useful as an operational amplifier.

5.12 *CE* STAGE WITH COLLECTOR FEEDBACK BYPASSED

The very low input impedance produced by collector feedback may be removed by placing a bypass capacitor at the midpoint of the bias resistor R_f, as shown in Fig. 5.8. The effect of the capacitor is to raise the voltage gain by 20 to 40 dB while keeping the stability and Q-point characteristics which are controlled by dc feedback. The effect of adding the capacitor is to change the stage current gain from S to β, and the voltage gain of an iterated stage increases to approximately β. The value of the capacitor C_f is best determined experimentally.

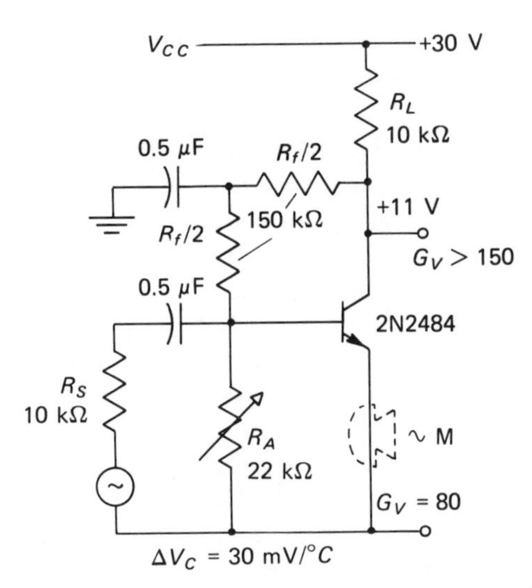

Figure 5.8. Collector-feedback stage (bypassed and biased).

In a low S-factor design the capacitor may increase the gain by a factor of 100, 40 dB, which is equivalent to an additional stage. The fact that the ac gain depends on β, and thus varies with the temperature and the transistor, may not be a disadvantage. The Q-point is under the control of dc feedback, whereas the ac gain is controlled better by feedback over two or three stages.

The dashed additions in Fig. 5.8 show how collector-feedback biasing may be used in a *CB* stage with a low impedance source. When connected between the emitter and ground, a carbon microphone is activated by the dc emitter current and has an impedance between 100 and 500 Ω. The source impedance R_S is not required, but the base should be connected to ground by a capacitor and the feedback capacitor is no longer necessary. The microphone amplifier has a voltage gain of 80 from the 100-Ω source. With a 100-Ω dynamic microphone replacing the 10-kΩ R_S, the signal amplification is 600.

5.13 Q-POINT ADJUSTMENT WITH COLLECTOR FEEDBACK

The Q-point of a collector-feedback amplifier tends to be too low for large-signal applications, and with a low supply voltage may be so close to the 2-V lower limit that the amplifier is impractical. As we have seen, significant feedback cannot be used unless the supply voltage exceeds 8 V. The difficulty is simply that a Q-point suitable for large signals and a reasonable Q-point shift may require an S-factor that is too large for significant feedback.

Experiment shows that a bias resistor R_A connected as shown in Fig. 5.8 generally raises the Q-point enough to permit large signals without otherwise affecting the circuit relations previously described. Unless the S-factor is quite low, values of R_A that are from two to ten times the value of R_L will raise the collector Q-point voltage to as high as one half the collector supply voltage. (Sometimes an amplifier will tolerate a higher output signal if the resistor is connected across the base-emitter junction.)

The use of a bias resistor has distinct advantages with low supply voltages where, otherwise, it is impossible to use significant feedback. The bias resistor makes possible the use of an S-factor of 30, as shown in Fig. 5.8, while keeping a Q-point voltage that permits higher signal levels and a wider temperature range than is possible with the stage shown in Fig. 5.4. A 12-V supply can be used with a 6-V Q-point when S is reduced to 8 and R_A = 10 kΩ. Thus, whether or not R_f is bypassed, the addition of R_A considerably improves the collector-feedback stage and eliminates the advantage that the emitter-feedback circuits have generally had with low supply voltages. A disadvantage of the bias resistor is that it tends to increase the temperature sensitivity of the collector Q-point. When using the resistor, the temperature shifts of a collector-feedback stage and an emitter-feedback stage are almost the same if both have the same S-factor and the same base resistor. The collector-feedback stage has the advantage of twice the voltage gain, and increasing the bias resistor reduces the temperature shift.

5.14 COMBINED COLLECTOR AND EMITTER FEEDBACK

The advantages of bypassing the collector feedback may be exploited differently by using emitter feedback to stabilize the ac gain and to increase the input impedance. This combination of dc collector feedback and ac emitter feedback may be the best all-purpose single-stage amplifier design. A circuit having the combined collector and emitter feedback is shown in Fig. 5.9.

The TG-IR representing the combined-feedback amplifier takes the form

$$G_v = \frac{\beta R_L}{R_S + \beta R_E} \tag{5.29}$$

To reduce the dependence of G_v on β, the input impedance βR_E must be made large compared with the source impedance R_S. If the gain is reduced by a factor of 4, a stage has significant feedback, and the gain is approximately $3R_L/4R_E$. In other words, if R_E is increased until the base signal is three-fourths the Thévenin equivalent source signal, the stage has significant ac feedback. Emitter feedback which reduces the stage gain by a factor of 4 makes the voltage gain less dependent on β by making the βR_E term in equation (5.29) equal to three times the R_S term. Thus, significant feedback reduces the voltage gain by a factor of 4 and reduces the gain change to one-fourth the change produced by a β change without feedback.

The amplifier shown in Fig. 5.9 has an S-factor of 30, which reduces the collector voltage to about 4 V, but a 20-kΩ bias resistor raises the collector voltage to 14 V, which permits a 26-V peak-to-peak output signal at room temperatures. The 130-Ω emitter resistor is selected to provide significant ac feedback.

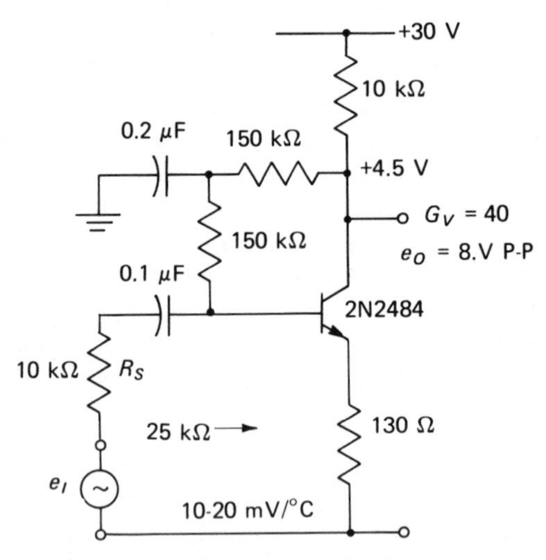

Figure 5.9. Collector-emitter feedback stage.

The collector-emitter feedback amplifier shown in Fig. 5.10 has a relatively high voltage gain of 33 and a 10-kΩ input impedance. The emitter feedback increases the input impedance ten to fifteen times with only a 2 to 1 gain reduction. The amplifier may be used with supply voltages from 5 to 50 V, is relatively insensitive to temperature, and permits peak-to-peak output signals exceeding one half the supply voltage. The gain may be reduced 40 dB by

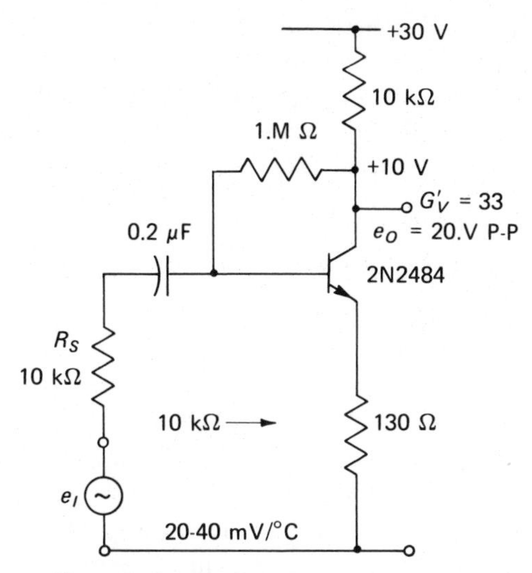

Figure 5.10. Direct-coupled iterated stage.

increasing the emitter resistor. This amplifier has the advantages of simplicity, few components, high gain, and better performance characteristics than a stage with emitter feedback biasing.

5.15 ADVANTAGES OF COMBINED FEEDBACKS

A combined-feedback amplifier has many advantages over the emitter-feedback amplifier so commonly used by industrial designers. Collector feedback is a better way to stabilize a Q-point, and the ac and dc feedbacks are separated by a capacitor C_f that is small compared with an emitter capacitor. High ac gains are obtained by using an emitter resistor that is between $R_L/30$ and $R_L/100$, and these resistor values do not appreciably change the Q-point. The analysis of combined feedback is simpler and the bias adjustment is not critical, as with emitter feedback.

The gain of a stage may be varied by using a variable collector-to-base feedback resistor or a variable emitter resistor. The gain may be varied over a 30- to 40-dB range with a shunt feedback resistor that is varied from $3R_f$ to βR_L. This gain adjustment is satisfactory in a stage that does not have emitter feedback. The adjustment has advantages over a series input gain control and does not appreciably affect the Q-point.

A variable emitter resistor is usually a better way to vary the gain, because increasing R_E tends to raise the Q-point voltage and to permit larger output signals than with the collector gain control. Increasing R_E by a factor of 100 (above the value for significant feedback) decreases the voltage gain by 40 dB and increases the input impedance.

An important advantage of emitter feedback is that a collector load tends to increase the feedback and reduce the effects of the load as seen by the preceding stage. With collector feedback the load tends to decrease the feedback, and a load capacitance may at high frequencies cause ringing and frequency distortion in an amplifier that has overall feedback.

5.16 COLLECTOR- AND EMITTER-FEEDBACK SUMMARY

Collector feedback is usually the most satisfactory way to stabilize the *Q*-point. The low ac input impedance caused by shunt feedback may be prevented by connecting a bypass capacitor at the midpoint of the feedback resistor. The addition of a small emitter resistor provides ac feedback and raises the input impedance without changing the *Q*-point. A base-to-ground bias resistor permits an adjustment of the *Q*-point as desired. A stage having the combined feedback offers several important advantages:

1. The impedance and gain characteristics of a stage with significant emitter feedback may be determined by inspection of the circuit components. With significant ac feedback the input impedance is three times the source impedance, and the stage voltage gain is one fourth the transistor β value. However, significant feedback reduces the gain dependence on β by approximately a factor of 4.
2. A stage with combined collector and emitter feedback operates over a wide range of the collector supply voltage, the *Q*-point is easily adjusted with a bias resistor, and the *Q*-point is not too adversely affected by temperature. The emitter feedback has the advantages that the input impedance is increased and load variations are not so easily reflected back to the input. The combined-feedback stage provides higher gain and less difficulty with ringing and frequency distortion.
3. The combined-feedback stage is generally more easily designed than an emitter-feedback stage and has a higher iterated gain. Except when the bias resistor is small, the temperature drift is the voltage change produced by the temperature change of β in equation (5.12). When a bias resistor is used, the collector voltage change increases to about 25 mV/°C, which is the temperature effect found in a low-gain emitter-feedback stage.
4. Approximately 40 dB of gain control may be introduced by making the emitter resistor variable and 100 times larger than the resistor used for significant feedback.
5. At the lowest frequency of interest the series input capacitor should have a reactance that is small compared with $4R_S$. Sometimes a stage may be coupled through the bias resistor, and the coupling capacitor is thus eliminated. High-frequency compensation may be provided by shunting a small capacitor across the emitter resistor.

6. A single stage using a high-β transistor with combined collector and emitter feedback gives 30 dB of iterated gain and excellent Q-point stability. With early low-β transistors and many designs currently in use, two stages are needed to give an equivalent performance. Smaller coupling and bypass capacitors are required with combined feedback than with emitter feedback. These advantages tend to make the combined-feedback stage economical and easy to test and service.

5.17 THE EMITTER-FEEDBACK AMPLIFIER

The best understood and most popular form of resistance-coupled amplifier is the emitter-feedback amplifier shown in Fig. 5.11. The characteristics of the emitter-feedback stage have been described in detail elsewhere (see reference 5-1), and only the results needed for design are reviewed here.

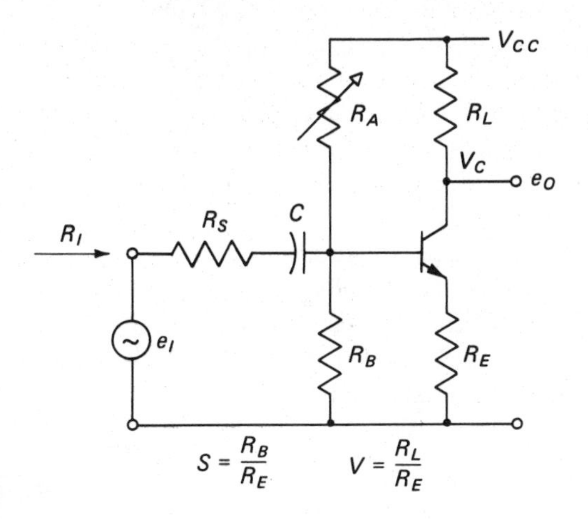

Figure 5.11. Emitter-feedback stage.

The dc current gain of the stage is fixed by the S-factor

$$S = \frac{R_B}{R_E} \tag{5.30}$$

and with adequate feedback the base-to-collector voltage gain is the V-factor

$$V = \frac{R_L}{R_E} \tag{5.31}$$

The impedance seen by the signal source is

$$R_I = R_S + R_B \tag{5.32}$$

By substituting S and R_I in the TG-IR, we find that the iterated voltage gain is

$$G_v' = V\frac{R_B}{R_S + R_B} \tag{5.33}$$

Equation (5.33) suggests that if all the resistors are fixed except R_B, the maximum gain is obtained by making R_B large compared with R_S. However, a study of the iterated designs used in industry shows that designers tend to make both the S-factor and the V-factor equal to 10. These values of S and V make $R_S = R_B = R_L$, and the voltage gain by equation (5.33) is 5. By making $S = 20$, R_B may be doubled, and the voltage gain is 6.7, but this is not much improvement. Moreover, any study of the emitter-feedback stage soon reveals that increasing either the S-factor or the V-factor above 10 makes the stage more sensitive to the value of the bias resistor, the temperature, and the supply voltage. A large-signal design may have S and V as high as 20, where an adjustment of bias and regulation of the supply voltage are permitted. Even so, the iterated gain is then only 10, which is small compared with the gain of 25 to 50 that may be attained with a collector-feedback stage offering comparable problems with bias and drift.

The foregoing paragraphs are summarized by noting that, unless a small increase of iterated gain must be obtained at any cost, the problems of design and the loss of Q-point stability usually do not justify the use of S-factors or V-factors over 10. When S and V are both 10, all the resistors except R_E are equal and the iterated stage gain is only 5.

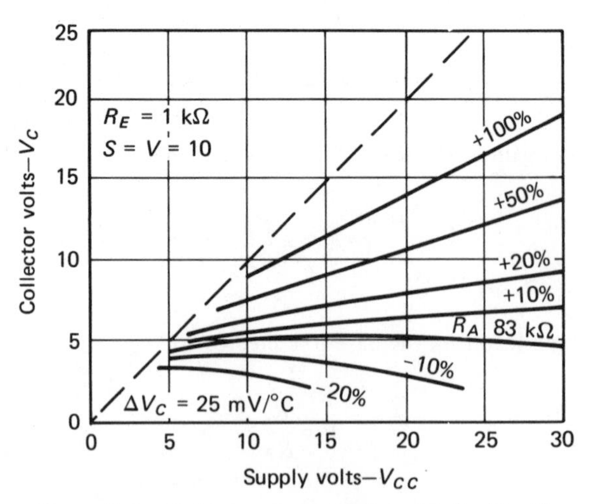

Figure 5.12. Collector characteristics with $S = V = 10$.

5.18 EMITTER-FEEDBACK-STAGE DESIGN AND BIASING

The characteristics of an emitter-feedback stage are illustrated by the curves shown in Figs. 5.12, 5.13, and 5.14, and the amplifier used to obtain the data is shown in Fig. 5.15. The curves, with the bias resistors given as the parameter, show how the collector Q-point voltage varies with the supply voltage. The amplifiers differ in the choice of the emitter resistor, so the figures indicate the effects of increasing S- and V-factors. Figure 5.12 is for $S = V = 10$; Fig. 5.13 is for $S = V = 15$; and Fig. 5.14 is for $S = V = 25$.

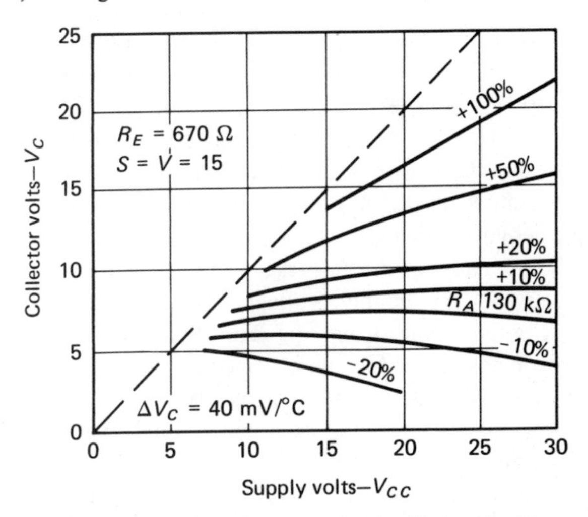

Figure 5.13. Collector characteristics with $S = V = 15$.

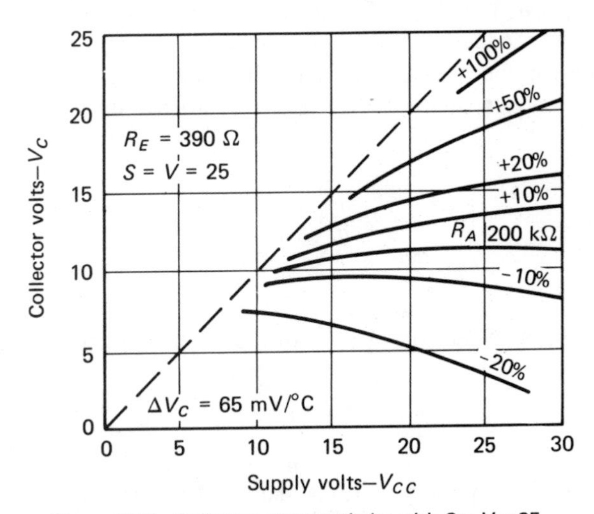

Figure 5.14. Collector characteristics with $S = V = 25$.

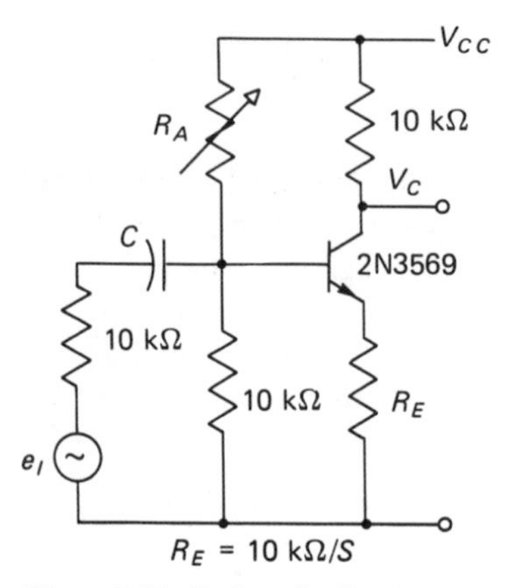

Figure 5.15. Emitter-feedback stage with $S = V$.

The curves in the figures extend over the range of Q-point conditions that tolerate a peak-to-peak signal greater than 1.4 V. Over parts of the curves the signal is too small for many applications, and to the left of the curves the amplifier is cut off and inoperative. When the curves are examined for Q-points that make V_C approximately one half of V_{CC}, we find that the collector supply voltage should be numerically equal to or greater than the amplifier V-factor. This fact means that while a CE stage is capable of a voltage gain of $[25\ V_{CC}]$, the dc voltage gain with feedback is limited to no more than $[V_{CC}]$. Because the ac gain is increased by adding an emitter capacitor, the ac gain is not so limited as the dc gain; but the dc limitation does suggest that there are major design problems.

For relatively low values of the bias resistor, the collector voltage tends to fall with an increase of the supply voltage. The biasing difficulties of an emitter-feedback stage come from the fact that a voltage change in the base-emitter loop tends to produce an exponential increase of the base current and, therefore, a decreasing collector voltage. Because a Q-point on a falling curve is generally unsatisfactory, the designer should bias for a Q-point that increases with the supply voltage.

The Q-point curves in Figs. 5.12, 5.13, and 5.14 reveal the interesting fact that a bias resistor can be selected to make the Q-point voltage independent of the supply voltage. If V_C is independent of V_{CC}, small changes of the supply voltage will not produce a signal at the collector, as with an inadequate hum filter. Analysis of the bias resistor value that makes V_C independent of V_{CC} reveals that the value of the four resistors of a stage must stand as a simple proportion:

$$\frac{R_A}{R_B} = \frac{R_L}{R_E} \tag{5.34}$$

Solving equation (5.34) for R_A gives the convenient design relation

$$R_A = SR_L \tag{5.35}$$

As the curves show, equation (5.35) actually gives a higher, and better, Q-point at which V_C increases with V_{CC}. If the S-factor is several times the V-factor, the bias resistor should be 0.8 times the value given by equation (5.35). The curves show also that when the stage is biased to make the collector Q-point voltage independent of the supply voltage, the Q-point voltage is approximately one half the V-factor. This relation holds for V-factors not too different from the S-factor.

When the S-factor is three or more times larger than the V-factor, two difficulties appear. The Q-point voltage is significantly lower and the Q-point temperature shift is approximately doubled. As the emitter-feedback stage is often direct coupled or used as a power-stage driver, where Q-point stability is required, these difficulties make high-S-factor designs less desirable than low-S-factor designs. The advantage of a high S-factor is a higher input impedance and only 25 per cent more voltage gain. If the S-factor is n times the V-factor, the curves in Figs. 5.12, 5.13, and 5.14 may be used with sufficient accuracy for design by using n times the indicated bias resistor. The times n rule may be applied with S-factors up to $\beta/3$, and the resulting curves have about the same position with a slightly steeper slope.

The temperature drift of the collector voltage is indicated in Figs. 5.12, 5.13, and 5.14. These values of drift show that the Q-point shift is caused mainly by the 2 mV/°C decrease of V_{BE}. The temperature drift of a collector-feedback stage is about the same when a bias resistor is used, and is perhaps one third or one half smaller without a resistor. The emitter-feedback stage has up to twice the drift indicated in the figures when S is between three and six times the V-factor.

A direct-coupled two-stage amplifier in which each stage has $V = S = 10$, as in Fig. 5.12, should be expected to have a second-stage collector voltage shift of 250 mV/°C. This temperature shift is the 25 mV/°C of the first stage increased by the voltage gain of the second stage. This calculation neglects the second-stage temperature effects. Figure 5.12 shows that if $V_{CC} = 20$ V and the second stage is biased for large-signal operation with $V_C = 10$ V, then a 2-V decrease of V_C may be about the maximum tolerable Q-point shift with increased temperature. This estimate shows that a pair of direct-coupled *CE* stages must be operated at room temperatures, although a single stage may be operated to about 100°C.

5.19 THE EMITTER-BYPASS CAPACITOR

When an emitter-feedback stage is driven by a zero-impedance source, the gain is maximized (see reference 5-1) by making the reactance of the capacitor small compared with h of the transistor. Sometimes the capacitor must be surprisingly large in order to provide a constant gain at low frequencies. However, with a high-impedance source the gain is maximized by making the reactance of the capacitor small compared with R_B/β, which may be several times larger than h.

The effect of fully bypassing the emitter resistor is equivalent to replacing S by β in the TG-IR. There are at least three reasons a designer may not want a large bypass capacitor. The capacitor makes the circuit characteristics vary with β, makes the input impedance very low, and eliminates the ac feedback.

A better design is obtained by using a resistor R_H in series with the emitter capacitor, as shown in Fig. 5.16. The resistor is proportioned to provide significant ac feedback, or more, as required, and the reactance of the capacitor C_H at the lowest frequency of interest should then equal the ohmic value of R_H.

The simplest and safest procedure for selecting the coupling and bypass capacitors is to examine the circuit experimentally and to avoid the errors of a calculation. To prevent the need for a supply of very large capacitors, a circuit may be examined at 10 times the lowest frequency of interest, and, when selected, the capacitors can be installed with 10 times larger capacitance.

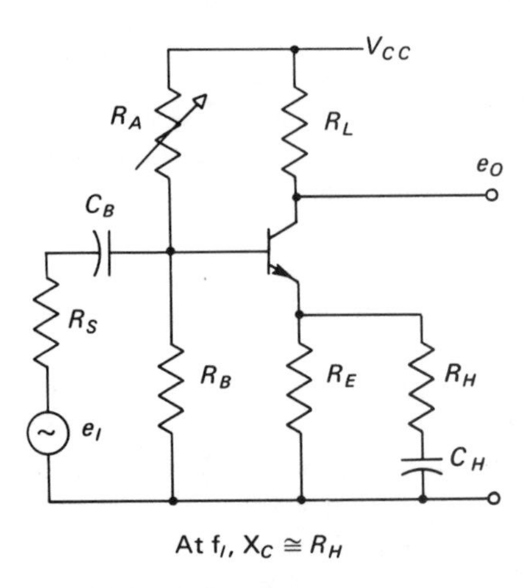

Figure 5.16. Emitter-feedback stage with bypass.

The procedure for selecting the capacitors is to begin by fully bypassing the emitter and selecting a coupling capacitor that gives a 3-dB loss at the low cutoff frequency. With these capacitors the gain at the midband frequency should be checked and found consistent with an iterated stage gain of β. The series emitter resistor is now inserted and adjusted to provide the desired amount of feedback. If significant feedback is required, the gain is reduced to one fourth the midfrequency gain.

By inserting the emitter resistor the cutoff frequency of the coupling capacitor is moved to a lower frequency, but the capacitor should not be changed. Reducing the emitter capacitor to reset the cutoff frequency minimizes the total capacitance, because the coupling capacitor is generally much smaller than the emitter capacitor. Finally, increasing both capacitors by the factor of 10, as mentioned above, should give the desired low-frequency response and the emitter feedback.

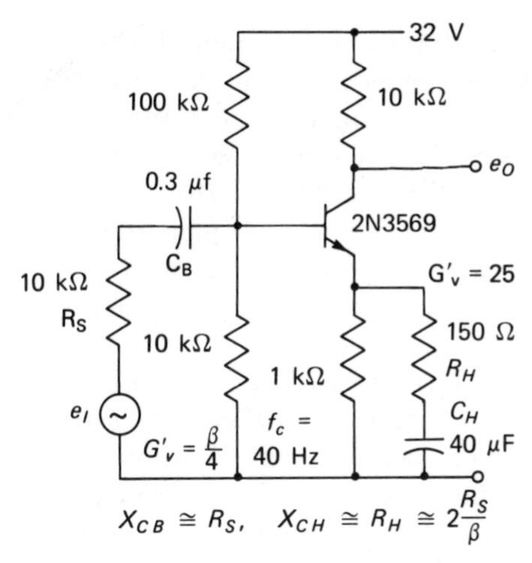

Figure 5.17. *CE* stage with significant emitter feedback.

The amplifier in Fig. 5.17 is an example of a *CE* stage with $S = V = 10$ and with significant emitter feedback. The voltage gain with the emitter fully bypassed is 100, and the partial bypass reduces the voltage gain to 25. The capacitors are sized to place the 3-dB cutoff at 40 Hz.

In summary, an iterated stage having significant feedback may be expected to have an overall voltage gain G_v' of approximately $\beta/4$. Without an emitter-bypass capacitor the voltage gain is about 5, so the increased gain is approximately a factor of 5 and is equivalent to an additional stage. If $\beta = 30$, an emitter-feedback stage with $S = 10$ has no more than significant feedback, and the emitter should not be bypassed. This fact explains why the partially bypassed emitter did not offer any real advantage until transistors became readily available with minimum βs exceeding 30. The base signal voltage with significant feedback is about 0.8 times the no-load base voltage. Thus, the existence of significant feedback may be tested by comparing the base voltage with the Thévenin equivalent voltage.

5.20 SUMMARY OF THE EMITTER-FEEDBACK STAGE

The *CE* stage having emitter feedback has for long been the most commonly used transistor amplifier. The amplifier offers an interesting combination of advantages and disadvantages.

1. The impedance and gain characteristics of a stage with a non-bypassed emitter are readily determined by inspection of the circuit diagram. The *S*-factor gives the current gain, the *V*-factor gives the base-to-collector voltage gain, and the input impedance is the resistance of the base resistor (corrected for the equivalent shunting of the bias resistor R_A).

2. The collector Q-point is set by a bias resistor that does not generally affect the gain and impedance parameters of the stage. The bias resistor usually must be set to within a 10 per cent tolerance when the amplifier is constructed, because the resistor value depends markedly on the supply voltage and the transistor parameters. The resistor value, which makes the component resistors stand in a simple proportion, gives a Q-point voltage that is approximately independent of the supply voltage. A higher value of the bias resistor may be used when a maximum peak-to-peak output voltage is required.
3. The S-factor of the emitter-feedback stage is usually between 5 and 20, with a value of 10 generally preferred by designers. The V-factor is similarly found to be between 5 and 30, with a value of 10 preferred. The V-factor should not exceed $[V_{CC}]$. The close limits placed on S and V come from the fact that the Q-point is controlled by the base current, which is exponentially related to the base voltage. These characteristics make the emitter-feedback stage sensitive to a change of circuit parameters and supply voltage.

   Whenever the S-factor and the V-factor exceed 10, the bias resistor must be carefully set with concern for the Q-point shift caused by ambient temperature changes.
4. The iterated gain of the emitter-feedback stage is only $V/2$. The input impedance with adequate feedback is R_B. A fully bypassed emitter resistor raises the iterated stage gain to β and makes the input impedance a short circuit. For significant ac feedback the emitter may be bypassed by a capacitor and series resistor, which reduce the iterated gain to $\beta/4$. The series emitter resistor is approximately $2/\beta$ times the equivalent source resistance seen by the transistor. With significant feedback the transistor input impedance is three times the equivalent source resistance. The existence of significant feedback may be tested by comparing the ac emitter voltage with the base voltage.
5. At the lowest frequency of interest the reactance of the input capacitor C_B should approximately equal the series equivalent of $R_S + R_B$, and the reactance of the bypass capacitor C_H should approximately equal R_H. Both capacitors may need to be much larger when there is less than significant feedback. The value of C_H is usually about β times the value of C_B.
6. The availability of transistors with β values exceeding 100 makes possible the use of a single stage with an ac gain of 25 to 100, which formerly required two stages. While a single stage with a low V-factor may be operated over a wide temperature range, a direct-coupled two-stage amplifier must be restricted to room temperatures.

   We observe, however, that a combined-feedback stage offers the same ac gain, uses a smaller bypass capacitor, and usually tolerates a wider range of parameter and V_{CC} values. The temperature drift of the collector-feedback-stage Q-point is usually about one third the drift in an emitter-feedback stage that has comparable *gain* characteristics.

7. The *CE* stage with emitter feedback may be recommended for low-gain applications requiring stable input and output characteristics. The amplifier is best used as a resistance-coupled or transformer-coupled class A power-output or driver stage.

PROBLEMS

5-1. Refer to Fig. 5.4 and assume that R_f is 540 kΩ. (a) Find the corrected S-factor, the collector voltage, and the voltage gain. (b) What are the approximate base-to-emitter input impedance and the intrinsic voltage gain?

5-2. Repeat Problem 5-1, using Fig. 5.5 with $R_f = 200$ kΩ and the switch open.

5-3. Show your calculations for the voltage gains given for Fig. 5.7. Explain why the Q-point temperature drift can be reduced to zero.

5-4. Confirm by calculations the voltage gains given for the amplifier in Fig. 5.8 when used as a microphone amplifier. Assume that the base is bypassed.

5-5. For the amplifier shown in Fig. 5.9 the text says that the S-factor is 30 and the voltage gain is 40. (a) Explain and show your calculations for the expected voltage gain. (b) To what extent does a 20-kΩ bias resistor lower the voltage gain?

5-6. The emitter-feedback stage in Fig. 5.15 is designed with $V_{CC} = 20$ V, $R_A = 200$ kΩ, and $S = 25$. (a) What are the dc emitter voltage, the base voltage, and the collector voltage? (b) If the design is changed to make $S = 10$ and the transistor is biased for the same collector voltage, what are R_A, the dc emitter voltage, and the base voltage? (c) What are the advantages and disadvantages of each design?

5-7. Consider the amplifiers represented by the curves in Fig. 5.12, 5.13, and 5.14. With a 20-V collector supply for each, what bias resistor should you use to obtain the smallest Q-point change if the bias resistor changes ±20 per cent?

DESIGNS

5-1. Design a collector-feedback stage to operate on a 20-V supply with a 5-V Q-point. The collector load impedance is 5 kΩ and the source impedance is 1 kΩ. Describe the characteristics of your design.

5-2. Design a combined collector-emitter feedback stage to operate on 20 V with $S = 10$, $V_C = 5$ V, $R_L = 20$ kΩ, and $R_S = 2$ kΩ. Describe the performance characteristics of your design.

5-3. In the laboratory observe the characteristics of the amplifier shown in Fig. 5.9. Compare those characteristics with the ones obtained with bias and with the emitter resistor removed. Repeat with C_f removed.

5-4. Design and evaluate a collector-emitter feedback stage that is suitable as a phonograph preamplifier with equalization for a magnetic pickup.

5-5. Design an input stage for an intercom that uses a loudspeaker as a microphone.

5-6. Modify the circuit shown in Fig. 5.7 for use with a 10-kΩ source and a 30-V power supply.

REFERENCES

5-1. Cowles, L. G., *Transistor Circuits and Applications*, Chapter 3 and Chapter 4, Sections 4.8 through 4.10. Englewood Cliffs, N.J.: Prentice-Hall, Inc., 1968.

5-2. *RCA Transistor Manual*, SC-13, "Biasing," pp. 22-26. Harrison, N.J.: Radio Corporation of America, 1967.

5-3. *GE Transistor Manual*, 7th ed., "Biasing," pp. 95-106. Electronics Park, Syracuse, N.Y.: General Electric Co., 1964.

6

TRANSISTOR PAIRS

Well-designed low-frequency amplifiers should have the transistors associated in direct-coupled pairs for circuit economy. A direct-coupled pair uses fewer components and provides higher gain than a pair of resistor-capacitor coupled stages. The direct-coupled pair allows one transistor to be used for voltage gain and the other for impedance matching, while the bias and S-factor control are considered only once. Direct-coupled pairs are usually designed with some form of overall feedback. Feedback over a pair is more easily applied than over three or more stages and is more effective than when applied as local feedback.

This chapter includes examples of the most commonly used direct-coupled-pair amplifiers. Some examples are designed for iterated circuits and others include an impedance change. With these amplifiers the designer can scale the impedance levels up or down, or change the feedback to meet the requirements of a particular application. With a few well-designed pairs from which to work the designer is able to proceed quickly toward a solution of each new problem.

6.1 DIRECT-COUPLED PAIRS

A pair of transistors may be series connected in nine different ways. As indicated in Table II, the first stage may be *CB*, *CE*, or *CC*, followed by a second stage in any one of the three configurations. The TG-IR shows that maximum amplification between equal input and output impedances is produced by providing a maximum current gain in each stage of the amplifier. Because a *CB* stage has a current gain of only 1, five of the nine arrangements are of interest only when current gain can be sacrificed for some other performance characteristic that requires a *CB* stage. Of the four combinations of *CE* and *CC* stages, the *CC-CC* pair does not provide voltage gain, but this pair does give current gain with an impedance step-down. The main characteristics of the transistor pairs are shown in Table II.

TABLE II
Characteristics of Transistor Pairs

First Stage	Second Stage		
	CE	*CC*	*CB*
CE	High G_V, iterated, low V_{CC}	Medium G_V, low R_L	Iterated video, low Miller effect
CC	Medium G_V, high R_I	Darlington step-down, $R_I = S^n R_L$	Emitter-coupled and special applications
CB	Impedance step-up, low Miller effect	Iterated video, low Miller effect	Rarely used, $G_i = 1$

Transistor pairs of the forms *CE-CE*, *CC-CE*, and *CE-CC* have broadly similar characteristics, and the main choice between them may be decided by the method selected to provide feedback biasing. A pair that is required to have exceptionally high voltage gain with a low collector supply voltage may be limited by the voltage gain attainable in a single *CE* stage, according to equation (3.15). Similarly, a *CC* stage has emitter feedback, so that a high load impedance may reduce the current gain below the gain attainable in a *CE* stage. For these reasons there is always the possibility that a given design requirement can only be satisfied by one of the three pairs. In general, the *CE-CE* pair seems to offer the most flexible and best combination for the application of feedback. A pair using a *CC* stage is best suited for applications in which the output impedance is a factor of 10 lower than the input impedance.

6.2 CHOOSING THE BEST PAIR

One of the pleasures of design is the challenge given by the many choices open to the designer for meeting the design requirements. Of all the choices there may be only a fairly limited range within which the desired objectives can actually be met. As an example of the choices presented by a direct-coupled pair, consider the following problem:

Assume that the designer is asked to supply an amplifier that has a voltage gain of 56 dB with an input impedance of 1000 Ω when the load impedance is 1000 Ω. Realizing that the final design may not be 100 per cent efficient, the designer may decide to use a voltage gain of 900, 59 dB, as a beginning objective. The gain and impedance relations of an *n*-stage amplifier are represented by the TG-IR

$$G_v = S^n \frac{R_L}{R_I} \tag{6.1}$$

By inserting the given gain and impedance values in equation (6.1), we have

$$900 = S^n \frac{1000}{1000} \tag{6.2}$$

Equation (6.2) shows that the product of the *S*-factors must equal 900, and we must decide how many stages are needed. A two-stage amplifier that uses transistors having an *S*-factor of 30 in each stage is the logical choice. The choice of stage types is the next decision to be settled.

Consider first a *CC-CE* direct-coupled-pair amplifier. The *CC* stage has a voltage gain of 1; hence, the entire voltage gain will have to be obtained in the second stage. For the second stage we have

$$900 = 30 \frac{1000}{R_{I2}} \tag{6.3}$$

so that R_{I2} must be 33 Ω, and $h = 1.1$ Ω.

The first stage may be a *CC* amplifier, as shown by the fact that

$$1 = 30 \frac{33}{1000} \tag{6.4}$$

The *CC-CE* amplifier that meets these design requirements is shown in Fig. 6.1, where the emitter current of the second stage is made 24 mA so that *h* is 1.1 Ω.

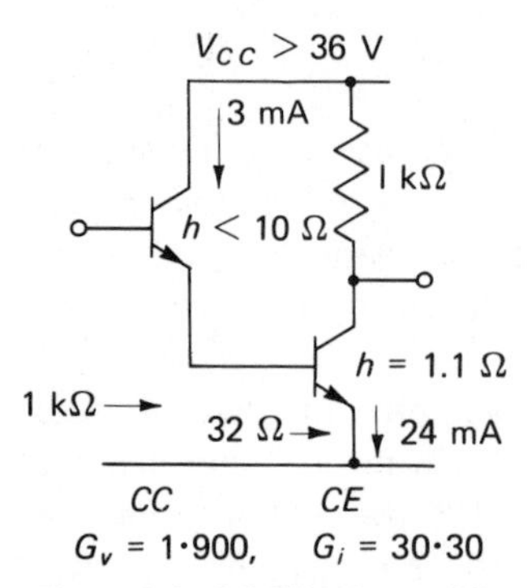

Figure 6.1. *CC-CE* pair amplifier.

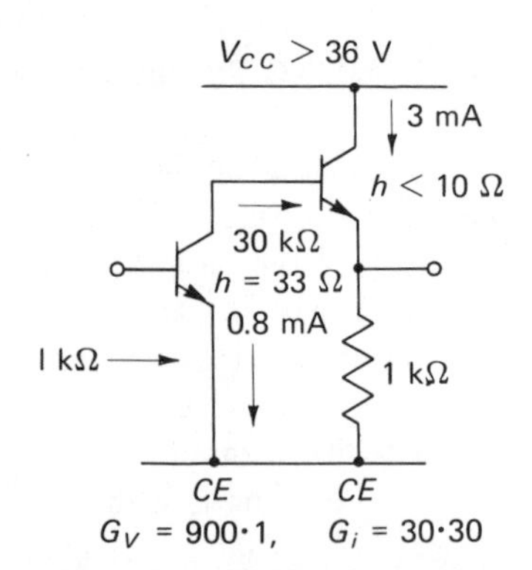

Figure 6.2. *CE-CE* pair amplifier.

The first stage requires an h that is small compared with the 33-Ω input impedance of the second stage. Hence, the first-stage emitter current must be about 3 mA. The entire voltage gain is placed in the second stage, so, by equation (3.15), we find that the collector supply voltage must be at least $\frac{900}{25} = 36$ V.

From the preceding trial design we might decide to raise the interstage impedance level by making the last stage the *CC* amplifier in order to reduce the emitter currents. Proceeding as before, we can meet the given specifications by the trial design shown in Fig. 6.2, and the TG-IR for the first stage is

$$900 = 30\,\frac{30{,}000}{30 \cdot h} \tag{6.5}$$

The entire voltage gain is in the first stage, so the supply voltage must still be about 40 V. The *CE* stage now requires that $h = 33$ Ω, so the emitter current drops to 0.8 mA and the total emitter current is about 3.8 mA.

As a last choice for the trial design, consider the effect of using the *CE-CE* pair with a voltage gain of 30 in each stage. The result of using this *CE-CE* pair is shown in Fig. 6.3, and the TG-IR for each stage is

$$30 = 30\,\frac{1000}{30 \cdot h} \tag{6.6}$$

Observe that both emitter currents are 0.8 mA and that equation (3.15) shows that the collector supply voltage may be as low as 1 V. These results indicate that the design requirements are more easily met by using a *CE-CE* pair and that the required dc power is minimal. The stages are iterated and alike, and this configuration evidently will give considerable flexibility as the design proceeds.

An examination of the design choices that can be made when the input-to-output impedance ratio is 100 with a voltage gain of 9 shows that the

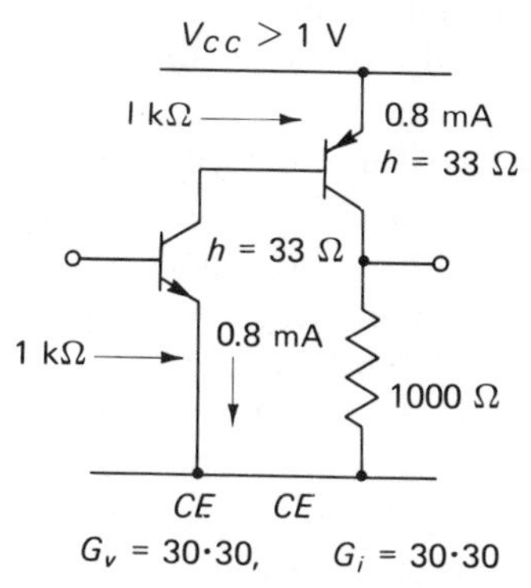

Figure 6.3. *CE-CE* pair amplifier.

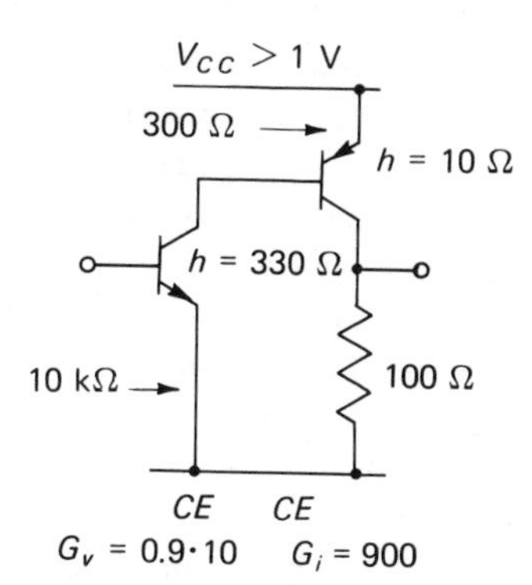

Figure 6.4. *CE-CE* impedance step-down pair.

CE-CE pair in Fig. 6.4 retains a marked advantage in power consumption. The pairs using *CC* stages do not require a high collector supply voltage because the overall voltage gain is only 9. Generally, the principal advantage in using a *CC* stage comes either from the simplicity of the *CC*-stage biasing and the circuit design or from the lack of signal inversion (change of phase). A *CC* input stage may have advantages when the highest practical input impedance is required, and a *CC* output stage may give a low output impedance without reflecting the load to the input, as with overall feedback. Except for such reasons, the *CE-CE* pair is generally the designer's best first choice.

6.3 THE CHOICE OF TRANSISTOR TYPES

The transistors that are used in a pair may be both *npn*, both *pnp*, or of mixed types. The choice of the transistor type for the first stage is usually determined by the polarity of the collector supply, because the emitter should be connected to the grounded side. This choice of the emitter return generally reduces the filtering required in the collector supply. With a positive collector supply the first transistor should be *npn*. If both polarities are available, the input noise, or cost and availability, may determine whether the transistor is an *npn* or *pnp*. The selection of the second-stage transistor probably is decided by the circuit configuration. Whenever the last transistor in an amplifier must deliver power, that transistor type is most likely to be a germanium *pnp* when cost or efficiency are important, or a silicon *npn* when high junction temperatures and high supply voltages are required.

A common reason for alternating *npn* and *pnp* transistors is to avoid stacking the collector voltages. The second-collector *Q*-point can be operated closer to ground with alternate types. With the availability of both positive and negative collector supplies, the second collector may be brought to ground or below, if desired. For this reason a three-stage amplifier is almost certain to use either alternate types or an output transistor of the opposite type. Moreover, when the second transistor is of the opposite type, the second collector is nearer the emitter voltage of the first stage, which simplifies the application of collector-to-emitter feedback. A disadvantage of using a pair of the same type of transistors is that the first collector has to be operated at a low Q-point voltage or the second emitter resistor will have a high voltage drop and require a large bypass capacitor. Generally, the main considerations in selecting the transistor types are the circuit configuration and the way feedback is to be applied.

When transistors are direct coupled, as in a pair, the dc changes of the first-stage collector are amplified by the gain of the second stage. This increase of the overall dc gain usually forces the designer to consider the effects of temperature on the output stage Q-point. The study of single-stage amplifiers indicated that temperature drift and the Q-point sensitivity to supply voltage are decreased by collector-feedback biasing. This consideration suggests, correctly, that a pair will have similar advantages when using shunt-feedback biasing. The double phase reversal from the first base to the second collector makes it impossible to obtain the first-stage bias from the second collector, but the second-stage base or emitter has the required phase turnover.

6.4 THE *CE-CE* TRANSISTOR PAIR

The design of a two-stage direct-coupled-pair amplifier is considerably more complicated than the design of a single-stage amplifier. There are more components to work with, and the direct coupling forces a more careful choice of the components. The amplifier shown in Fig. 6.5 is a carefully designed basic

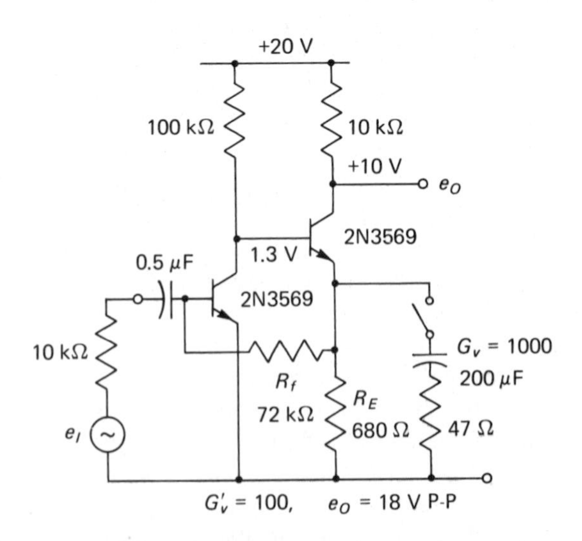

Figure 6.5. High-gain *CE-CE* amplifier.

two-stage dc amplifier having a form of shunt feedback that limits the overall current gain. The source and load resistors are both 10 kΩ, making this design an iterated one. Known as a current-feedback amplifier, the circuit offers flexibility in design and good Q-point tracking when the supply voltage varies.

Because two transistors are used, we might anticipate an open-feedback-loop current gain β^2 equal to about 10,000, and therefore make the ratio of R_f/R_E about 1000. When $R_E = 10$ kΩ and $R_f = 1.5$ MΩ, the voltage gain from the input *base* to the output *emitter* is experimentally observed to be only 450. This is considerably below the gain needed for significant feedback. However, the emitter follower cannot have a voltage gain greater than 1, and the first stage cannot have a voltage gain greater than $25[V_{CC}] = 500$. These facts mean that the loop current gain cannot exceed 500 and the measured voltage gain of 450 is a reasonable value. To place the current gain under feedback control, the ratio R_f/R_E should be about 100, which makes the overall voltage gain about 100.

Having decided on a value for the loop current gain, the designer may logically ask what happens when R_f and R_E are varied together while their ratio is fixed. In the circuit shown, the voltage gain to the second collector is 100 and unchanged for R_E (and R_f) varied from 330 Ω to 200 kΩ. However, the second-stage collector Q-point does change and is one half V_{CC} when $R_E = 680$ Ω. Moreover, as long as $R_E = 680$ Ω, the collector Q-point is essentially independent of R_f (and the loop current gain), as R_f is varied from 1 to 100 kΩ. This result means that the Q-point and the current gain of this *CE-CE* feedback pair may be independently adjusted.

The resistor in the collector of the first stage acts as a constant current source, supplying collector and base current for each stage respectively. The value of this resistor will depend on the transistors and is best determined experimentally. In the circuit shown in Fig. 6.5 the first-stage collector resistor should be about 10 times the second-stage collector resistor. The Q-point stability characteristic of this circuit comes from the low collector voltage 1.3 V, of the first stage. Any increase or decrease of this voltage tends to be offset by an almost exponential increase or decrease of the feedback current. For this reason the second-stage emitter voltage is independent of R_f and approximately independent of R_E. Hence, any change of R_E must produce a change of I_C and the second-stage Q-point.

The ac voltage gain of the *CE-CE* transistor pair may be increased about 20 dB, either by bypassing the second emitter or by bypassing the center point of the feedback resistor R_f. A 3.3-kΩ resistor in series with a 2-μF capacitor at the midpoint of R_f raises the overall voltage gain. The voltage gain of the first stage is increased by a factor of 6 to become 38, and the voltage gain of the output stage is 10,000/680 = 14. The net overall voltage gain is 500 (54 dB). When the feedback is bypassed, the 2-μF capacitor at the midpoint is sufficient to hold the 3-dB low-frequency cutoff at 30 Hz, but the Miller-effect cutoff in the first stage makes the high-frequency cutoff fall at 20 kHz. If the midpoint is bypassed by a large capacitor without the series resistor, the Miller-effect cutoff is at 3.5 kHz.

When the second-stage emitter is bypassed by 200 μF in series with 47 Ω, the overall voltage gain is raised to 1000 (60 dB). The emitter capacitor decreases the voltage gain in the first stage, and the Miller-effect cutoff is then in the second stage at 30 kHz. With the emitter fully bypassed by 1000 μF, the overall voltage gain is the expected 5000 (74 dB) with the upper cutoff frequency at 7 kHz.

With either type of ac feedback the amplifier performs very well, as the supply voltage is varied upward from 17 V. The high-gain amplifier shown in Fig. 6.5 is entirely satisfactory for room-temperature applications. Amplifiers in this form are designed for wide temperature range applications by reducing the dc voltage gain of the second stage until the *Q*-point drift is acceptable. With further adjustments, the first-stage drift can be made to compensate for the second-stage drift over a part of the temperature range.

Amplifiers in the *CE-CE* form are sometimes designed with the emitter fully bypassed and with ac feedback from the second collector to the first emitter. This combination of dc and ac feedback has essentially the same performance characteristics as the partially bypassed amplifier. The double-feedback amplifier is more complicated and is not known to offer any useful advantage in an iterated design.

6.5 *CE-CE* PAIR WITH OPPOSITE-TYPE TRANSISTORS

A *CE-CE* amplifier using opposite types of transistors, *npn-pnp*, does not perform as satisfactorily as a high-gain iterated amplifier. An amplifier of this type is shown in Fig. 6.6 for illustration, but not as a working example. The difficulty with the exampe as a shunt-feedback amplifier is that the second-stage emitter voltage is so much above the input base voltage that any value of R_f that sets a reasonably low current gain, R_f/R_E, also supplies so much base current that one or both stages are saturated. Without feedback the amplifier has such a high overall dc current gain that linear operation is possible only for a limited range of the supply voltage. The difficulty with series feedback to the first emitter is similar. The amplifier is very sensitive to a change of the collector voltage, and, with a voltage gain of 1000, the *npn-pnp* amplifier lacks enough feedback for bias control and satisfactory performance.

However, the *CE-CE* amplifier that uses opposite transistors does give satisfactory operation when the emitter feedback is increased until the overall voltage gain is less than 100. The input impedance is increased, and the amplifier has applications when some voltage gain is required with an input impedance that is 10 to 100 times larger than the load impedance. The amplifier also offers

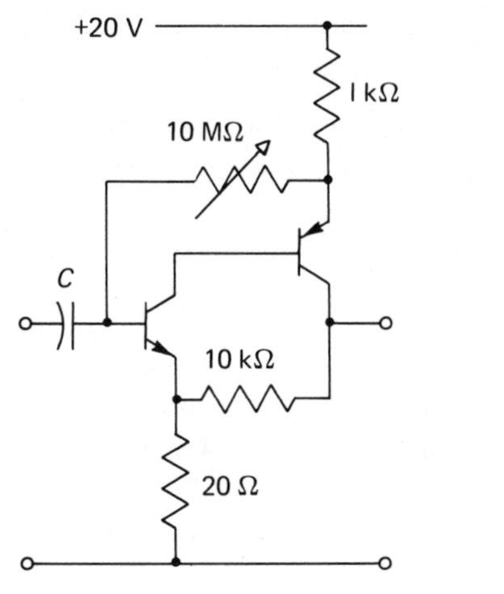

Figure 6.6. *CE-CE* amplifier with *npn-pnp* transistors.

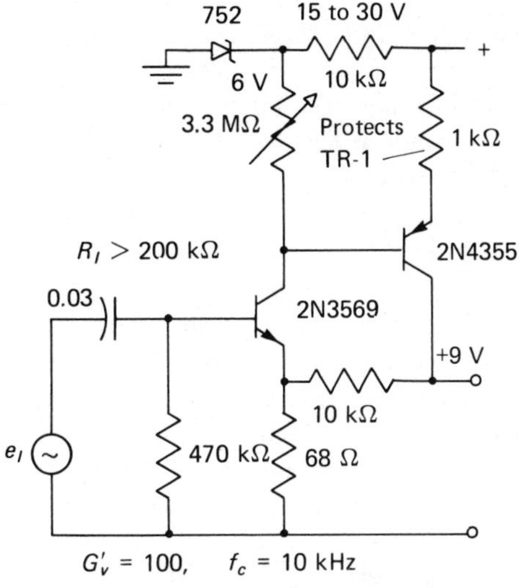

Figure 6.7. *CE-CE* amplifier with series feedback.

a low output impedance, although not simultaneously with high input impedance. Examples of this type of amplifier are shown in Figs. 6.7 and 6.8. In the first example the input impedance is over 200 kΩ with a voltage gain of 100. In the second example the input impedance is 5 MΩ with a voltage gain of 10. Both amplifiers operate satisfactorily for supply voltage changes from 15 to above 30 V, which shows an unusual insensitivity to the supply voltage. The success of the high-gain amplifier comes from the regulated bias, and of the low-gain amplifier from the feedback.

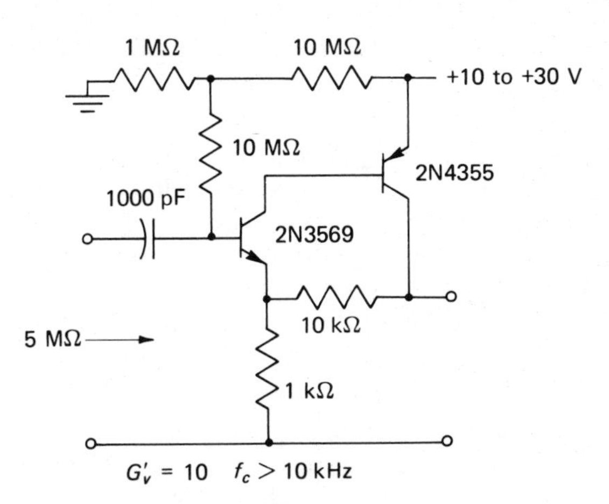

Figure 6.8. *CE-CE* high-input-impedance amplifier.

6.6 THE *CC-CC* PAIR

The familiar Darlington compound is a pair of emitter followers, which for many applications can be viewed as a single high-β transistor. A pair of transistors with an intrinsic current gain $\beta^2 = 10{,}000$ is usually operated with resistors that make the *S*-factor a factor of 10 or 100 times below the available current gain.

The circuit of a high-gain Darlington pair with *S*-factor control is shown in Fig. 6.9. The amplifier transfers the input signal at an impedance level of 0.8 MΩ to the load at 1 kΩ. The ac voltage loss in the amplifier is practically negligible, while the dc loss is 1 to 2 V, depending on whether the transistors are low-power devices, as in the figure, or power transistors, as in a power amplifier. The ac power gain in the amplifier is simply the ratio of the input impedance to the load resistance, which is 800, 29 dB. In a power amplifier in which the transistors are operated at high current levels and with lower *S*-factors, the power gain may be only 10 dB.

In former years the Darlington pair was often used to provide a high input impedance, a characteristic that is now more easily provided by an FET. Until recently, the difficulty of manufacturing an FET in an integrated circuit made the *CC-CC* pair the designer's choice for the input stage of high-impedance integrated circuits. Because the bias resistors and the collector-to-base leakage have limited the input impedance, the impedance is sometimes increased by a technique referred to as "bootstrapping." This scheme couples a point on the bias resistors through a capacitor to the load resistor. When this point on the bias resistor follows the input signal, there is no ac current in the resistor, and the bias resistor acts as a relatively high ac load. Bootstrapping usually raises the input impedance by a factor of approximately 10. The availability of the high-β planar transistors and FETs has made the complications of bootstrap circuits relatively obsolete, as a comparison of circuits will show. Two reasons for bootstrapping, however, are to increase the high cutoff frequency and to increase the input impedance of a power stage.

As in feedback amplifiers, there are problems with the Darlington compound at frequencies near the transistor cutoff frequency f_β. The high-frequency cutoff in circuits is always caused by reactances, which means there is an accompanying phase shift. At a frequency about a decade below the transistor f_T, the gain-bandwidth product, where the transistor β is decreasing 6 dB/octave, the β-cutoff causes the load capacitance and resistance to reflect into the input as a negative resistance and an inductance, respectively. The effect of these reflected components is to cause instabilities or to produce an unintended oscillation.

Because the feedback circuit is completed through the source, the Darlington compound is open-circuit stable, which means that the stability may be improved by using high source impedances, by a series resistance to offset the negative resistance, by avoiding inductive source impedances, by reducing the overall S-factor, and by using transistors that have staggered f_T values. (For additional information about the β-cutoff instability, see reference 6-1.)

The design of a Darlington *CC-CC* amplifier is quite simple. The transistors should be selected to have a high current gain at the operating Q-point, and this means in many cases that the transistors should not be the same. If the second stage is required to dissipate several watts, the second transistor may be a germanium power device driven by a low-power *pnp* silicon planar transistor.

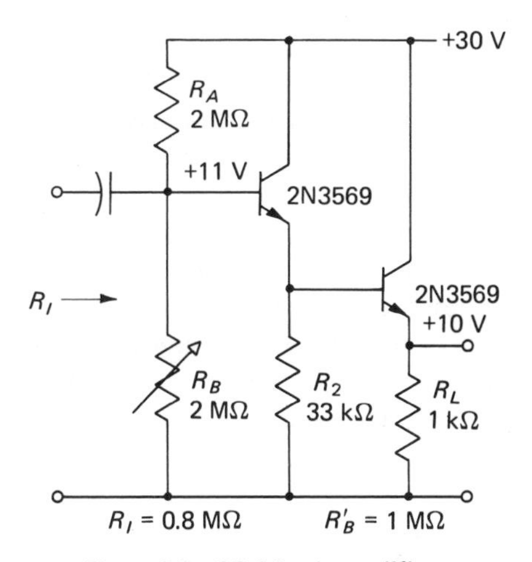

Figure 6.9. *CC-CC* pair amplifier.

The overall S-factor of a *CC-CC* pair, as in Fig. 6.9, is the resistor ratio R'_B/R_L, where R'_B is the parallel equivalent of R_A and R_B. The intermediate resistor R_2 ensures that each stage shares in the overall current gain, and the resistor also reduces the possibility of high-frequency and high-temperature instabilities. Generally, R_2 is made equal to $\sqrt{R'_B R_L}$, the mean value of R'_B and R_L. It is possible to use even a *pnp-npn* transistor sequence (see Fig. 6.10B) with the advantage that the offset voltage from input to output is the difference of two V_{BE} voltages instead of a sum. The Darlington pair with complementary transistors has the advantage also that the β temperature changes have opposing effects on the second-stage emitter current with the possibility of drift reduction, as in Fig. 5.7.

Because of the relatively high amount of emitter feedback, the Darlington pair is a linear amplifier, even at high signal levels. Generally, if the bias is adjusted to make the second emitter voltage half the supply voltage, a *CC-CC* amplifier will handle peak-to-peak signal levels to within 1 V of the supply voltage. With relatively high S-factors or with low supply voltages the resistor R_B may be several times larger than R_A. The stage input impedance is, of course, R'_B.

A Darlington pair is easily substituted for a single transistor almost any time a higher current gain can be used to advantage. Sometimes three or more stages of emitter followers are found in circuits of a decade ago. However, with today's devices, two high-β transistors or an FET stage usually provide a better way to obtain high input impedance.

6.7 THE *CC-CE* TRANSISTOR PAIR

The *CC-CE* transistor pair using like transistors, as shown in Fig. 6.10(a), is really a Darlington compound with the load in the collector. This pair may be treated as a high-β *CE* transistor, and the circuit may be designed by the methods already discussed. (For an amplifier that uses the *CC-CE* pair with collector feedback, see reference 6-2.)

If the pair has complementary transistors, as in Fig. 6.10(b), the amount of feedback can be controlled by connecting the collector to a tap on the emitter resistor. The interstage resistor R_2 supplies current to both transistors and is an essential component. Except for special applications, this compound with complementary transistors has not been shown to have particularly useful advantages over the pair with like transistors.

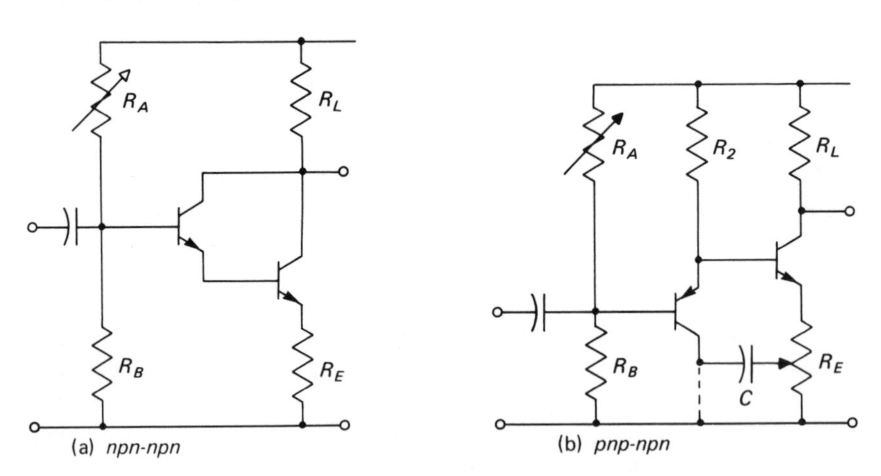

Figure 6.10. *CC-CE* pair amplifiers.

The amplifier shown in Fig. 6.11 is an example of a *CC-CE* pair used for high current gain and preceded by an FET for high input impedance. This amplifier delivers 1 W to a 150-Ω load with a power gain of 80 dB. The amplifier may be used to drive a low-impedance electrodynamic load or tape recorder from high-impedance transducers. For dc recording, the output capacitor may be eliminated by returning the load to an adjustable tap on the power supply.

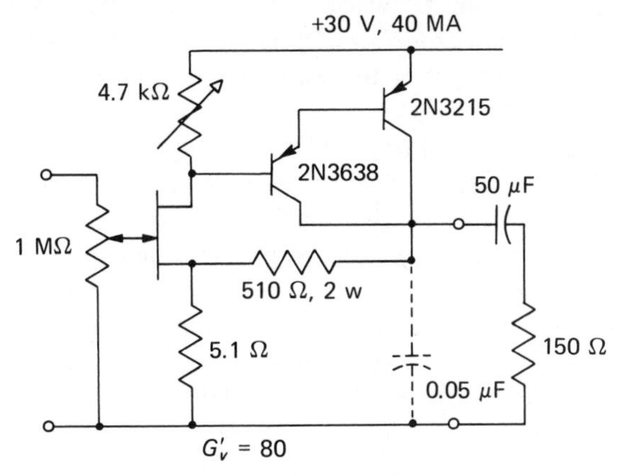

Figure 6.11. High-impedance FET-*CC-CE* amplifier.

6.8 THE *CE-CC* TRANSISTOR PAIR

The *CE-CC* transistor pair is equivalent to a single high-β transistor and may be used as one. The pair is commonly used when both impedance step-down and voltage gain are required. As with most *CE-CE* amplifiers, the load may be connected to an emitter for a low output impedance. With the load at the second emitter, as in Fig. 6.12, the pair gives a voltage gain of 10 with an impedance step-down exceeding 200. As shown in the figure, the amplifier has significant feedback, and the 2N1711 needs a clip-on heat sink to dissipate 1 W.

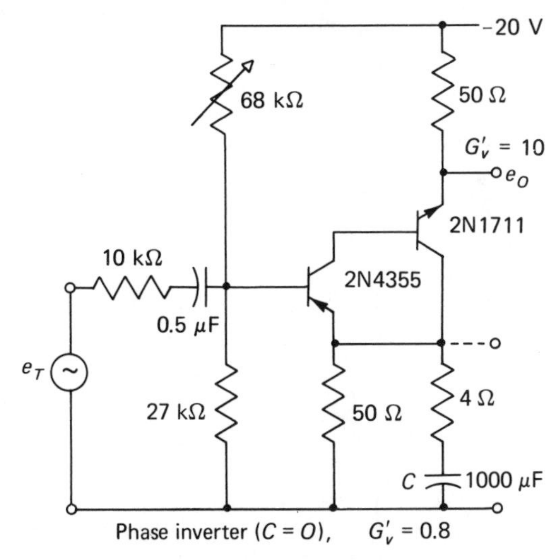

Figure 6.12. *CE-CC* pair amplifier or phase inverter.

Removing the capacitor C increases the overall feedback and the input impedance, and the amplifier is converted to a phase inverter with a gain slightly less than 1. A germanium power transistor may be used in the second stage to lower the impedance and increase the power output by replacing both transistors with complementary types and reversing the power polarity.

A compound that uses a silicon *npn* followed by a germanium power transistor is very much like a high-β silicon power transistor. The intermediate resistor for this compound may need to be connected to either the collector supply or to ground, depending on the transistor characteristics. (For additional application information, see the references given in reference 6-1.) If the load is moved to the collector, this pair becomes a Darlington compound that uses complementary transistors, as described in Section 6.7.

6.9 THE *CB-CC* AND *CE-CB* PAIRS

The pairs of this section have similar characteristics. One stage has current gain and the other, a *CB* stage, is used for the voltage gain. The use of a *CB* stage usually implies that current gain is sacrificed for an advantage that cannot otherwise be easily obtained. The main advantages of using a *CB* stage for one of a pair are to obtain a high gain-bandwidth by reducing the Miller effect, or to use the high breakdown voltage rating of the *CB* transistor. The stage used for current gain may at times be a *CE* stage in place of the *CC*, or vice versa.

An example of a *CB-CC* direct-coupled vido amplifier is shown in Fig. 4.13. In that amplifier with minor changes the load can be transferred to the second-stage collector for operation as a *CB-CE* pair, and some additional gain may be obtained without seriously reducing the bandwidth.

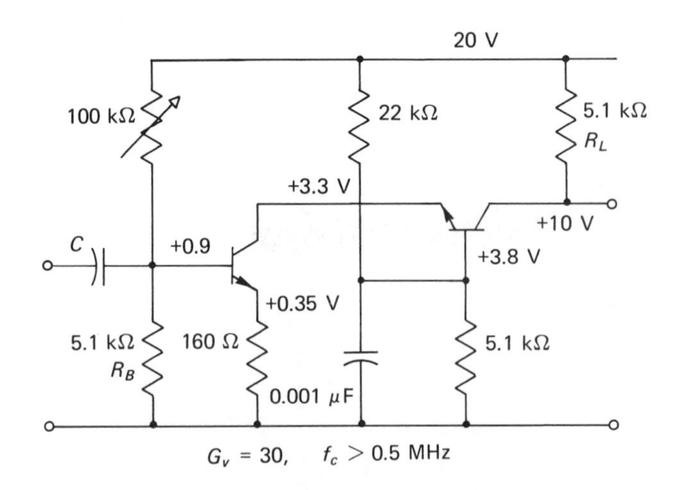

Figure 6.13. *CE-CB* wide-band amplifier.

The amplifier illustrated in Fig. 6.13 is a *CE-CB* cascode in which the *CB* stage is used to eliminate Miller-effect loading at the input. The amplifier is designed as an iterated stage with equal input and output resistors, R_B and R_L. The overall voltage gain is 30, numerically the *S*-factor. The current gain is provided by the first stage and voltage gain by the second. Because the input impedance of the *CB* stage effectively shorts the *CE* collector to ground, the *CE* stage has a voltage loss, but provides a current gain of 30 up to a frequency of nearly f_T/S. The low voltage gain essentially eliminates the first-stage Miller effect.

Although the voltage gain of the pair is entirely in the *CB* stage, the second-stage collector voltage is in phase with the emitter voltage, and there is again no Miller effect. The important advantage of a cascode is that interaction between the load and the input can be minimized, even up to ultrahigh frequencies.

The *CE-CB* cascode has a low sensitivity to V_{CC} changes, and the *CB*-stage biasing is relatively noncritical. Amplifiers of this type are used as oscilloscope preamplifiers and wide-band preamplifiers operating up to 250 MHz. With a tuned collector load a cascode is often used at low frequencies to take advantage of the high-breakdown-voltage characteristics of the *CB* output stage.

A cascode is usually designed with a single bias network supplying both stages. As illustrated in Fig. 6.14, the bias network is a series of three resistors. The design of the single-string bias network is left as a problem for the reader.

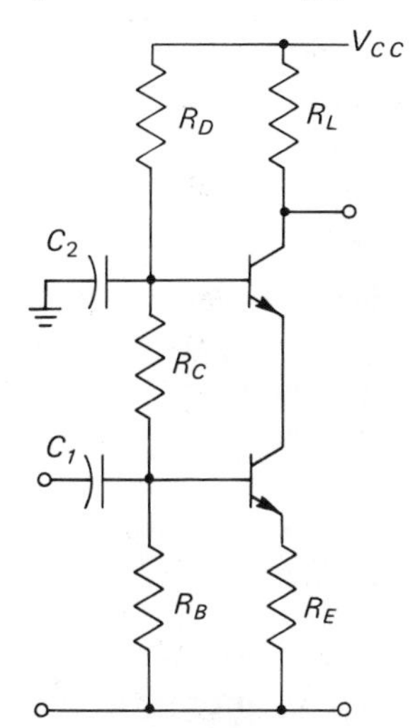

Figure 6.14. *CE-CB* cascode with common bias string.

6.10 THE *CE-CB* EMITTER-COUPLED PAIR

The amplifier shown in Fig. 6.15 has identical *CE* and *CB* stages that are direct-coupled at the emitters. The signal is applied at the base of the first stage and returns to ground through the second-stage base. Because of the relatively high value of the emitter resistor R_E, the current leaving one emitter enters the other and forces the second stage to follow the first with the same signal amplitude but with reversed polarity. The amplifier responds as if the input signal were applied from base to base, so there is no signal current in the emitter resistor. The voltage gain on each side is R_L/h, where R_L is the load impedance on one side. The signal measured on one side is, therefore, $R_L/2h$ times the input signal e_I, and the collector-to-collector signal is R_L/h times the input signal.

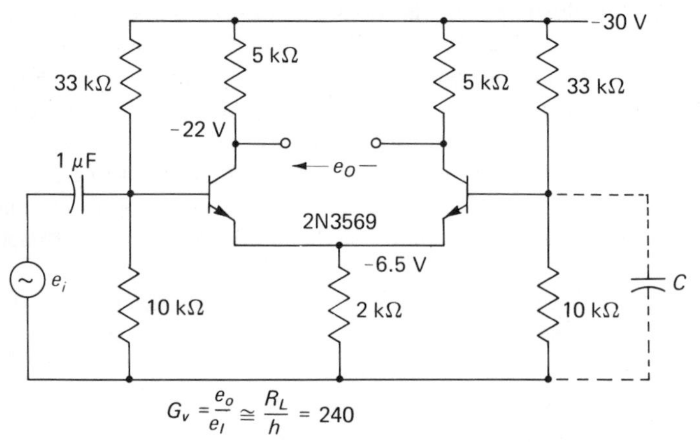

$$G_v = \frac{e_o}{e_i} \cong \frac{R_L}{h} = 240$$

Figure 6.15. Emitter-coupled amplifier.

The amplifier has the gain of a *CE* stage with the emitter bypassed and is used when high gain is required and an emitter bypass is not permitted. The amplifier has the advantage of a *CE* stage without phase turnover or may be used as a high-gain phase-inverter with signals of both polarities available at the collectors. With the input stage collector connected to the power supply without the resistor, the amplifier is a form of cascode that gives high gain without the Miller feedback. In many respects the emitter-coupled pair is a remarkably versatile amplifier.

6.11 COMPLEMENTARY SYMMETRY

The term *complementary symmetry* designates a circuit in which opposite types of transistors are connected in parallel for operation on alternate half-cycles, as in a class-B amplifier. The advantage of a complementary-symmetry circuit is that the signal circuits for alternate half-cycles are essentially alike. The transistors are conducting, push-push, so that the load is driven from a low impedance, and the amplifier is able to carry a higher load because the duty cycle of each transistor is 50 per cent ON and 50 per cent OFF.

The complementary emitter follower shown in Fig. 6.16 lacks true symmetry but is used to improve high-speed switching and to drive capacitor loads. With the capacitor connected, the upper transistor charges the capacitor and the lower transistor provides a low-impedance circuit for discharging the capacitor. If the lower transistor is removed, the emitter of the *npn* transistor is reverse biased whenever the base voltage is less than the emitter voltage, so the capacitor cannot discharge on the alternate half-cycles. A demonstration of the circuit in the figure provides a convincing example of the reasons for using a complementary emitter follower.

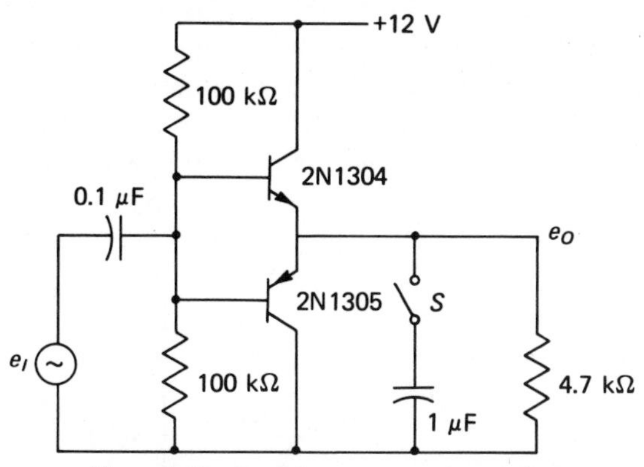

Figure 6.16. Complementary emitter follower.

The complementary emitter follower in Fig. 6.17 uses silicon transistors that require a small forward bias to eliminate crossover distortion at low signal levels. The circuit shows one diode and a series resistor to provide an adjustment of the forward bias. In many cases the resistor is replaced by a second diode. The 50-Ω resistors in series with the collectors protect the transistors from second breakdown by limiting the peak collector currents. These resistors reduce the output signal very little.

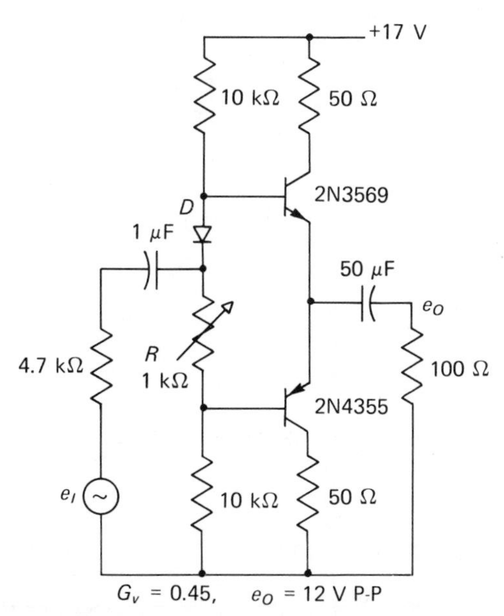

Figure 6.17. Complementary emitter follower with crossover diode (D).

The two-stage amplifier in Fig. 6.18 is Sziklai's complementary-symmetry class-B compound. This double compound is a remarkable two-stage, push-push, class-B amplifier with overall feedback. The amplifier can be constructed entirely of transistors and is available as an integrated circuit. In the form shown, the amplifier has unity voltage gain and low distortion and requires positive and negative power supplies.

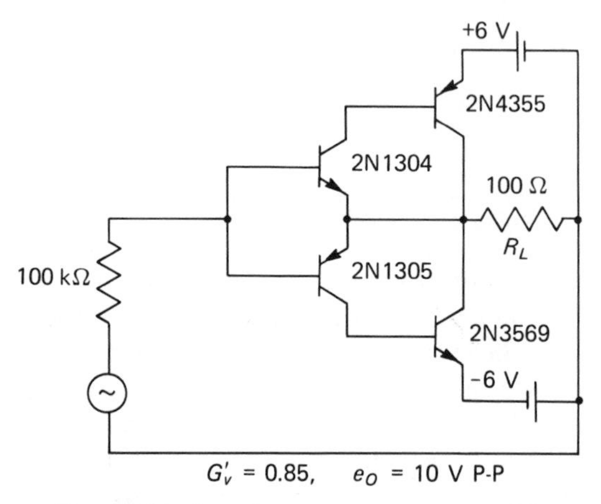

Figure 6.18. Complementary-symmetry amplifier.

A complementary-symmetry class-B audio amplifier is illustrated in Fig. 6.19.

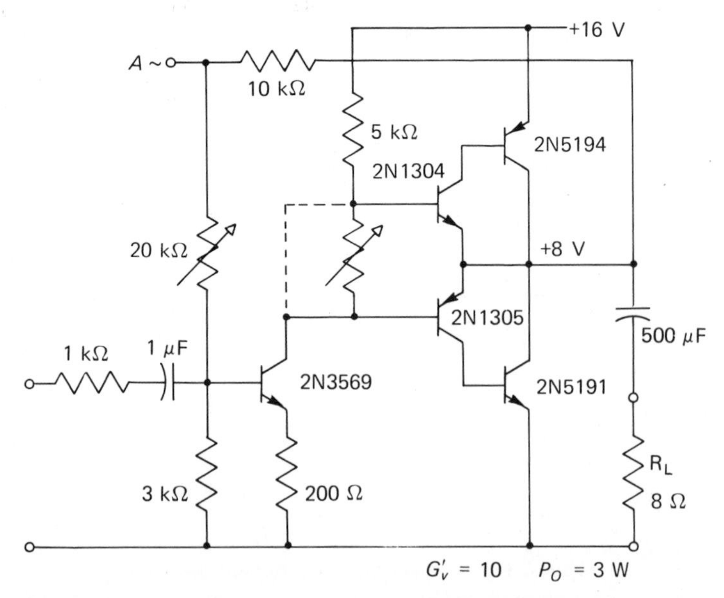

Figure 6.19. Complementary-symmetry class-B amplifier and driver stage.

The output section is a two-stage *CE-CE* amplifier with collector-to-emitter feedback. The upper and lower sides of the amplifier use complementary transistors and operate class B. The output section operates as a high-gain emitter follower with unit voltage gain. The input stage provides voltage gain, which is used partly for overall feedback.

The bias for a complementary-symmetry amplifier is adjusted to make the dc voltage at the output terminal just half the supply voltage. Because the bias resistor supplies both dc and ac feedback, the amount of feedback may be measured as the increase of gain that is produced when point *A* is connected to ground through a large capacitor. The amplifier in Fig. 6.19 has few components and provides an interesting example for design studies.

6.12 QUASI-COMPLEMENTARY-SYMMETRY AMPLIFIERS

Disadvantages of the complementary-symmetry amplifier are the difficulty of manufacturing complementary transistors in an integrated circuit and the expense or nonavailability of complementary power transistors. For these reasons the class-B power amplifiers are often constructed as quasi-complementary-symmetry amplifiers, having the basic form shown in Fig. 6.20. These amplifiers use complementary low-power input transistors with the bases connected in parallel. The output stage uses like power transistors connected in series. One side is a two-stage emitter follower, and the other is a complementary two-stage *CE-CE* collector follower. The bias battery, which is shown for simplicity, supplies negligible current and may have a high internal impedance. If preferred, the bias may be supplied as in Fig. 6.17.

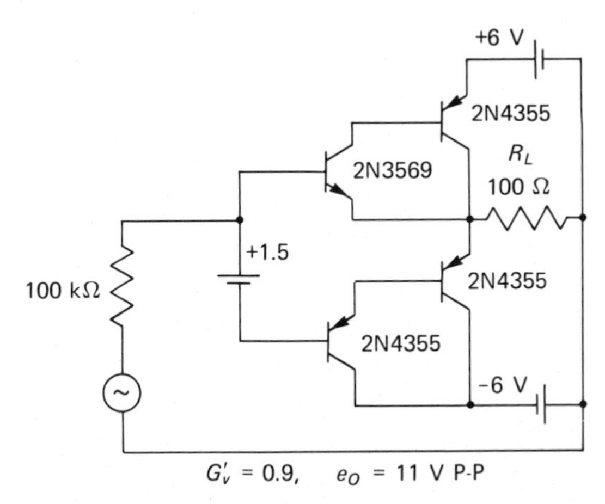

$G'_v = 0.9, \quad e_O = 11$ V P-P

Figure 6.20. Quasi-complementary-symmetry amplifier.

Although the quasi-complementary-symmetry circuit is not so simple as a true complementary circuit, the circuits perform equally well, and the design procedures are practically identical. Generally, one form of amplifier can easily be converted to the other.

The amplifiers shown in Fig. 6.18 and 6.20 require neither a push-pull drive nor an output transformer and may be used as direct-coupled power amplifiers. The main advantage of the complementary-symmetry amplifiers is that heavy and expensive transformers are not required, as in a push-pull amplifier. Because the phase shift of the transformer is eliminated, these amplifiers can be designed with considerable feedback. Generally, if the feedback is not too close to the singing point, the transistors do not need to be matched, as some writers suggest.

A 5-W quasi-complementary-symmetry amplifier with metal tabs for attaching the heat sink is available as an integrated circuit. (General Electric Co., type PA-234.)

6.13 TEMPERATURE PROBLEMS

Direct-coupled amplifiers have the disadvantage that temperature-engendered drifts in the first stage are amplified by succeeding stages and may produce *Q*-point drifts in the last stage large enough to make the amplifier inoperative at some high and low temperatures. The *Q*-point drift in direct-coupled pairs is usually controlled by reducing the dc gain until an acceptable performance is established. Occasionally, simple compensating tecnhiques are used. When a considerable amount of dc gain is required, the amplifier is more easily designed as a differential amplifier.

Generally, a single-sided direct-coupled two-stage amplifier operates satisfactorily over the temperature range experienced by electronic equipment in everyday use, as in an automobile, or in equipment in which nearby components produce considerable heat. Unless one or more stages are *CC* stages that do not magnify the voltage changes of the prior stages, a three-stage amplifier may present design or operating difficulties. A good practical rule is that a single-sided amplifier should have no more than two *CE* stages.

6.14 SUMMARY

Transistors are easily direct-coupled to form pairs that can be treated in design much like a single transistor with enhanced characteristics. Transistor pairs provide more gain and simpler circuits and, generally, simplify the application of overall feedback.

The TG-IR shows that the main difference between a pair and a single transistor is that the current gain is increased to β^2. The choice between one pair and another is often a matter of convenience or of the manner in which feedback is applied. On the other hand, we have shown that in an iterated design with a low supply voltage or low available dc power the pair should be *CE-CE*.

The Darlington *CC-CC* pair is particularly suited for impedance step-down at the input or output of an amplifier. The complementary *CE-CE* pair has similar characteristics and may be used as a substitute for a single transistor to provide increased current gain.

The *CE-CE* pair comprised of like transistors has outstanding characteristics as a high-gain shunt-feedback amplifier. This pair is widely used in intermediate-Frequency (IF) and radio-Frequency (RF) applications because the bias circuit has the advantages of collector feedback without requiring a collector resistor.

The series feedback *CE-CE* pair is best suited for low-gain, high-input-impedance applications. This pair has the characteristic of the single-stage *CE* emitter-feedback stage in that a stable *Q*-point is obtained only by using a low dc voltage gain.

Transistor pairs that have one *CB* stage are generally used for wide-band amplifiers when Miller-effect feedback cannot be tolerated. The emitter-coupled pair has similar characteristics and is widely used, particularly because high gain may be obtained without an emitter capacitor.

Complementary-symmetry amplifiers are mainly used in class-B audio-frequency amplifiers where the advantages of direct coupling eliminate the need for interstage and output transformers and permit increased feedback.

PROBLEMS

6-1. (a) Show that the gain and impedance values given for Figs. 6.9 and 6.13 are consistent with the TG-IR. (b) Give your estimate for the emitter currents and *h*-parameter for each stage in each amplifier.

6-2. Repeat Problem 6-1, using Figs. 6.12 and 6.15.

6-3. (a) Show that the voltage gains given in Fig. 6.5 are approximately correct. (b) Does the ac input signal at the input base increase or decrease when the switch is closed at the second emitter? (c) Explain.

6-4. Compare Figs. 6.13 and 6.14 and decide resistor values that should be tried in Fig. 6.14. Examine the amplifier experimentally and explain what may have been overlooked in your initial design.

6-5. A three-stage amplifier is required to have R_I = 22,000 Ω, G_v = 10, and an 8-Ω load. The transistor current gains are 30. If two *CC* stages and one *CE* stage are used without feedback, what values of *h* are required in each stage for each sequence given? (a) *CE-CC-CC*; (b) *CC-CE-CC*. (c) *CC-CC-CE*. (d) Which sequence requires the lowest emitter current in the amplifier?

6-6. (a) Repeat the analysis as given in Fig. 6.2, assuming that the input impedance is 500Ω and the load is 2000Ω. (b) Repeat similarly for Fig. 6.3. (c) Show how Fig. 6.4 is changed to give 45-dB gain when R_I = 5 kΩ and R_L = 200Ω. The second-stage emitter current is to be 30 times the first-stage value.

DESIGNS

6-1. Design an amplifier that meets the requirements of equation (6.4).

6-2. Design a *CE-CE-CC* amplifier that has a voltage gain of 500 with a 10-kΩ source and a 240-Ω load (which may carry transistor *Q*-point current).

6-3. Draw your circuit for an amplifier obtained by modifying Fig. 6.13 to meet the following specifications. The second stage is to use a *pnp* power transistor and a positive supply. The load is to be 10Ω and the source impedance 1 kΩ. Specify all components and the expected voltage gain. Estimate the maximum class-A power output and the current required of the dc supply.

6-4. In the laboratory study the design and performance of a class-B power amplifier of your own choice. Avoid a complicated circuit and avoid power outputs exceeding 7 W. Good examples are available in the references.

REFERENCES

6-1. Cowles, L. G., *Analysis and Design of Transistor Circuits*, Chapter 6, "Direct-Coupled Pairs." New York: Van Nostrand Reinhold Company, 1966.

6-2. Cowles, L. G., *Transistor Circuits and Applications*. Englewood Cliffs, N.J.: Prentice-Hall, Inc., 1968.

6-3. *GE Transistor Manual*, 7th ed. Electronics Park, Syracuse, N.Y.: General Electric Co., 1964. Chapters 9, 10, and 11 have examples of direct-coupled-pair amplifiers. See also the *RCA Transistor Manuals*.

7

TRANSFORMER-COUPLED CLASS-A AMPLIFIERS

This chapter describes the design of class-A amplifiers in which the collector load is connected to the amplifier by a transformer. Transformers with magnetic cores are used at audio frequencies when the designer needs to obtain either increased gain by approximately matching a high-impedance amplifier to a low-impedance load or to obtain increased power in a low-impedance load.

Transformer-coupled amplifiers are commonly used for the driver stages of class-A and class-B amplifiers and for low-power audio and instrument amplifiers. A transformer output stage is generally required whenever an amplifier is a part of a larger system, as in communication, servo, and control applications. In this chapter we describe the characteristics and limitations of a class-A amplifier and show how the power output is determined by the Q-point. We also show how the physical size of the transformer is determined by the amplifier and load and how the designer finds a suitable transformer.

7.1 TRANSFORMER COUPLING AT AUDIO FREQUENCIES

Consider the stage illustrated in Fig. 7.1 in which a transformer replaces the collector resistor of a resistance-coupled stage. Because the transformer has a primary-to-secondary turns ratio n, the collector load impedance is n^2R. The voltage gain of the stage is

$$G_v = \frac{n^2R}{h} \cdot \frac{1}{n} = \frac{nR}{h} \tag{7.1}$$

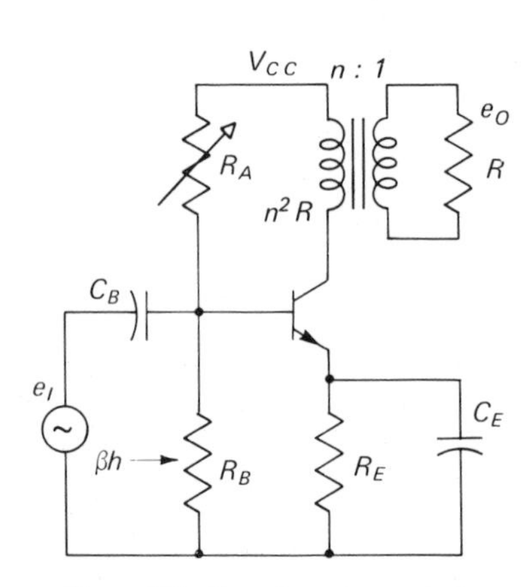

Figure 7.1. Transformer-coupled audio amplifier.

The transistor is biased like an R-C coupled stage except that the transformer makes the collector voltage equal to the supply voltage. With the collector voltage independent of the emitter current, the designer may begin by selecting an emitter current as high as practical in order to maximize the voltage gain. However, the collector power dissipation and the Q-point temperature drift are important considerations that usually make it desirable to use a low S-factor and to limit the collector temperature rise. Generally, the temperature rise may be neglected if the collector power is less than 50 mW. If the power is to be limited to 50 mW with a 10-V supply voltage, the designer should use an emitter current that is less than 5 mA. Observe that the bias resistor R_A should not be set so low that the transistor current or power ratings are exceeded.

The emitter resistor is generally selected to make the dc emitter voltage about 5 per cent of the supply voltage, but for adequate dc feedback the emitter voltage should be at least 1 V. However, the main Q-point consideration is the emitter current, not the voltage, because the current determines the collector power dissipation and the voltage gain.

Whenever transformer coupling is used, we generally find either an emitter capacitor or a grounded emitter. The advantage of using the emitter resistor and capacitor is that the bias adjustment and the emitter current are relatively independent of temperature and the transistor. The voltage gain, equation (7.1), is then independent of the transistor and temperature. If the emitter is grounded to eliminate the cost of R_E and C_E, the gain is somewhat more variable and the manufacturer may need to select the transistor or to adjust R_A.

7.2 THE TRANSFORMER GAIN ADVANTAGE

If the load R is placed in the collector circuit, as with R-C coupling, the stage gain is R/h, so the advantage of using the transformer is proportional to n, the turns ratio. Theoretically, the stage voltage gain is increased by coupling any load through a step-down transformer as long as the high-winding impedance is small compared with the internal impedance of the transistor. In practice the difficulty of handling small-diameter wire generally limits the high-winding impedance of an audio transformer to about a 10-kΩ maximum. The gain advantage of the transformer is, therefore, only available for load impedances less than 10 kΩ. If the load is 10 Ω, the gain advantage is approximately the square root of the impedance ratio, or 30.

In an iterated stage the load impedance is βh, and the voltage gain advantage is $\sqrt{10{,}000/\beta h}$. If $\beta = 100$, the advantage is $\sqrt{100/h}$. We see that there is no advantage unless h is less than 100 Ω, which means that in a practical situation transformer coupling is useful only when the emitter current exceeds 0.25 mA and the unbypassed emitter resistor is small compared with 100 Ω. These facts make transformer coupling of dubious value for low-power iterated designs, as in a hearing aid. The principal application of transformer-coupled stages is in power amplifiers, where the load impedance is small compared with the transistor input impedance, and the emitter currents are relatively high.

7.3 A LOW-POWER AMPLIFIER DESIGN

The circuit of a typical low-power transformer-coupled stage is shown in Fig. 7.2. The amplifier is intended to operate from a source impedance of 10 kΩ to a

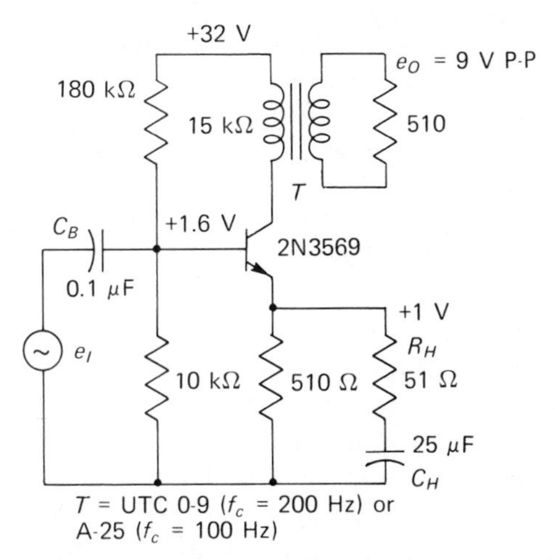

Figure 7.2. Low-power audio amplifier (to 500 Ω line).

500-Ω line. The stage operates class A and is designed to provide a maximum gain without regard to the power output.

The designer of the amplifier begins by selecting a suitable transformer. In this case the amplifier is intended for voice-frequency applications, so a transformer with a small core is satisfactory. With a 15-kΩ input impedance and a 500-Ω line impedance, the turns ratio is 5.5. The smaller transformer will tolerate 4-mA dc, which is twice the collector current.

The designer decided to use a dc S-factor of 20 and a partially bypassed emitter as a means of adjusting the gain. The required input impedance determines the 10-kΩ value for R_B, and $R_E = 10{,}000/20 = 500\ \Omega$. If $V_E = 1$ V, $V_B = 1.6$ V, and the emitter current is 2 mA. The supply voltage is 20 times the desired base voltage, so R_A must be about 20 times R_B, or 200 kΩ. The TG-IR shows that we may expect a voltage gain of

$$G_v = Sn\frac{R_L}{R_I} = 20(5.5)\frac{500}{10{,}000} = 5.5 \tag{7.2}$$

We measure 4.8.

A minimum value for the emitter resistor R_H is calculated as follows: If $\beta = 200$, the input impedance is lowered to 5000 Ω when $R_H = 10{,}000/\beta = 50$ Ω. For voice frequencies let $X_C = 50\ \Omega$ at 150 Hz. The base-to-collector voltage gain is $15{,}000/(50 + h) = 240$, and the transformer step-down is $\sqrt{30}$. The calculated voltage gain is 43, and we measure 38. By changing R_H, the voltage gain may be varied from 4.8 to 38, 18 dB. We note that if the line can be direct-coupled to the collector, the voltage gain is only

$$G_v = 200\frac{500}{10{,}000} = 10 \tag{7.3}$$

However, equation (7.3) neglects the input and output coupling losses, so the voltage gain would be something less than 7 without the transformer. Aside from the fact that the transformer is needed when coupling an amplifier into a line, the transformer increases the available gain by n, a factor of 5.5, or 15 dB.

7.4 POWER-AMPLIFIER DESIGN

The formulas commonly used for the design of power amplifiers are tools for the convenience of the designer that cannot be used as a substitute for an understanding of the real objectives and problems of the design. The design objectives are usually given as an amount of approximately undistorted power required in a particular load resistance. The designer begins by assuming that the distortion requirement can be met by an amplifier that produces a few decibels of excess power at the overload limits of the amplifier. The distortion specification is then met by operating at reduced output power and adding overall feedback.

The designer sees his problem as an exercise of his technical skill in obtaining the required power from transistors that have known limitations. Sometimes the desired amount of power cannot be obtained from a pair of any transistors that are available or economically useful. More often, the objective is to obtain the desired amount of power from a given and conveniently available transistor type. The problem is, therefore, one of obtaining power from a transistor without exceeding one or more practical limitations of the device. A designer must know the transistor limitations that are important in a power design and how these limitations relate to the specified power and load.

The primary limitations of transistors in power amplifiers are:

1. The transistor must dissipate the heat developed by the dc collector power without exceeding the junction temperature specified by the manufacturer. The designer meets this limitation by conducting the heat away from the transistor, selecting a transistor having a higher dissipation rating, or increasing the efficiency of power conversion.
2. The transistor must have a useful current gain β at the maximum peak collector current. The current gain of a transistor falls off rapidly at a collector-current value that may be only one tenth the manufacturer's "maximum collector current" rating. The maximum instantaneous collector current needed by the design is calculated by a formula that relates this current to the supply voltage and the ac power output.
3. The transistor must be able to withstand the maximum collector voltage impressed during the cycle. The maximum safe voltage cannot be simply specified because its value depends on the load current, the reactive nature of the load, the junction temperature, and on the time duration of the maximum voltage. Except when the voltage breakdown characteristic is discussed in more detail, it will be assumed that the transistor breakdown voltage is the transistor BV_{CEO} rating. In a power amplifier the peak collector voltage is usually twice the supply voltage. A transistor BV_{CEO} rating equal to $4V_{CC}$ is not too conservative for class-A operation when the transistor is operated near its peak current rating. The phenomenon, known as *second breakdown*, is discussed later.

In summary, a power amplifier is designed by the process of selecting transistors that will perform adequately at the stress limits imposed by the circuit and the load. For each type of power amplifier the designer needs to know the amount of power that must be dissipated by the transistor, the maximum collector current at which the transistor must have a useful current gain, and the maximum instantaneous collector voltage. Our objective here is to present the formulas that relate the stress parameters to the design objectives. The detailed derivation of these relations is avoided because derivations are a favorite topic in design literature, and it is more important that we emphasize how a design is actually accomplished.

7.5 PUSH-PULL AMPLIFIERS

The power output of an amplifier may be increased by using two transistors instead of one in the output stage. One way of connecting two transistors is by paralleling all three leads and lowering the circuit values so that all currents are doubled. Transistors share the load only if equalizing resistors are used, but these resistors consume valuable power. Parallel-connected transistors are mainly used at high frequencies and only rarely at audio frequencies.

A push-pull amplifier has a pair of amplifiers connected to share an input and output transformer, as shown in Fig. 7.3. The transistors are phased in such a way that they carry the load peaks alternately and are not required to divide the load equally, as with parallel transistors. Because the transistors carry the load alternately, it is possible to bias the amplifier so that dc collector power flows only when there is a signal. Amplifiers biased and operated with the collector power supplied at a fixed continuous value, regardless of the signal level, are known as *class-A amplifiers*. An amplifier that is biased so that the dc collector power flows only when the signal is applied, and then for the entire cycle, is known as a *class-B amplifier*.

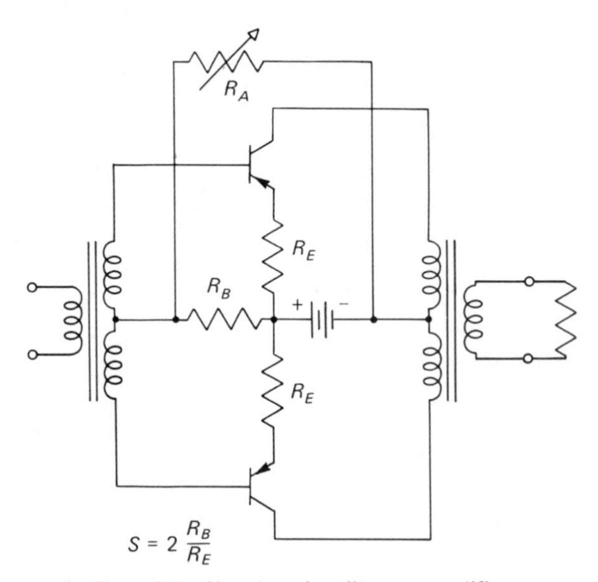

Figure 7.3. Class-A push-pull power amplifier.

Push-pull class-A amplifiers have the advantage that the dc collector current flows in opposite directions in the transformer and tends to cancel the dc core magnetization. They have the disadvantages that considerable heat must be dissipated by the transistors, and the dc power loss is a maximum under no-signal standby. A class-B amplifier requires very little dc input power under standby conditions and has, under all operating conditions, a higher power conversion efficiency than a class-A amplifier.

Push-pull transistor amplifiers operate so well near cutoff that there is little reason to use class-A push-pull amplifiers. An amplifier could be biased with vacuum tubes to reduce the dc input power in class-B service and obtain more than twice the power obtainable from a single tube. However, the difficulty of maintaining sufficient balance between tubes and the nonlinear characteristic of tubes near cutoff have made the class-B *vacuum-tube amplifier* generally unacceptable for linear (sound) systems. With reasonable care transistor amplifiers can be designed to remain balanced at distortion levels 1 or 2 orders of magnitude lower than with tubes. The advantages of the class-B amplifier are its low standby power and the small heat sink required in music and speech amplifiers. The class-A amplifier has 6-dB higher power gain and presents a constant load on the power supply.

7.6 OUTPUT TRANSFORMERS

A power amplifier may need an output transformer either to isolate the load from the amplifier or to provide an efficient impedance match. When a transformer is required, a major problem for the designer may be that of selecting and obtaining a suitable output transformer. An industrial designer is often forced to adapt an available transformer to his purposes, or he may have to design and build the transformer.

As with an ac power transformer, the physical size of an output transformer depends on the power that is transmitted and on the lowest signal frequency. For class-A amplifier service the primary winding of the transformer carries direct current, and the power capacity of the transformer may be less than one tenth the VA rating of an ac power transformer. Because the dc current tends to magnetize the core and to reduce the effective inductance, the core in an output transformer must be surprisingly large and should have an air gap.

As a practical guide for single-sided class-A amplifiers, the minimum area of the core may be obtained from the formula

$$A_c > \frac{150}{f_l} \sqrt{I_{CC} V_{CC}} \tag{7.4}$$

The core area is in square centimeters, f_l is the 3-dB low-frequency cutoff, I_{CC} is the dc current in the winding, and V_{CC} is the dc supply voltage. For high-fidelity applications, f_l is 30 to 50 Hz, and for portable battery-operated radios and sound systems (intercoms), f_l may be as high as 300 Hz. The formula gives a minimum value for the core area because a low-loss or a low-distortion specification may indicate the need to provide more than twice the area given by the formula.

The minimum core area for a push-pull class-A or class-B design is given by the formula

$$A_c > \frac{50}{f_l}\sqrt{P_O} \tag{7.5}$$

In equation (7.5) the output power P_O is in watts and the remaining units are the same as for equation (7.4).

A comparison of equations (7.4) and (7.5) shows that for the same power output the core area required for a single-sided class-A amplifier is from four to six times the area required for a push-pull amplifier. The size, weight, and cost of the transformer tend to limit the use of transformer-coupled-class-A amplifiers to about 1 W, which requires at least a 5-cm^2 core area.

The load impedance for which a transformer winding is intended is usually given by the manufacturer. (For additional information concerning the characteristics of transformers, see reference 7-1.)

7.7 CLASS-A AMPLIFIER DESIGN

Amplifiers are referred to as power amplifiers when the design requirements specify a certain amount of power that must be delivered to a given load resistance. A single-sided class-A amplifier is generally used when the required output power is 1 W, or less, and when for economy the designer wishes to use a single transistor. As an example, a small radio operating on an 80-V collector supply usually requires an output transformer to couple the 4- or 8-Ω speaker load to the high-impedance collector, and a single transistor will supply the required power without needing a large heat sink. For higher power outputs the designer would use a pair of transistors, and the transformer may be eliminated by changing to a low-voltage power supply. The circuit of a typical class-A 1-W power amplifier is shown in Fig. 7.4.

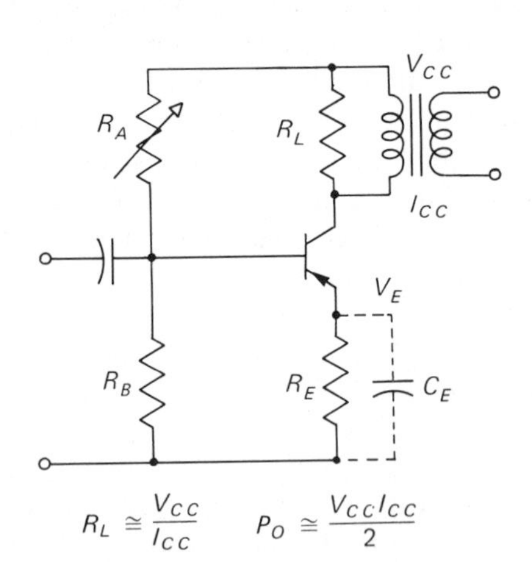

Figure 7.4. Class-A single-sided audio amplifier.

A class-A power amplifier is characterized by an efficiency of nearly 50 per cent, and the dc input power $V_{CC}I_{CC}$ does not change with the signal. These facts mean that, regardless of the transformer impedance, the bias should be adjusted to make the collector dc input power about twice the required power output P_O. To simplify calculations, we neglect the emitter voltage V_E, as compared with V_{CC}, and neglect the power V_EI_E.

A designer begins by selecting a transistor that will dissipate the total input power $2P_O$ and by confirming that space can be provided for the heat sink. Tentative values of the supply current and voltage are selected to fit the relation

$$I_{CC}V_{CC} = 2P_O \tag{7.6}$$

By referring to the collector characteristics shown in Fig. 7.5, consider the instantaneous peak collector current and voltage values that exist when the signal is a sine wave. When the signal has the value that gives the maximum power output, a high efficiency of dc to ac power conversion is obtained by making the instantaneous collector current increase until the collector voltage is driven down to zero volts. Similarly, on the other half-cycle the collector voltage is increased until the instantaneous collector current is zero. For linear operation the signal peaks on both halves of the cycle must be alike. Hence, the instantaneous current must increase to $2I_{CC}$ when the peak voltage is zero, and the voltage must increase to $2V_{CC}$ when the peak current is zero.

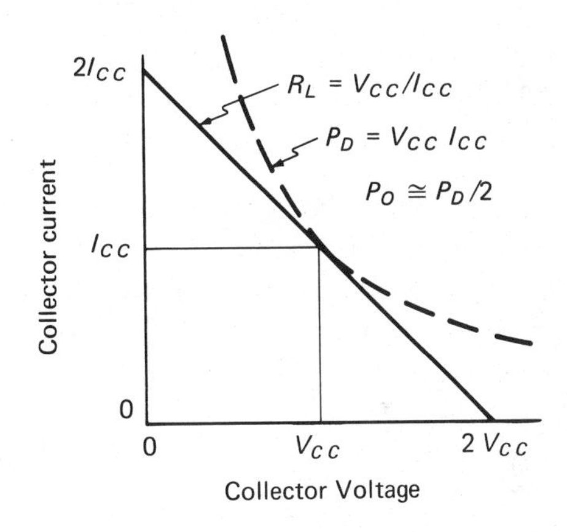

Figure 7.5. Class-A load line ($V_E \ll V_{CC}$).

The sinusoidal signal power output may be calculated by converting the peak-to-peak values to rms values, whence

$$P_O = \frac{2I_{CC}}{2\sqrt{2}} \cdot \frac{2V_{CC}}{2\sqrt{2}} = \frac{I_{CC}V_{CC}}{2} \tag{7.7}$$

The designer of a class-A power amplifier proceeds by selecting a transistor that can withstand the maximum instantaneous voltage

$$V_M = 2V_{CC} \tag{7.8}$$

while maintaining adequate current gain at the maximum instantaneous collector current

$$I_M = 2I_{CC} = \frac{4P_O}{V_{CC}} \tag{7.9}$$

To avoid collector voltage breakdown, the Q-point voltage V_{CC} is usually limited to about one fourth BV_{CBO}, a severe restriction. Because the power gain depends on the transistor current gain, low distortion at maximum output power requires that the transistor have a current gain several times the S-factor used in the design. A difficulty in meeting the required peak voltage or current may sometimes be met by changing both, as long as the product that controls P_O remains constant.

As soon as the transistor is decided upon, the transformer primary impedance may be calculated from the coordinates of the Q-point

$$R_L = \frac{V_{CC}}{I_{CC}} \tag{7.10}$$

Observe that the power output [equation (7.7)] and the transformer collector-to-collector impedance are both fixed by the coordinates of the collector Q-point. If the collector supply voltage is decided upon, the load impedance is inversely proportional to the power output

$$R_L = \frac{V_{CC}^2}{2P_O} \tag{7.11}$$

Because the output transformer consumes a part of the power developed by the amplifier, the transformer impedance is usually lowered enough to offset this power loss. Small transformers are quite inefficient, so it may be necessary to make P_O in equations (7.9) and (7.11) almost twice the desired load power. The secondary impedance is fixed, of course, by the given load impedance.

A disadvantage of the single-sided class-A amplifier is that the Q-point current tends to saturate the transformer core unless the core is made relatively large, heavy, and expensive. Together, the core and the heat-sink problems tend to make the push-pull class-B amplifier relatively more attractive when the required output power exceeds about 0.1 W.

7.8 AMPLIFIER CHARACTERISTICS

For proper operation a class-A amplifier must be operated with the Q-point close to the intended values. The temperature of the collector junction varies with the ambient temperature via the heat sink, with the signal, and with the length of time the amplifier has been operating after turn-on. Because the Q-point of a power amplifier is relatively more sensitive to temperature than a small-signal amplifier, special care should be given to the use of adequate feedback for Q-point stabilization. Satisfactory stability may usually be secured by providing emitter feedback with an S-factor of 10, or less. Because feedback increases the power required of the preceding stage, the driver, an emitter capacitor is usually used to prevent feedback at the signal frequencies.

A designer always considers the possibility of voltage breakdown of the transistors, and, when possible, will keep the peak voltage below about one half the transistor breakdown rating, BV_{CEO}. Second breakdown is not usually a problem in a small class-A design, but a nonlinear element may be needed to prevent damage by the transient voltage spikes caused by a sudden open circuiting of the transformer primary when the transistor is driven into cutoff.

A power amplifier is usually expected to provide a distortion-free output signal, and a harmonic content of less than 1 per cent of the signal voltage is considered low distortion. For low distortion a transistor should be driven by a current source, provided β is a constant, but a current source generally implies that the input circuit is operated at a low power efficiency. The power gain of a power amplifier without feedback is proportional to β, and the variability of β may be offset by emitter feedback. However, acceptably low distortion may be obtained by driving the input from a source having a finite impedance such that the input current-per-volt of signal increases with the signal enough to offset the β falloff on the signal peaks.

The small signal distortion may be reduced by overall feedback, but the abrupt rise of distortion at the maximum output level is fixed mainly by the design. The abrupt rise above three to five per cent generally begins at the signal level where the transistor ceases to have adequate current gain. Sometimes the overload is caused by transformer saturation or an inadequate power supply. The overload distortion cannot be reduced by feedback.

Careful testing of a class-A design is really the only way to locate potential problems and to decide that the design meets the requirements of an intended application. Transistor failures are an opportunity for the study of potential problems in the field, and a breakdown must never be dismissed as an accident of "early failure." Failures have a bad habit of becoming persistent when least expected and most inconvenient. A good design depends in large measure on the ability of the designer to recognize the early warnings that are present during the performance tests.

7.9 LINE-OPERATED 1-W AMPLIFIER

A 1-W amplifier that operates on the 120-V power line without an ac power transformer is illustrated in Fig. 7.6. Commonly used as the output stage of a

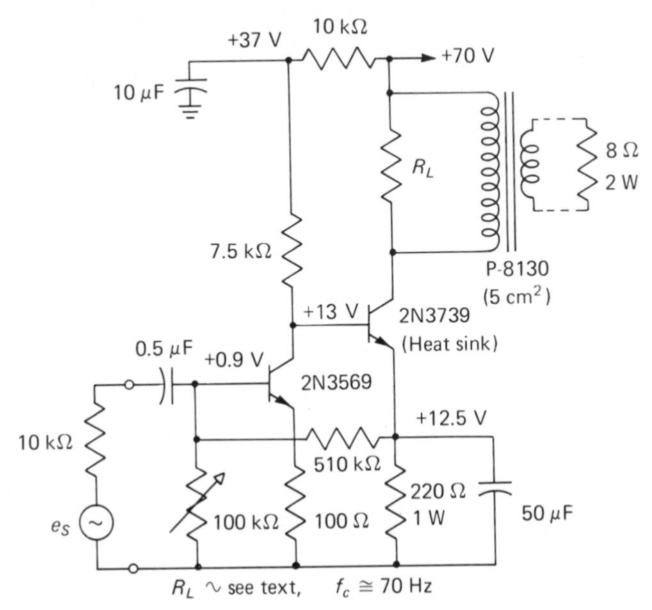

Figure 7.6. One-watt amplifier.

transistor TV, this amplifier provides an interesting study of the typical class-A power amplifier.

For dc power this amplifier uses a half-wave rectifier with a series resistor and a shunt capacitor that are not shown in the figure. The series resistor is selected to make the collector supply voltage 70 V, and the capacitor is made large enough to provide the required reduction of ac hum. Additional filtering for the driver stage is provided by connecting a low-voltage capacitor at a point on the collector resistor.

The amplifier is a direct-coupled *CE-CE* pair with current feedback from the second emitter to limit the overall *S*-factor. The procedure for design is approximately as follows:

1. The amplifier has a conversion efficiency of about 35 per cent, so with 1-W output the collector dissipation is 3 W. If the dc emitter voltage is 10 V and the voltage drop in the transformer is 3 V, the collector-to-emitter voltage is only 70 − 13 = 57 V. For 1-W output the collector dissipation is made 3 W. The collector voltage is 57 V, so the collector current will be 53 mA.
2. At high signal levels the 2N3739 power transistor may have a current gain as low as 20. Hence, the maximum required base drive current is about 2.5 mA. When the driver transistor is full OFF, the second stage is driven by the current in the first-stage collector resistor. These estimates suggest using a collector resistor of about 20 kΩ.

3. The emitter resistor should be as high as practical in order to ensure thermal stability of the second stage and to prevent runaway damage. Moreover, the emitter resistor should be as large as is permitted by the collector-circuit voltage loss. The usual design practice is to make the emitter voltage about 10 V.
4. The input-stage emitter resistor is selected by trial and error and serves largely to increase the input impedance level. As shown in Figure 7.6, the amplifier has an input impedance of about 15 kΩ without feedback. The emitter resistor also provides a convenient place to introduce feedback. When the 8-Ω voice-coil winding of the output transformer is connected to the first emitter through 1000 Ω, the overall voltage gain is reduced to 10. This connection raises the input impedance to 30 kΩ and tends to make the distortion independent of the input source impedance.
5. The designer selects an output transformer after finding the desired turns ratio and the required core area. The dc collector-to-emitter voltage calculated in step 1 is 57 V. This voltage with the expected 53-mA emitter current suggests a 1000-Ω load impedance [equation (7.10)]. To match the 8-Ω voice-coil impedance, the transformer should have an 11:1 step-down turns ratio. The core area of the transformer is found by use of equation (7.4). Substituting $V_{CC} = 57$ V, $I_{CC} = 0.05$ A, and $f_l = 75$ Hz gives $A_c = 3.5$ cm^2. Transformers for this application are not usually available as cataloged items. The output transformers intended for 70- or 25-V lines make satisfactory class-A output transformers for low-power applications. Generally, the cores of these transformers are too small to tolerate the dc current of a 1-W amplifier. In some cases a small transformer may be capacitor-coupled to the transistor with the dc collector current supplied through a resistor, as with resistance coupling.
6. If as in this example, a satisfactory transformer is not readily available, the designer looks for a substitute so that the design may be examined experimentally. The fact that the peak ac voltage on the primary winding is to be 57 V suggests that a 60-Hz filament transformer may be satisfactory for a test. Power transformers, especially when designed for 50 to 400 Hz, are suitable for voice frequency and similar applications. A 120- to 12-V transformer has the correct step-down ratio, and the 60-Hz operating frequency suggests that the winding inductance may be high enough for the intended application. The transformer shown in Fig. 7.6 has a core with more than the required area but is a power transformer that does not have enough inductance or air-gap for a 75 Hz class A amplifier. However, even with these limitations, the transformer can be used to test the amplifier at voice frequencies with the design of a more suitable transformer left until later.
7. Whether an output transformer actually has the correct load impedance for the transistor is easily determined. The load check is made by

measuring the ac power delivered to several different values of load impedances connected across the primary. After finding the optimum primary load impedance and power, the designer compares that load power with the power delivered to the equivalent load on the secondary. Because of the loss in the transformer, the power into the secondary load is expected to be 85 per cent of the primary input power. For the load measurements the power is calculated from the peak-to-peak voltage and the load resistance, using the equation

$$P_O = \frac{(V_{p\text{-}p})^2}{8R_L} \tag{7.12}$$

8. Because the power transistor may be damaged by transient spikes if the amplifier is operated without a resistance load, the transformer primary should be shunted by a Thyrite varistor. Thyrite varistors are rated by a clipping voltage at which the current increases as the fourth power of the voltage and by the power that can be dissipated. The varistor for the 1-W amplifier should be rated for 60 to 100 V and $\frac{3}{4}$ W. The amplifier may be safely tested without the varistor if the load is connected in a way that prevents accidental removal.

7.10 CHARACTERISTICS OF THE 1-WATT AMPLIFIER

The 1-W amplifier was assembled and tested, using the 12-V filament transformer to couple a 10-Ω load. Full output of 1 W is obtained with 0.35-V peak-to-peak input signal from a 10-kΩ source. The small-signal voltage gain from the base to the load is 40 without feedback, and is 25 when referred to the no-load voltage of a 10-kΩ source. The conversion efficiency of the transistor is 40 to 45 per cent, depending on the accepted distortion, and the transformer loss reduces the efficiency to about 33 per cent. When the $\frac{3}{4}$-W power loss in the emitter resistor is included, the overall efficiency is only 25 per cent.

A single-sided class-A amplifier has the disadvantage that the signal is clipped on one side only at the maximum power output. Increasing the load impedance to about twice the calculated impedance decreases the peak collector current and eliminates most of the odd harmonic distortion. The maximum power output is decreased about 10 per cent by this change, and the efficiency is reduced.

The experimentally observed characteristics of the 1-W amplifier suggest that the 12-V transformer may be loaded by 16 Ω on the full winding or by 4 Ω on the 6-V center tap. An improved low-frequency response may be obtained by doubling the turns of both windings and by inserting an air gap in the core.

The power gain of the amplifier is calculated from

$$G_p = G_v G_i = G_v^2 \frac{R_I}{R_L} \tag{7.13}$$

Inserting the base-to-collector voltage gain gives

$$G_p = (40)^2 \ \frac{15{,}000}{10} = 2.3 \times 10^6 \text{, 64 dB} \tag{7.14}$$

The fact that a power gain of 30 dB per stage is reasonable for this design may be found by converting the observed voltage gain to an overall current gain. Thus, by the *TG-IR*

$$G_i = G_v \frac{R_I}{R_L} \tag{7.15}$$

and

$$G_i = 40 \ \frac{15{,}000}{10} = 60{,}000 \tag{7.16}$$

The overall current gain includes a 10:1 current gain in the output transformer. Hence, the open-loop two-stage current gain is 6000. If we assume that $\beta = 100$ for the input transistor, the power transistor appears to have a reasonable $S = 60$. This result with the observed 1-W output confirms the success of the design. The main limitation of the class-A design is illustrated by the large transformer and by the poor low-frequency response with full power output.

7.11 THE TRANSISTOR HEAT SINK

The power transistor needs a small heat sink to dissipate 4 W, although it may be operated for a few minutes without a heat sink. The thermal resistance of the transistor itself is small enough to be neglected in comparison with that of the heat sink. If we assume a 100°C temperature rise with 4-W input, the heat sink may have a thermal resistance as high as 25°C/W. However, the designer should consider the problems of mounting a heat sink that is 100°C above the ambient and the problems of protecting nearby components from the heat. A larger heat sink with a thermal resistance of 10°C/W, or less, would limit the heat-sink temperature to a more reasonable 40°C above the ambient.

Small heat sinks have a thermal resistance of about 500°C/W for each square centimeter exposed to ambient. With 50 cm^2 (in parallel) the thermal resistance is 10°C/W. A 5- by 5-cm aluminum plate exposed on both sides, or a 7- by 7-cm plate exposed on one side, makes a satisfactory heat sink. The required area may be confirmed by referring to the more accurate chart giving heat-sink dimensions in Appendix A.3.

7.12 SUMMARY

Class-A transformer-coupled amplifiers are generally found in low-power applications where the transformer is used to isolate the load from the amplifier. If a step-down transformer is used, the transformer effectively increases the voltage gain of the stage by n, the primary-to-secondary turns ratio. However, in a low-power iterated amplifier, as in a hearing aid, the gain advantage is available only if the emitter current exceeds approximately 0.25 mA.

The transformer for an audio amplifier must have an iron core and close coupling for an efficient transfer of power throughout the relatively wide frequency band. The dc collector current of a class-A amplifier tends to saturate the core and make it necessary to use impractically large transformers when the output power exceeds 1 W. The core area required for a push-pull amplifier is from four to six times smaller than for a single-sided class-A amplifier with the same power rating.

The designer of a class-A power amplifier begins by selecting a supply voltage and current that give an input power three to four times the required output power. A transistor is selected that has current gain at a peak collector current approximately twice the Q-point current and a voltage rating that is two to four times the supply voltage. If a suitable transistor cannot be found, the terms of equations (7.8) and (7.9) must be adjusted to accommodate the ratings of the available transistors. The class-A impedance is approximately the ratio of the dc collector voltage to the collector current. For a maximum gain in low-power applications the primary impedance should be as high as practical.

A class-A amplifier must have a heat sink that will dissipate the Q-point power input. Q-point changes with temperature and thermal runaway are both controlled by using a low S-factor and an adequate heat sink. In some cases a designer may use temperature-sensitive resistors for temperature compensation in the bias network.

Difficulty in obtaining a satisfactory transformer for a power amplifier is an everyday problem for the designer. A manufacturer of radios, for example, may be able to design and build a special transformer, but the designer of a prototype for small-quantity production must be prepared to find substitutes or to modify his design. The designer may be forced to accept a reduced power output or to design and build a simple transformer. These problems are a continual source of frustration and are time-consuming. However, each new problem leads the designer into a new area for study and contributes to the broad field of practical experience that in time is most rewarding.

PROBLEMS

7-1. (a) Draw a circuit showing the changes you would make if the amplifier shown in Fig. 7.2 is to be operated on a −12-V supply with a 50-Ω load. (b) What voltage gain and peak output signal do you expect from your design?

7-2. A single-sided class-A amplifier is designed to operate on a 17-V supply using an output transformer that has a 3:1 step-down turns ratio. The load has 11Ω. (a) Draw the load-line characteristic for this amplifier with current and voltage scales. (b) What is the power output?

7-3. (a) Construct a load-line characteristic for the output stage illustrated in Fig. 7.6. (b) Show how the power output of this amplifier is calculated.

7-4. (a) Consider the 1-W amplifier shown in Fig. 7.6. Calculate the expected power output if the collector voltage is lowered to 42 V when all other Q-point values remain the same. (b) To what value can the collector current be reduced by biasing and obtain the power output calculated in (a)?

7-5. Explain the differences between the designs represented by Figs. 7.2 and 7.4.

7-6. Find a circuit with component values for a single-ended amplifier similar to that shown in Fig. 7.4. Specify suitable current and voltage ratings for the power transistors, compute the expected power output, and describe the bias circuit.

DESIGNS

7-1. Design a single-sided class-A amplifier to provide 0.25 W when operating on a 30-V supply.

7-2. Construct the amplifier shown in Fig. 7.2 and examine the possibility of obtaining greater power output by using a different output transformer.

7-3. Design a class-A push-pull power amplifier that delivers 1 W to an 8-Ω load when operated on a 12-V supply.

REFERENCES

7-1. Cowles, L. G., *Transistor Circuits and Applications*. Englewood Cliffs, N.J. Prentice-Hall, Inc., 1968.

7-2. Snyder, G. E., and E. L. Haas, *A Four Transistor Line-Operated Radio Receiver*, Application Note No. 90.26, 4 pp. Schenectady, N.Y.: General Electric Co., 1965.

7-3. *RCA Transistor 40264 Data Sheet*. Harrison, N.J.: Radio Corporation of America, 1964.

8

CLASS-B POWER AMPLIFIERS

A class-B power amplifier offers high efficiency and has low standby power requirements and low distortion. Because voice and music are highly intermittent signals, high-power output may be obtained from a pair of relatively low-power transistors.

Transistors are remarkably well adapted for class-B operation. When biased slightly ON for class-B operation, a transistor has high gain and has an essentially linear response. Moreover, substantial feedback may be used in direct-coupled transistor amplifiers. With these contributing characteristics, the crossover distortion of the class-B amplifier can be reduced to a negligible level. Because power transistors are high-current devices, many low-impedance loads—e.g., voice coils—can be driven without the output transformer that has been a source of distortion and low-frequency cutoff in vacuum-tube amplifiers.

Class-B power amplifiers are easily designed with the use of discrete components and are becoming available as integrated circuits for low-power applications. This chapter concludes with the design of a 7-W class-B amplifier and preamplifier suitable for a stereo phonograph amplifier.

8.1 PUSH-PULL CLASS-B AMPLIFIER

A class-B amplifier is biased to make the collector current nearly zero when no signal is applied. With a signal, one transistor amplifies the positive side and the other transistor amplifies the negative side of the signal. The wave form of the primary current is a series of half-cycles, as shown in Fig. 8.1. In a push-pull class-B amplifier these signals are combined by the output transformer to restore the original wave form.

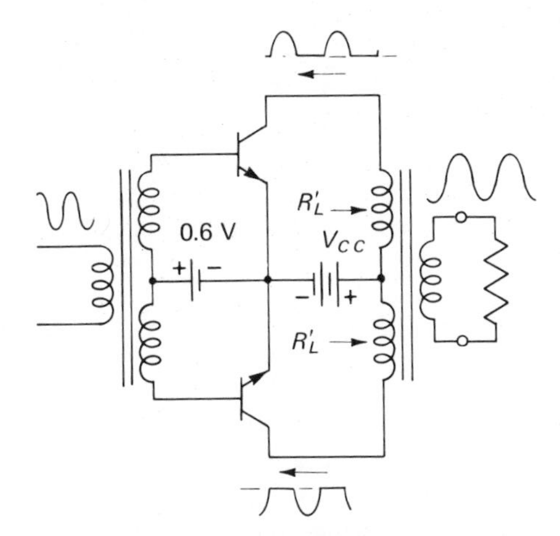

Figure 8.1. Class-B amplifier.

A class-B amplifier cannot be biased exactly at cutoff because cutoff biasing removes small signals and causes crossover distortion. However, when a bias is used that barely turns on both sides of the amplifier, most of the small-signal distortion is removed. Because the class-B amplifier has high gain at the bias point, the relatively small nonlinearity is easily removed by feedback. Thus, it is possible to design class-B amplifiers having exceedingly low distortion. By eliminating the output transformer and direct coupling the driver stages, a class-B amplifier may be given 40-dB feedback with a broad frequency band and a high degree of stability. These characteristics have brought transistor hi-fi amplifiers to a perfection well beyond that attained with vacuum tubes.

8.2 CLASS-B AMPLIFIER DESIGN

A class-B amplifier may be designed easily when the designer recognizes that the transformer carries the signal in one side on alternate half-cycles. The amplifier has a theoretical efficiency of nearly 78 per cent because the dc power input increases with the signal and is very low on standby. The maximum dc to ac conversion efficiency is achieved by driving the collector to zero volts at the peak of the maximum output signal. The peak signal voltage on one side of the transformer is V_{CC} across the side impedance R_L'. The power output is developed on alternate sides of the transformer on alternate half-cycles; hence, the full-cycle power is P_O, where

$$P_O = \left(\frac{V_{CC}}{\sqrt{2}}\right)^2 \cdot \frac{1}{R_L'} \tag{8.1}$$

When solved for the load impedance on one side of the output transformer, equation (8.1) gives

$$R_L' = \frac{V_{CC}^2}{2P_O} \tag{8.2}$$

and the nominal primary winding impedance is four times the side impedance, or

$$R_{CC} = \frac{2V_{CC}^2}{P_O} \tag{8.3}$$

As equation (8.3) shows, the power output of a class-B push-pull amplifier is determined merely by the supply voltage V_{CC} and the impedance R_L' presented by the output transformer. Small class-B transformers, both input and output, have a high copper loss, which must be supplied in addition to the output load. For example, an amplifier supplying 1 W to a load may have to develop an output power of nearly 2 W. The collector load impedance of such a transformer must be sized for 2 W. These considerations reduce equation (8.3) to

$$R_{CC} \cong \frac{V_{CC}^2}{P_O} \tag{8.4}$$

as a more practical relation for class-B designs at the 1-W level, or below. The published nomograms are usually based on the last equation.

The load diagram for one side of the class-B amplifier is represented in Fig. 8.2. As shown there, the transistors are turned OFF when the maximum instantaneous voltage,

$$V_m = 2V_{CC} \tag{8.5}$$

is highest. With resistance loads the transistor voltage rating BV_{CEO} need be only a little above $2V_{CC}$. Hence, transistors may be operated class B with nearly twice the supply voltage that is used for class A.

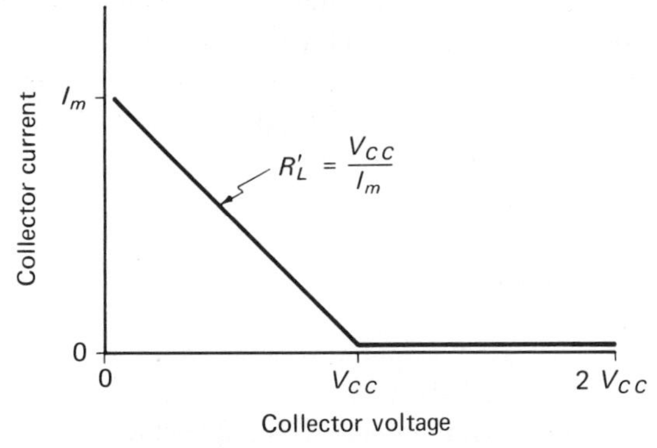

Figure 8.2. Class-B amplifier load line (one side).

The transistors should be selected to have adequate current gain when the transistor is turned full ON. The transistor connects the supply voltage V_{CC} across the load R_L', so the maximum instantaneous current is V_{CC}/R_L'. For design purposes the maximum current is expressed in the more useful form

$$I_m = \frac{2P_O}{V_{CC}} \tag{8.6}$$

As shown elsewhere (see reference 8-4), a class-B amplifier may use the same transformer as a push-pull class-A amplifier that is producing the same power output. The class-B transistors operate at one half the peak collector current of the equivalent class-A amplifier and closer to the transistor VB_{CEO} rating. These are significant advantages, although the class-B amplifier has one half the power gain of the class-A amplifier.

Because of the varying demand on the power supply, the output wave form and the maximum power output depend partly on the "stiffness" of the power supply. Good regulation is desirable and the cost of a stiff power supply is an important consideration in the design of a class-B system.

The efficiency of a class-B amplifier falls off at low signal levels. The maximum power dissipation is 40 per cent of the maximum power output and occurs at 40 per cent of the maximum output, where the efficiency is 50 per cent. However, the crossover bias generally increases the standby power to 10 per cent or more of the maximum dc input power. Hence, the maximum power dissipation may be 50 per cent of the maximum output power. Speech and music signals have an intermittent character that makes it possible to design audio amplifiers with the transistor heat sink only large enough to dissipate the standby power. Because an amplifier may operate accidentally with a sustained high-frequency feedback signal, better reliability is ensured by providing enough heat sink to dissipate 50 per cent of the maximum power. A small class-B amplifier is rarely damaged by thermal runaway and will operate normally after the transistors cool. Silicon transistors are often operated in class-B service without heat sinks when the power output rating is 1 W or less. When desired, the circuit can be designed to protect the transistors from damage by runaway. (For a description, see reference 8-5.)

8.3 DRIVER AMPLIFIERS

The power gain of a class-B amplifier is about 200 times (20 to 30 dB). The driver amplifier is usually a class-A stage, rated to supply about one twentieth the power output of the class-B amplifier. The interstage, or input, transformer cannot be easily specified theoretically and is best selected by trial and error. Transformer manufacturers usually offer sets of input and output transformers with a typical circuit. As a starting point, the base-to-base winding, the secondary, usually has the same nominal impedance as the following collector-to-collector winding. The power rating of the transformer should be about one twentieth the output stage power rating, and the primary impedance may be determined by equation (7.11) because the driver is a class-A stage.

Because the input impedance of a class-B amplifier varies with the signal level, some feedback systems (e.g., servo) operate better when the driver amplifier is loaded by a resistor connected across the primary of the driver transformer. This resistor lowers the loop gain with a low-level signal without affecting the gain at high signal levels. By this action the resistor may improve the stability of a tightly controlled feedback system.

The description of a 2-W, class-B, push-pull power amplifier may be found in reference 8-4. A 0.25-W, push-pull, class-B amplifier suitable for a small radio or intercom is illustrated in Fig. 8.3.

8.4 CLASS-B BIASING

For power economy and for distortion control, a class-B amplifier needs to be driven by an approximately constant voltage source. The low dc resistance of the base circuit makes the bias point sensitive to temperature and β. Because a class-B stage is biased near cutoff, even a small change of the V_{BE} voltage produces a large change of the crossover distortion or of the dc collector power. To stabilize the bias circuit, a class-B amplifier usually includes a thermistor to decrease the bias current with temperature, or a diode, as shown in Fig. 8.3. The diode should be mounted on the transistor heat sink so that its temperature follows that of the transistor. By this means the diode voltage drop tends to offset and to change with the V_{BE} voltage of the transistors.

Drift engendered by both β and temperature may be further reduced by

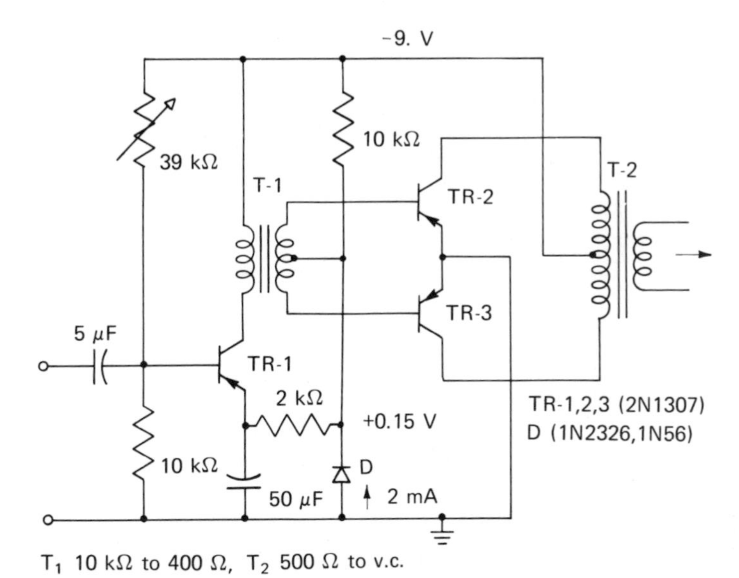

Figure 8.3. Push-pull 0.25-W class-B amplifier.

emitter feedback. The emitter resistors are selected to give 0.5- to 1-V voltage drop with the maximum output signal, and the dc resistance in the transformer should be limited in order that the *S*-factor will be 10 or lower. Because of the temperature problems, silicon transistors are usually preferred for class-B applications. Capacitors are not normally used to bypass the resistors in a class-B stage because the unidirectional nature of the currents tends to charge the capacitor and move the *Q*-point toward or beyond cutoff.

8.5 SINGLE-ENDED CLASS-B AMPLIFIERS

A single-ended class-B amplifier differs from a push-pull stage by having the power transistors connected in series, as in Fig. 8.4, with the load capacitor-coupled to the midpoint. This arrangement eliminates the output transformer, but, by keeping the driver transformer, has the advantage of permitting a low *S*-factor. (An example of a 15-W, single-ended, class-B amplifier is given in reference 8-4.) The main difference between a single-ended and a push-pull design is in the fact that the driver transformer must have two secondaries. In the example given, each secondary has the same impedance as the output load (8 Ω), and the *S*-factor is about 15.

The single-ended amplifier has advantages for applications that require high power output and good temperature stability. Good driver transformers are

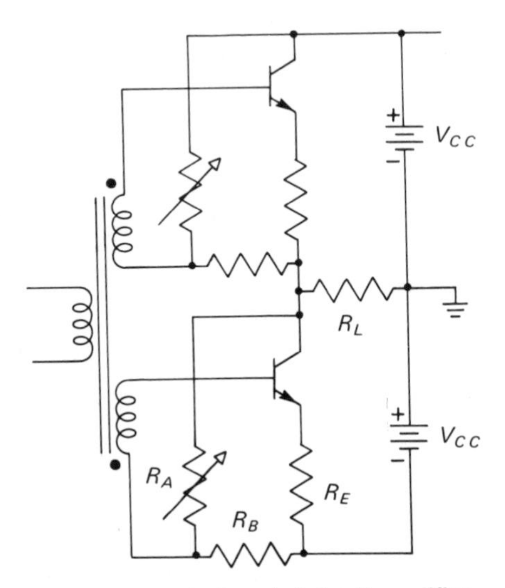

Figure. 8.4. Single-ended class-B amplifier.

more easily obtained and are much less expensive than output transformers, so this circuit is often used in inexpensive public-address systems. The transformer-coupled driver eliminates some of the temperature problems of a direct-coupled amplifier and may give more reliable operation under unfavorable operating conditions and improper maintenance. An amplifier having even one transformer presents difficulties in applying feedback. For this reason, the transformerless direct-coupled class-B amplifiers are much preferred for high-fidelity designs.

The single-ended and the transformerless (or complementary-symmetry) class-B amplifiers are designed with the aid of the same equations, since both circuits have the transistors in series and the load is connected at the center tap. Because the transistors are never subjected to more than the total supply voltage, an easier comparison with the push-pull relations is made by assuming that a transformerless amplifier is operated on twice the supply voltage of the push-pull amplifier. Twice the supply voltage may be easily represented by assuming that the supply is constructed to provide $+V_{CC}$ and $-V_{CC}$, as shown in Fig. 8.5.

The transistors are biased class B so that the center loadpoint is at ground potential at the center of the supply. The transistors alternately connect the load across one or the other side of the power supply, as indicated in Fig. 8.6. The maximum instantaneous collector current is

$$I_M = \frac{V_{CC}}{R_L} = \frac{2P_O}{V_{CC}} \tag{8.7}$$

and the maximum collector voltage is

$$V_M = 2V_{CC} \tag{8.8}$$

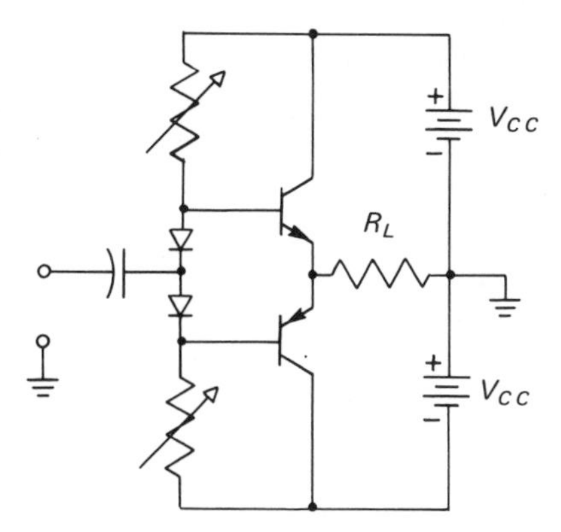

Figure 8.5. Single-ended class-B amplifier.

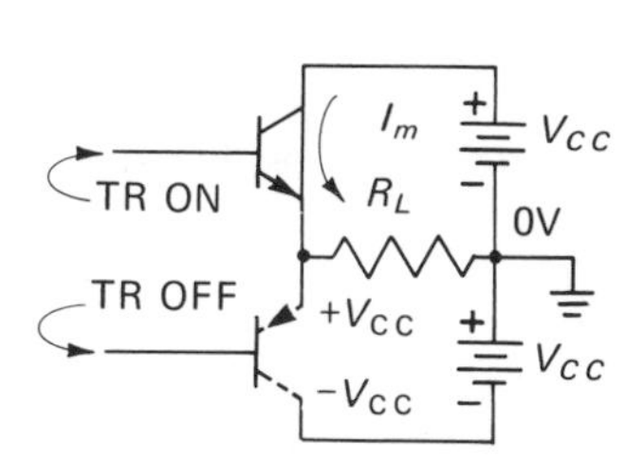

Figure 8.6. Peak signal relations in a class-B amplifier.

The power output is calculated by converting the peak voltage V_{CC} across the load to rms voltage, and we have

$$P_O = \left(\frac{V_{CC}}{\sqrt{2}}\right)^2 \cdot \frac{1}{R_L} = \frac{V_{CC}^2}{2R_L} \tag{8.9}$$

Observe that by assuming the double-voltage power supply, all the equations for the transformerless and the push-pull circuits are the same, with the load R_L taking the place of the single-side load R_L'. The similarity of the results means that for the same power output the transistors in the transformerless circuit operate with the same current and voltage ratings as with a transformer, provided the power supply has double the voltage, and, of course, half the current capacity.

The higher voltage power supply makes power-hum filtering a little easier. The principal advantage of the transformerless amplifier is that it is possible to use more feedback and obtain as much as 30 W with 0.1 per cent distortion from 20 Hz to 100 kHz. With an 8-Ω load, equation (8.9) shows that 10 W requires V_{CC} = 13. V. In actual practice the supply voltage must be about 30 per cent higher, or more, depending on the success that is achieved in driving the output transistors to full output and on the capacity of the power supply to maintain the rated voltage for the duration of the signal.

8.6 POWER SUPPLY FOR CLASS-B AMPLIFIERS

The dc power required to operate a class-B amplifier at maximum power output may be calculated from the power output and the expected conversion efficiency. If we assume a 70 per cent efficiency and no output transformer, the collector input power is $P_O/0.7$ W. The input power gives only the dc power that must be developed by the power supply. The designer needs to know the VA rating and voltage rating required for the ac power transformer. The VA rating (see reference 8-4) is 30 per cent larger than the dc load power when the rectifier is full wave.

The full-load voltage $2V_{CC}$ of the power supply is calculated from the peak signal voltage across the load R_L by solving equation (8.9) for V_{CC}, whence

$$2V_{CC} = \sqrt{8P_O R_L} \tag{8.10}$$

and the supply voltage required for each side is

$$V_{CC} = \sqrt{2P_O R_L} \tag{8.11}$$

With full-load output the dc collector input power is $2V_{CC}I_{CC}$, which is calculated from the power output and the estimated efficiency. Thus,

$$2V_{CC}I_{CC} \cong \frac{P_O}{0.7} \tag{8.12}$$

The full-load dc collector current is obtained by combining equations (8.10) and (8.12) to give

$$I_{CC} = \frac{1}{2}\sqrt{\frac{P_O}{R_L}} \tag{8.13}$$

As an example, suppose that we want 7.5 W across an 8-Ω load. Equation (8.13) gives as the side-to-side dc voltage of the supply

$$2V_{CC} = \sqrt{8(7.5)8} = 22 \text{ V} \tag{8.14}$$

and for each side

$$V_{CC} = 11 \text{ V} \tag{8.15}$$

From equation (8.13) the full-load dc collector current is

$$I_{CC} = \frac{1}{2}\sqrt{\frac{7.5}{8}} = 0.49 \text{ A} \tag{8.16}$$

and the VA rating of the transformer is approximately

$$1.3(2V_{CC}I_{CC}) = 1.3(22 \times 0.49) = 14 \text{ VA} \tag{8.17}$$

8.7 PRACTICAL DESIGN LIMITATIONS

The theoretical relations that have just been used in the example may be used as a guide, but not for a practical design. The difficulty is that a class-B power amplifier cannot be driven to make the peak output voltage equal to the supply voltage, as implied by equation (8.9). The problem is mainly that power transistors cannot be turned full ON, and, in circuits using an emitter follower as driver, the emitter voltage is at least 1 V less than the peak base voltage. In other words, the limitations of transistors and practical circuits make it impossible to use the entire supply voltage. These limitations affect the design of an amplifier operating on a low-voltage supply more than one on a high-voltage supply.

A number of factors increase the load that must be supplied by the power supply above the V_{CC} and I_{CC} values just calculated for the class-B amplifier. Three principal factors need to be considered:

1. The supply voltage must be increased to offset the V_{BE} voltage drop required to turn the driver transistors and power transistors full ON. If a Darlington pair of silicon transistors is used on one side of the amplifier, the voltage drop from the first base to the output emitter may be nearly 2 V at the peak of the signal. If the opposite side of the amplifier is a pair of *CE* stages, the output collector at the peak signal may be nearly 1 V below the opposite supply voltage. Taken together, these voltage drops force the designer to allow about 3 V more than the voltage $2V_{CC}$ calculated by equation (8.10). If emitter resistors are used to prevent short-circuit damage to the power transistors, the voltage drop in the resistors should be included also. If germanium transistors are used in the output stage, the V_{BE} voltage drops are about one half those assumed for the silicon transistors, but the allowance for the series resistors may be higher.
2. The regulation of the power supply itself is usually a significant factor in causing the supply voltage to fall during the signal peaks at full output. A line-operated amplifier usually has a capacitor input filter and a step-down transformer that is as physically small as practical. A small transformer is used to minimize the initial cost and the space required for the transformer. A small transformer usually has a 10 per cent voltage drop in the windings at full load, and the rectifiers have nearly 1 V each at full load. At full output the capacitor input filter usually has only three fourths of the no-load voltage, or about 1.1 times the transformer no-load rms voltage. More serious considerations for persons designing for

small-quantity applications are that custom-built transformers are expensive and the choice available in the cataloged voltage and VA ratings is quite limited. A transformer made by one manufacturer is usually identical to one made by another.

3. A final consideration entering the design of the power supply is the current load required for driver stages and associated amplifiers, relays, lamps, etc. This load with the current required for the crossover bias should be expected to be at least 25 per cent of the full-load current.

   Circuit design is never the simple straightforward process implied by the theoretical relations given in textbooks on design. Circuits are generally designed because they cannot be purchased as a ready-built package, and the designer has to face at least two practical problems. First, the circuit selected for the design has practical limitations, and the specifications of the design introduce others. Second, the components available at a reasonable cost usually impose quite severe limitations on the freedom available to the designer. For example, components may be available for a 7- and a 20-W amplifier but not for a 12-W amplifier. The difference between a 12- and a 7-W amplifier is really quite small, but the practical differences between a 7- and a 20-W amplifier are quite large in weight and cost. In adapting a design to available components or in keeping a design within space and weight limitations, the designer faces many difficult and tantalizing choices.

8.8 A 7-W HIGH-FIDELITY PHONO AMPLIFIER

The circuit of a practical 7-W high-fidelity phonograph amplifier is shown in Fig. 8.7. The first three stages of the amplifier are a preamplifier, and the last six transistors are in the power amplifier. The preamplifier serves to amplify and shape the signal received from a transducer, the pickup. The power amplifier is comprised of the single-sided amplifier TR-4 and TR-5, the complementary drivers TR-6 and TR-7, and the power transistors TR-8 and TR-9.

The first two stages of the preamplifier employ a direct-coupled FET-transistor pair for high input impedance and for an amplification of about 100. The feedback from the collector to the FET source reduces the frequency response, beginning at 60 and terminating at 500 Hz, as a part of the RIAA equalization for a magnetic pickup. The remaining RIAA cutoff from 1 kHz upward is provided in the high-frequency tone control R_3. By placing the cutoff in the tone control, the control may be adjusted to give a high-frequency boost or a loss.

The stage with TR-3 is a low-frequency boost amplifier. The *CE* stage is similar to the usual base-boost amplifiers except that only a part of the available gain is used for feedback, so the stage has a net gain.

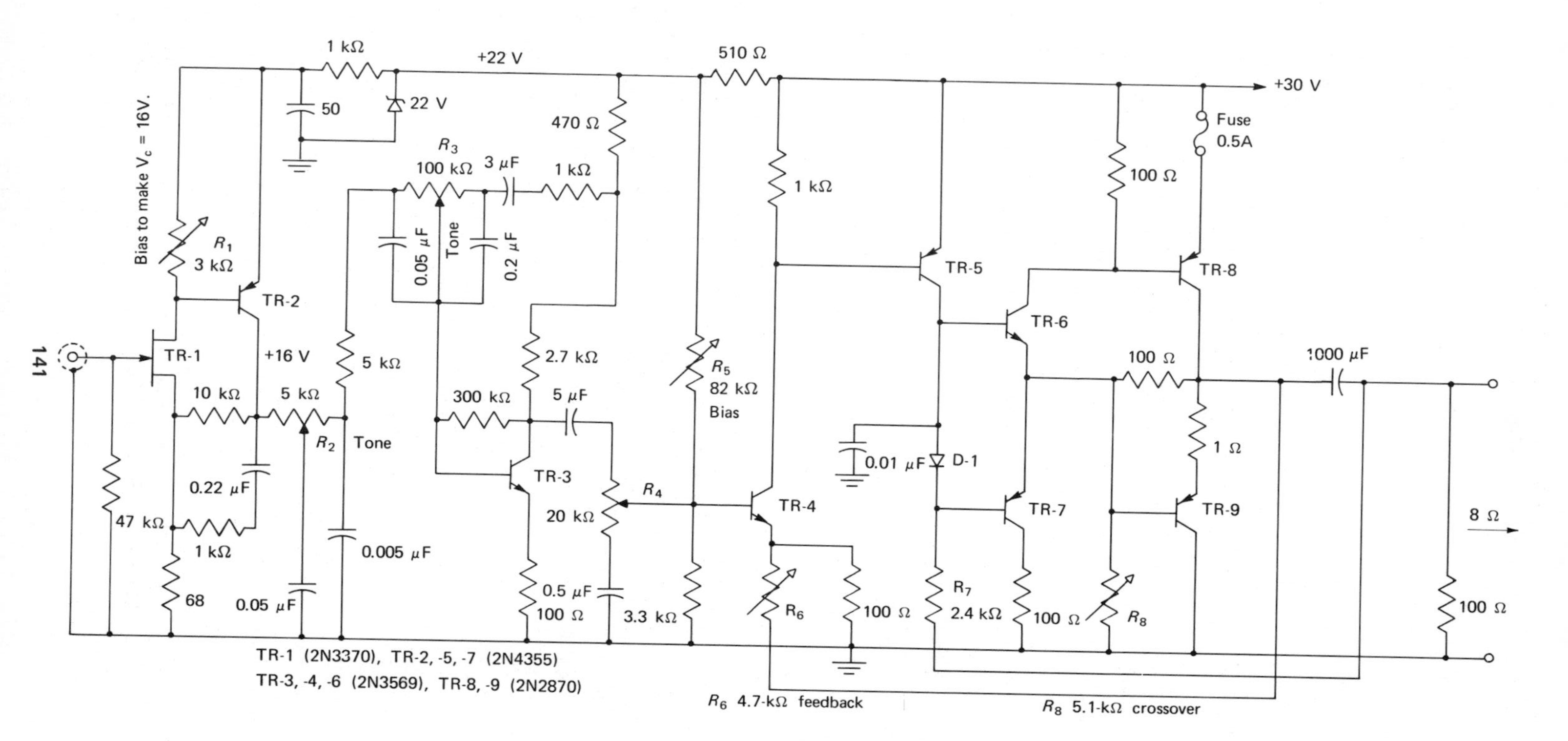

Figure 8.7. A 7-W high-fidelity phonograph amplifier.

The input stage of the power amplifier uses the CE configuration to minimize the interaction between the preamplifier and the power amplifier. The power amplifier has approximately 40-dB feedback, and interaction with a prior stage may make the power stage unstable. The amplifier gain control R_4 is between the preamplifier and the power amplifier, where there is no danger of signal distortion at low-gain settings of the control. The 0.5-μF capacitor provides an approximate Fletcher-Munson frequency compensation at low-gain settings.

The power section of the amplifier begins with a *CE-CE* pair for gain, followed by the complementary-symmetry drivers TR-6 and TR-7. The output stage uses a pair of *pnp* 2N2870 transistors that are series connected to complete the quasi-complementary-symmetry amplifier. The crossover distortion is adjusted by R_8, which has the effect of moving the distortion point to a high signal level, where the residual distortion is relatively unimportant.

A small increase or decrease of the overall gain may be obtained by changing R_6 and the amount of feedback. These changes generally require a readjustment of the bias resistor R_5, and the bias is correct when the TR-8 collector voltage is one half the supply voltage.

Although the power section has between 30- and 40-dB feedback, the amplifier is stable for all ordinary loads and does not require a feedback gain-step. The designer may wish to add the gain-step for an improved transient response.

The power amplifier is protected from a short-circuited load mainly by the 100-Ω resistor in the TR-7 collector. The transistors TR-4 through TR-7 are high-β types with 60 V, or greater, collector voltage ratings. The bias for TR-6 and TR-7 is supplied through a 2.4-kΩ resistor connected to the load. This connection, a simplification of the usual bootstrap, effectively removes the bias resistor as a load on TR-5 and increases both the amount of feedback and peak signal output.

The amplifier offers low noise and a low hum level, provided the high-current ground circuits are carefully separated from the low-level circuits. The output power may be increased by operating the amplifier on a higher-voltage higher-capacity power supply. Because of the simplicity of the circuit and the relatively small number of components, the amplifier is attractive for experimenters who wish to build their own stereo system.

8.9 DESIGN OF THE 7-W POWER AMPLIFIER

In planning the design of the 7-W amplifier, the designer knows that some power is lost in the emitter resistors, so he may select 8 W as the design level of power output. Substituting 8 Ω, the load impedance, and 8 W in equation (8.11) shows that the collector voltage on one side, V_{CC}, should be $8\sqrt{2} = 11.3$ V.

From equation (8.7) we find that the transistors should be selected for adequate current gain at $2 \times 8/11.3 = 1.4$ A. We know also that the transistors should have a favorable second-breakdown rating under short-circuit conditions. The 2N2870 transistors were found to have a minimum current gain of 50 at 1 A and a collector-to-emitter breakdown rating of 50 V with 0.6-A collector current. These germanium transistors are readily available and have proved satisfactory with a short-circuited load.

The remainder of the design is simply an application of the principles outlined in preceding chapters combined with a desire to simplify and improve those circuits found in the references given at the end of this chapter. At this stage of the process, design becomes a cut-and-try process guided by experience and an understanding of principles.

The design of the amplifier power supply is outlined in Section 8.7. The amplifier uses, except for the regulator, a 34-V power supply. (For a description see Chapter 7, reference 8-4.) The power supply uses a readily available power transformer, and higher output power may be obtained by connecting two of these power supplies in series. A transformer with a higher VA rating should provide better regulation and higher output.

8.10 SUMMARY

Transistor class-B amplifiers are widely used in applications requiring high output power, high efficiency, and low harmonic content. The transformer-coupled class-B amplifiers tend to be simpler, more efficient, and are better suited for industrial applications. The complementary-symmetry amplifiers have the particular advantage of not requiring transformers and, therefore, of permitting considerable feedback.

The power capability of a class-B power amplifier is determined by the load impedance and the available supply voltage, as shown by equations (8.2) and (8.9). A given transistor in a class-B amplifier can deliver four to five times more output power than in a class-A power amplifier. With music and voice signals, a class-B amplifier dissipates less than one tenth the peak power output, so small heat sinks are generally satisfactory.

The power capability of a class-B amplifier depends considerably on the use of an adequate dc power supply. Depending on the needs of the amplifier, the VA capacity of the dc supply should be from three to five times the peak power output required of the amplifier. Good regulation and hum reduction are particularly desirable for a class-B amplifier. The physical size and cost of the power supply is appreciably increased by the use of the transformerless output circuits.

Design procedures are outlined for several forms of the class-B amplifier, and the design of a 7-W complementary-symmetry amplifier is given as a practical example of a class-B design.

The output transistors in a class-B amplifier must have significant current gain at the peak load current V_{CC}/R_L and a voltage-breakdown rating of about $2V_{CC}$. However, a more severe restriction, the second-breakdown rating, must be observed unless there is circuit protection from a short-circuiting load.

PROBLEMS

8-1. Describe the advantages of operating an amplifier in class-B as compared with class A.

8-2. (a) Explain why diodes are used in the bias circuit of a class-B amplifier. (b) How can you tell when the bias is set correctly? (c) How does the normal bias affect the efficiency?

8-3. (a) What are the peak current and voltage ratings required of a transistor operating push-pull class-B on 12 V dc and delivering 5 W to a 4-Ω load? (b) What are the ratings if the supply voltage is 24 V and the power output is 10 W?

8-4. A single-ended class-B audio amplifier is to be operated on a 50-V-dc power supply. (a) Specify maximum voltage and current ratings required of the power transistors if the load resistance is 8 Ω. (b) What is the full-load current required of the power supply? (c) Why is the transistor peak current rating approximately three times the average collector supply current?

8-5. A transformerless class-B amplifier is needed to drive an 8-Ω load. The dc power is to be obtained by means of a 12-V transformer. (a) Find the expected full-load dc voltage. (b) Specify transistor ratings. (c) What is the VA rating required of the transformer?

REFERENCES

8-1. *RCA Silicon Power Circuits Manual*, SP-50. Harrison, N.J.: Radio Corporation of America, 1967.

8-2. *GE Transistor Manual*, 7th ed. Electronics Park, Syracuse, N.Y.: General Electric Co., 1964.

8-3. "15 Watt Stereo Amplifier," Application Note No. 31. Kokomo, Ind.: Delco Radio Division, General Motors Corporation, May 1965. For higher power see Application Notes Nos. 35 and 36.

8-4. Cowles, L. G., *Transistor Circuits and Applications*. Englewood Cliffs, N.J.: Prentice-Hall, Inc., 1968.

8-5. Cowles, L. G., *Analysis and Design of Transistor Circuits*. New York: Van Nostrand Reinhold Company, 1966.

DESIGNS

8-1. Design a transformerless class-B power amplifier that will supply a maximum power to an 8-Ω speaker using a 12-V supply, as in an automobile.

8-2. Change the design illustrated in Fig. 8.7 so that the power output may be increased to approximately 25 W.

9

FEEDBACK–A DESIGN TOOL

The coming of the transistor, bringing high gain with the nonlinearity of a diode, has made feedback of primary importance in contemporary design. Feedback is now used in almost any one- or two-stage transistor amplifier to control impedances, to improve linearity, and to stabilize gain. Feedback was occasionally used in vacuum-tube amplifiers, and the problems of design were relatively simple, but it is undoubtedly true that feedback has become a vital element in transistor-circuit design. Some amplifiers have four or more kinds of feedback within a single overall feedback loop.

The concepts required to understand feedback are simple enough, but the mathematical analysis is complicated by algebraic terms that are retained through a reluctance of writers to make practical approximations. The use of approximations is fully justified, because the simpler equations make feedback a *tool* for the designer. Feedback is truly the key to most of today's circuits, and the concepts in this chapter are needed by everyone concerned with applications or design.

9.1 FEEDBACK TERMS AND FEEDBACK USES

One problem with feedback is that a variety of terms is required to describe the many ways of using it. An amplifier is said to have negative feedback when a voltage derived from the amplifier output is added to the input signal in such a way as to oppose the input signal. The feedback within a single stage is called *local feedback*, whereas the feedback over several stages is called *overall feedback*. Local feedback is easily applied and understood, but it makes an uneconomical use of gain. Overall feedback may reduce the gain of an amplifier by a factor of 10 to 100 and may be difficult to apply while keeping the amplifier stable. Overall feedback is preferred, because a given gain reduction is used where it is most needed. Internal feedback is the feedback within a transistor that is usually seen as a source of high-frequency problems.

The terms *shunt* and *series feedback* distinguish whether the feedback is applied in shunt or in series with the input, because this feedback affects the input impedance. Shunt feedback tends to lower the input impedance, and series feedback raises the input impedance.

The terms *voltage* and *current feedback* are used when the source of the feedback at the output may have an effect on the output impedance. Voltage feedback that is returned from the output of an amplifier reduces the output impedance, and current feedback increases the output impedance. The effect of feedback on the output impedance is sometimes of interest, as in a hi-fi amplifier in which a low impedance is desired in order to damp out the speaker transients.

Generally, feedback amplifiers can be designed without concern for the effect of feedback on the internal output impedance. However, the effect of the load impedance on the feedback must always be considered. With voltage feedback the load may shunt and thereby reduce or eliminate feedback that is intended to control a gain, an impedance, or a frequency response. A reactance load may affect the response or the stability of an amplifier.

Significant feedback, as explained later (in Chapter 16), exists when the feedback is sufficient to mke a significant change in the amplifier performance. A 4 to 1 change of gain is defined as significant. *Adequate feedback* is a term used to indicate that the 1 in $(1 + AB)$ terms of formulas may be neglected. Loop feedback exists when a signal path, or loop, may be traced from the amplifier input, through the amplifier, back through the feedback network, and into the amplifier input. As we shall show, loop feedback has the effect of making the output signal more like the input signal. With loop feedback the characteristics of an amplfier may be improved, either to make the amplifier less sensitive to the ambient temperature and the supply voltage, or to change the frequency response and the impedance characteristics, or both. Loop feedback is not significant unless the net gain around the loop is 3, 10 dB. Adequate loop feedback generally requires a careful design of the loop amplitude-phase characteristics in order to control the transient response or the stability of the amplifier under closed-loop conditions. The stability problem is the concern of Chapter 10.

Feedback is an exchange of power gain for an improved performance, and many performance characteristics cannot be obtained without feedback. Because feedback reduces the gain, amplifiers must be designed with additional stages. Transistor circuits are so much simpler than vacuum-tube circuits that with the help of feedback better performance can be obtained than was formerly considered practical.

9.2 AMPLIFIER GAINS AND OVERALL SERIES FEEDBACK

With adequate local feedback the voltage gain and the current gain of a stage are fixed by the resistors of the circuit and not by the transistors. In a multistage amplifier with adequate overall feedback, the overall gains are fixed in a similar way by the resistors in the feedback network. For ordinary combinations of local and overall feedback, the gain of an amplifier may be easily evaluated by finding the feedback resistors and making simple calculations of the kind used to evaluate a stage that has local feedback.

As examples of series feedback, consider the following comparison of the familiar CE stage having local feedback, as shown in Fig. 9.1, with the amplifier shown in Fig. 9.2. Both amplifiers are driven by a current source and have an input resistor R_B, a feedback resistor R_E, and a load resistor R_L. In both examples the current in the load resistor is substantially the same as the current in the feedback resistor. The similarity between the single-stage and the multistage amplifier suggests that the characteristics of the amplifier can be derived from the known gain and impedance characteristics of an emitter-feedback stage.

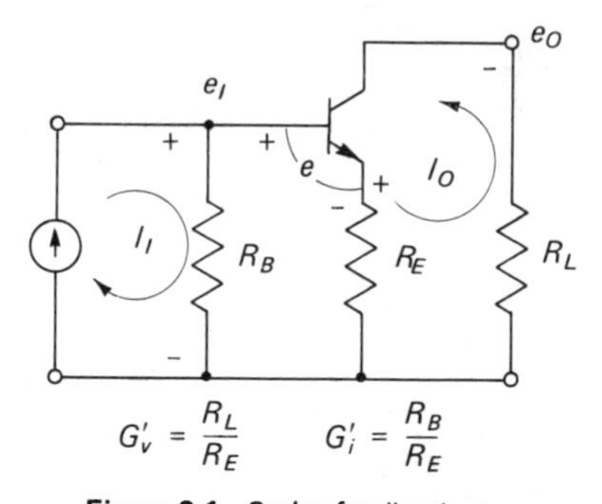

Figure 9.1. Series-feedback stage.

If there is adequate feedback from the output loop back to the input loop for the amplifier shown in Fig. 9.2, the amplifier current gain is R_B/R_E, the voltage gain is R_L/R_E, and the input impedance is R_B. Observe that nothing is said of the amplifier except that the feedback is adequate, meaning the amplifier has enough gain so that the gain and impedance relations are fixed by the feedback network. The feedback is adequate if the amplifier voltage gain without

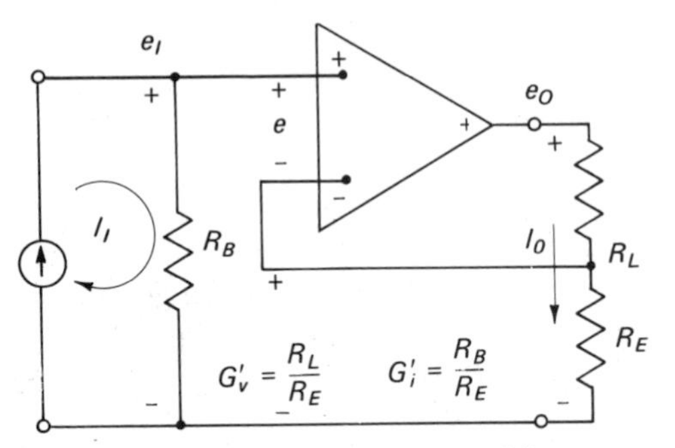

Figure 9.2. Series-feedback amplifier.

feedback is large compared to the resistance ratio R_L/R_E, and the current gain without feedback is large compared to the ratio R_B/R_E.

There is a difference between the voltage feedback and current feedback that we must recognize. The voltage gain with feedback e_O/e_I is fixed by the resistor ratio R_L/R_E regardless of R_B or the source impedance. However, the current gain with feedback is fixed by the resistor ratio R_B/R_E only if there are both adequate current feedback and adequate voltage feedback. If both feedbacks are adequate, the input impedance seen by the source is R_B, and the amplifier input voltage e may be considered negligible compared with the voltages across R_B and R_E.

An important practical difference between single-stage and multistage feedback is that the feedback in a single-stage amplifier may be significant but is not often adequate. The feedback in a two-stage amplifier probably can be made adequate without instability problems. A multistage amplifier usually needs special attention to the design of the loop gain and phase characteristics to prevent instability when adequate feedback is required.

If an amplifier is found by measurements to have gains and impedances that conform to the values predicted by the circuit resistors, we may assume that the feedback is adequate for ordinary design purposes. Conceivably, the original design of an amplifier may require more than adequate feedback, as the term is used here, and there is no easy way to know from the gain relations and measurements that the excess of feedback actually exists. Whether there is enough feedback to ensure a required stability of gain or a required reduction of distortion has to be evaluated by measuring the loop gain or by other measurements, such as a stability test or a distortion measurement.

9.3 AMPLIFIER GAINS AND OVERALL SHUNT FEEDBACK

As examples of overall shunt feedback, consider the amplifiers shown in Fig. 9.3 and 9.4. The amplifier in Fig. 9.3 is the familiar single-stage form with

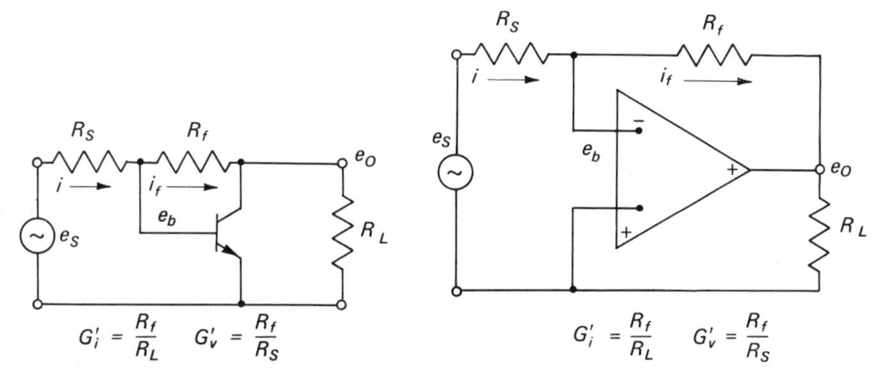

Figure 9.3. Shunt-feedback stage.

Figure 9.4. Shunt-feedback amplifier.

collector-to-base feedback. The multistage amplifier in Fig. 9.4 has shunt feedback as used with operational amplifiers. We examine the characteristics of these amplifiers in detail because the characteristics of shunt feedback are not so well known as those of series feedback.

Shunt feedback is applied through the resistor R_f, which is assumed large compared with the load R_L. The signal is connected through the resistor R_S, which is sized to make the signal e_S large compared with the amplifier input signal e_b. This choice of R_S makes the amplifier voltage gain without feedback large compared with the voltage gain with feedback. To the extent that e_S is large compared with e_b, we may say that the voltage across R_S is e_S, and the input impedance seen by the source is R_S. Also, to be useful as a feedback amplifier, the output voltage e_O must be large compared with the input voltage e_b hence, the voltage across R_f is e_O.

From the foregoing relations the voltage gain with feedback G_v' is

$$G_v' = \frac{e_O}{e_I} = \frac{i_f R_f}{i R_S} \tag{9.1}$$

We assume adequate feedback, which was shown in Section 5.7, to mean that the S-factor is small compared with β, and that the amplifier input impedance is large compared with the Miller equivalent input impedance. For this reason the current input to the amplifier is negligible compared with i and i_f, so that $i = i_f$, and equation (9.1) reduces to

$$G_v' = \frac{R_f}{R_S} \tag{9.2}$$

The current gain of an amplifier with shunt feedback is easily found by inserting the gain and impedance values just calculated into the TG-IR:

$$G_v' = \frac{R_f}{R_S} = G_i \frac{R_L}{R_S} \tag{9.3}$$

Solving for the current gain gives

$$G_i = \frac{R_f}{R_L} \cong S \tag{9.4}$$

In summary, both the current gain and the voltage gain of a shunt-feedback amplifier are determined by the feedback resistors if the R_f/R_L ratio is small compared with the amplifier current gain without feedback, and if the R_f/R_S ratio is small compared with the amplifier voltage gain without feedback. The voltage gain is controlled by the resistor ratio only if both the current and voltage feedbacks are adequate. When both feedbacks are adequate, the input impedance is R_S, and the amplifier input current is negligible compared with the current in R_S and R_f.

One reason for using high-gain integrated circuits and severely reducing the overall gain by feedback is that, according to equation (9.2), a voltage gain may be very precisely and easily adjusted by only two resistors. Similarly, using complex impedances in the feedback network of operational amplifiers, the amplifier is said to operate on a signal by changing the response according to the mathematical relations represented by the impedances in the feedback network. By similar means the frequency response of an amplifier may be improved or shaped by a feedback network.

9.4 LOOP FEEDBACK

The extent by which feedback reduces internal noise and nonlinearities is determined by the signal that is returned to reenter the amplifier. The feedback returned into the amplifier is referred to as *loop feedback* and is measured by the net gain experienced by a signal making one round trip around the loop.

Loop feedback is established by combining signals at the input of an amplifier, as shown in Fig. 9.5. The amplifier shown as driven by a source e_S', which has an internal impedance R_S, supplies an output voltage e_O' across the load R_L. The feedback resistor is assumed small compared with the load resistor, and the ratio R_E/R_L is called the *voltage loss factor* B_v. Gains and signals with feedback are designated by primed symbols, and gains and signals without feedback do not have primes. The input signals are normalized (adjusted) for discussion purposes to make the amplifier input voltage exactly 1 V. Because the amplifier gain is A, the output signal is A volts.

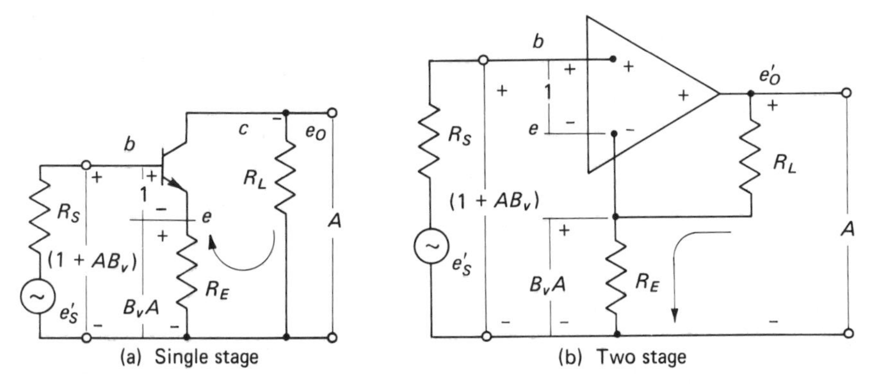

Figure 9.5. Series-feedback voltage relations.

The feedback resistor R_E is shared by the output and the signal circuits. If the amplifier input impedance R_I is large compared with $R_S + R_E$, the voltage loss in R_S may be neglected, and the drive signal required to sustain the indicated voltages is $e'_S = (1 + B_vA)$. The voltage gain without feedback, or the ratio of the output voltage A to the input 1, is G_v, where

$$G_v = A \tag{9.5}$$

and the voltage gain with feedback, the ratio of A to e'_S, is G'_v, where

$$G'_v = \frac{A}{1 + B_vA} \tag{9.6}$$

The feedback is negative because it opposes the input signal and reduces the overall gain. The complete circuit from the input through the amplifier and the B_v network and back to the input *b-e* is the feedback loop. The voltage gain B_vA is the loop gain G_l, and the magnitude $|1 + B_vA|$ is called the *feedback factor.*

The effectiveness of loop feedback is measured by the amount by which the overall gain is reduced when the feedback is applied. Significant loop feedback generally exists when the amplifier voltage gain and current gain are both reduced by factors of 6. The difficulty in estimating the loop gain is that the amplifier input impedance may not be known, so the return loss cannot be calculated.

Generally, it is supposed that the loop gain can be measured by opening the loop. If opened for a measurement, the loop must be carefully terminated so that the gain and impedances are not changed. When the feedback supplies bias, the ac gain may have to be measured by a method that maintains the dc feedback while removing the ac feedback. Sometimes the amplifier is too unstable with the loop opened to permit gain measurements. In practice, the difficulties of measuring the loop gain usually require more effort and time than can be justified or permitted.

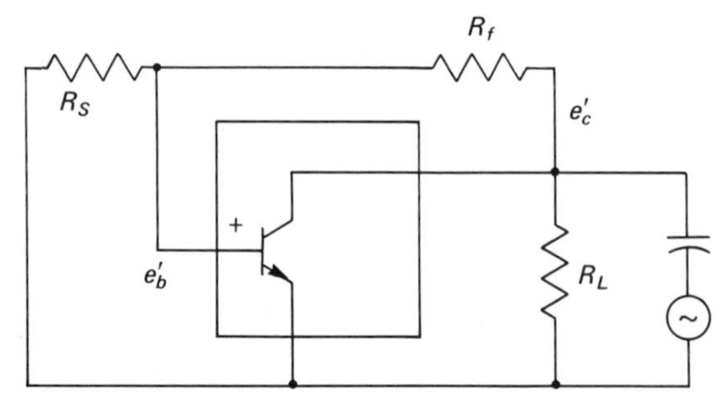

Figure 9.6. Loop-gain test with shunt feedback.

With shunt feedback an approximated value of the loop gain can be obtained by the method indicated in Fig. 9.6. A signal is applied at the output so that the return loss e_b'/e_c' can be measured. The forward voltage gain

$$G_v = \frac{e_c}{e_b} \tag{9.7}$$

is measured with the signal applied through R_S in the usual way. The loop gain is approximately

$$G_l = \frac{e_b'}{e_c'} G_v \tag{9.8}$$

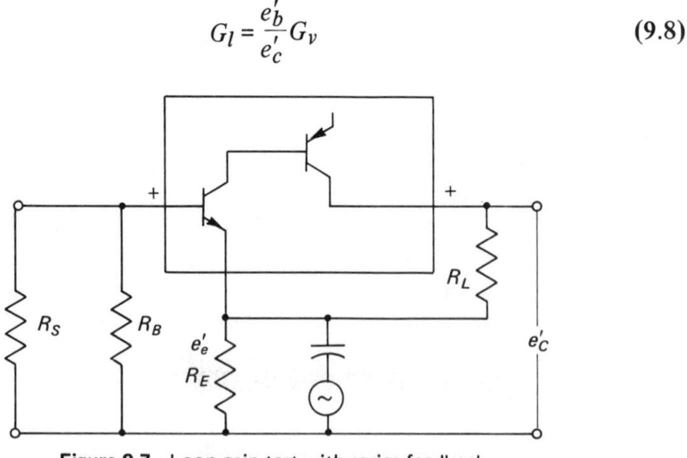

Figure 9.7. Loop-gain test with series feedback.

With series feedback the loop gain may be measured by the method indicated in Fig. 9.7. The ac feedback is removed by introducing a signal at the emitter, and we measure the signal ratio e_c'/e_e'. The loop gain is approximately

$$G_l = \frac{e_c'}{e_e'} \frac{R_E}{R_L} \tag{9.9}$$

This measurement does not give the dc gain, which may, of course, be different from the ac gain, and the formula has several approximations. In some cases the loop current gain, which has the same value as the loop voltage gain, may be measured with an ac clip-on current probe.

For the present, the reader is cautioned that the loop gain may not be significant if the source impedance is so high that the feedback signal is reduced on its return through the source. An amplifier designed to provide a high input impedance may have loop feedback when driven from a low-impedance source, and may not have the advantages of loop feedback when driven from a high impedance source. In summary, any purpose for which loop feedback is intended may be offset by the impedance characteristics of the source or the load.

9.5 USES OF LOOP FEEDBACK

An examination of equation (9.6) shows that when the feedback factor is large, the voltage gain is the reciprocal of the loss factor B_v. Thus

$$G_v' = \frac{1}{B_v} \tag{9.10}$$

When the loop gain is approximately 10 or more, the 1 in the denominator of equation (9.6) may be neglected, and the feedback is said to be adequate. Equation (9.10) states that, as long as the feedback is adequate, the voltage gain is $1/B_v$, independent of the amplifier itself, and is fixed by the resistors that make up the feedback network. This remarkable behavior exists because the 1 V applied to the amplifier input (Fig. 9.5) represents a small difference between the relatively large feedback voltage and the drive signal. In other words, when the feedback is large compared with the 1-V input, the signal within the amplifier itself is the correction required to make the feedback voltage follow the input signal. In effect, the amplifier is used merely to correct any difference between the feedback and the drive. We see that feedback makes an amplifier improve the quality of the output signal, provided the B_v network is comprised of linear components. Similarly, the frequency characteristics of an amplifier may be improved or shaped by the B_v network.

Suppose, as shown in Fig. 9.8, that an amplifier has a frequency response that is constant between the frequencies f_l and f_h, and the amplifier has an excess of gain. By applying feedback the excess of gain may be exchanged for an increased

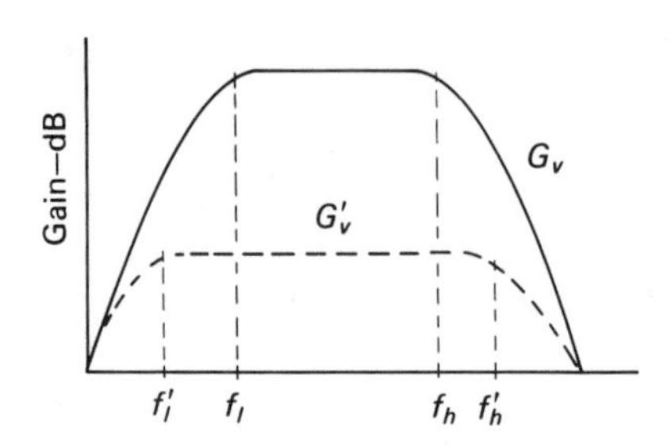

Figure 9.8. Amplifier gain-frequency curves.

frequency band. If the feedback network is comprised of resistors, the gain is reduced to a maximum gain $G'_V = 1/B_V$, as shown in Fig. 9.8, and the frequency response with feedback is constant from f'_l to f'_h. This technique is often used to improve the frequency response at small-signal levels, but the technique may not be effective when the amplifier is operated at full power output.

For example, suppose that a 10-W amplifier, shown in Fig. 9.9, can only deliver 1 W at 20 Hz because 20 Hz is below the low-frequency cutoff of the output transformer. The addition of feedback can make the amplifier show a flat frequency response with small signals down to 20 Hz, but the amplifier still cannot deliver more than 1 W at 20 Hz. An output exceeding 1 W at this frequency would require more than 10-W input to the transformer, and feedback cannot increase the basic power capability of the power stage.

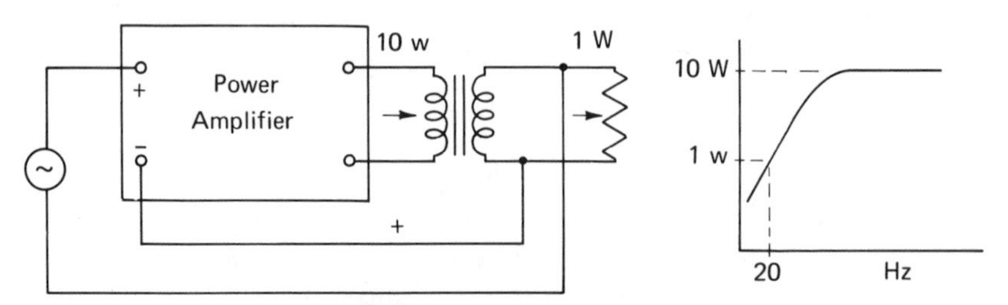

Figure 9.9. Amplifier power-frequency characteristics.

Feedback that has been used to reduce the gain of an amplifier may be removed in a restricted frequency range to compensate for a signal loss somewhere else in a transmission system. A small capacitor in shunt with the feedback resistor R_E reduces the feedback at high frequencies, as shown in Fig. 9.10, and may restore the gain to the value that exists without feedback. Observe that the gain is not increased above the no-feedback curve, although this is possible if the feedback signal becomes in phase, as with positive feedback.

In this example the capacitor reduces the input impedance, so the gain may not increase as expected. The gain increases at high frequencies only if R_S is small, so the amplifier input signal e_I is not changed by the addition of the capacitor. Sometimes the feedback can be made to serve two such purposes by increasing the loop gain, which means, probably, that the amplifier gain must be increased. The advantages of feedback are available only when the amplifier has enough excess gain to be divided between the required improvements.

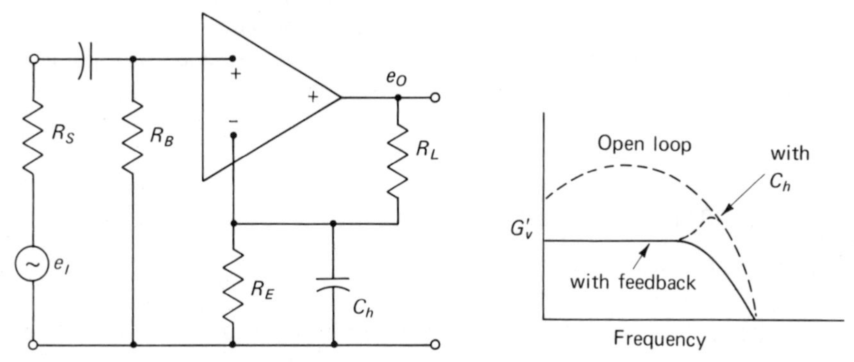

Figure 9.10. Feedback frequency compensation.

9.6 GAIN CHANGES WITH LOOP FEEDBACK

As an example of the advantages of loop feedback, consider an amplifier having a voltage gain A of 400 (52 dB) without feedback, and assume that the amplifier is modified by adding significant loop feedback. With significant feedback the loop gain B_vA is 3, which makes the denominator of equation (9.6) equal to 4, and the gain without feedback is reduced by 4 (12 dB). We say that the amplifier has 12 dB of feedback and the gain with feedback is 100 (40 dB). Suppose now that the amplifier gain falls off, for any reason whatever, to only 200 (46 dB), meaning that the gain A drops 6 dB (52 − 46). Substituting in equation (9.6), we now have

$$G'_v = \frac{200}{1 + 1.5} \tag{9.11}$$

Observe that the numerator and the second term in the denominator decreased 2 to 1 because the gain A decreased 2 to 1, and the overall gain is now 80 (38 dB). We find that the significant feedback has reduced a 6-dB gain change in the amplifier to only a 2-dB (40 − 38) change. By constructing the amplifier with 12 dB of excess gain and reducing the gain 12 dB by feedback, the 6 dB gain change has been reduced to a 2 dB change. Observe further that the 3 to 1 improvement is approximately equal to the feedback factor 4.

The calculation just above has assumed that the feedback remained exactly in phase, as required for negative feedback. When the gain changes are produced as a function of frequency by reactive components, as with high-frequency cutoff, the relative phase of the B_vA term changes so that the feedback is not necessarily negative feedback. For small amplitude-frequency changes, the voltage gain may change less than calculated above, and for a large change the phase angle of the feedback signal may shift enough to cause an increase of the overall gain with feedback. Because of this characteristic of amplitude-frequency changes, the design of a feedback amplifier with an improved frequency response is somewhat more difficult than for improved gain stability. When the phase of the feedback signal shifts 180°, the feedback signal is in phase with the input signal, and the gain may increase enough to produce an unstable amplifier or an oscillator.

9.7 DISTORTION AND HUM REDUCTION BY FEEDBACK

Distortion and power-supply hum are reduced by negative feedback in very much the same way that amplitude changes are reduced. A distortion signal that appears in the output is fed back and compared with the input signal, just as the main signal components are compared. However, the distortion is subsequently offset at the output only if the distortion components are amplified linearly. When an amplifier is operated at the point of overload, the loop gain falls off so much that the feedback is ineffective. For this reason, even with a considerable amount of feedback, the maximum output power cannot be significantly increased. A distortion, like crossover distortion, which occurs about 20 dB below the maximum output level, may be reduced in proportion to the amount of feedback.

Hum that appears in the output of an amplifier may be reduced by feedback as long as the hum that is fed back to the input is proportional to and in phase with the output hum. Because the ground side of the feedback signal circuit may be different from the ground side of the output circuit, the hum in these circuits may also differ. This difference between the output and the feedback circuits is particularly troublesome in push-pull amplifier when the feedback is returned from the primary side of the transformer. Whenever possible, feedback should be derived from the output terminals of a power amplifier. Hum, noise, and spurious signals that enter an amplifier *with* the input signal are treated as an input signal and are not reduced by feedback.

9.8 NOISE REDUCTION BY FEEDBACK

The amplifier noise inherent in high-gain amplifiers, e.g., thermal and shot noise, cannot be reduced by feedback as long as a fixed gain is required. An 80-dB amplifier without feedback has the same noise output as an amplifier having 80 dB of gain with feedback. Noise and hum that are produced in an amplifier are reduced in the same way they would be reduced by a loss network in the output, except that feedback does not reduce the power output capability of the amplifier. Spurious signals and noise inherent in the input stages of high-gain amplifiers are only reduced by the careful selection of components or by circuit design that eliminates the basic sources of noise.

9.9 THE MANY USES OF FEEDBACK

The simplest use of feedback is the reduction of the gain of an amplifier to control the Q-point biasing or the Q-point drift. Transistor amplifiers usually use both dc and ac feedback to reduce the effects of β, which otherwise would make many circuit designs highly impractical. The feedback used with high-gain amplifiers forces the amplifier to respond in a very precisely determined manner, as is required for operational or instrument amplifiers.

Feedback is used to shape the frequency response of phonograph preamplifiers, transducer amplifiers, active filters, and wide-band amplifiers. Industrial controls and servos are forms of low-frequency feedback amplifiers, which may include machines as components of the system. The overall gain of an RF amplifier is generally controlled by AGC feedback bias developed by the signal. The AGC circuit is a feedback loop that is carefully designed to control rapid signal changes while avoiding an unstable feedback characteristic.

The stability of a complicated feedback system is an important consideration for the designer. A feedback loop always has reactances that introduce phase shift so that at some frequency the feedback becomes in phase. Unless the loop loss at the inphase frequency is 10 dB or more, a feedback system is unstable and may oscillate or have an undesirable transient response. The problem of designing a stable feedback amplifier is the topic of Chapter 10.

9.10 SUMMARY

1. Negative feedback improves the performance of an amplifier by comparing the output signal with the input signal. The price paid for the improved performance is reduced gain, which is offset by providing an excess of amplifier gain. Feedback is introduced in an amplifier by coupling the output signal to the input through a resistor network, so the characteristics of the amplifier depend on the feedback resistors. If the feedback is adequate, the gain depends mainly on the characteristics of the feedback network and is relatively independent of the amplifier and transistors.
2. Local feedback simplifies biasing, is easier to use than overall feedback, and makes the stage less dependent on the transistor and temperature. A stage with significant feedback has enough feedback to reduce the gain by a factor of 4, and the stage is approximately a factor of 4 less sensitive to a change of the active element or the circuit components.
3. Series feedback increases the input impedance of an amplifier and reduces the voltage gain. Shunt feedback reduces the input impedance and lowers the current gain. These changes of the input impedance are the means by which the feedback improves the amplifier performance. The effect of

series feedback in controlling the amplifier voltage gain is offset by using a high source impedance. Similarly, a reduction of series feedback to increase amplifier gain reduces the input impedance. The impedance and gain characteristics of the shunt-feedback amplifier are similarly interrelated.

4. Loop feedback makes both the voltage gain and the current gain of an amplifier depend on the feedback network. With series feedback, loop feedback is obtained only by using a source impedance that makes the overall current gain small compared with the open-loop current gain. Similarly, with shunt feedback loop feedback exists only when the source impedance makes the overall voltage gain small compared with the open-loop voltage gain.
5. With adequate loop feedback the gain characteristics of an amplifier are determined by the external feedback resistors. Adequate feedback usually exists when the current and voltage gains with feedback are both small compared with the open-loop current and voltage gains. The amount of loop feedback, which controls the quality of an amplifier, is evaluated by measuring or calculating the gain in one round trip around the loop. The loop gain is often difficult to measure without disturbing the amplifier and making the measurements of questionable value.
6. Loop feedback improves the frequency response and reduces the distortion and power-supply hum of an amplifier. The noise in later stages is reduced by loop feedback, but the noise in the input signal or in the input-circuit components is not reduced relative to the signal. Loop feedback reduces, by the amount of feedback, any gain changes that arise within the amplifier itself.

Whether or not an amplifier is stable with loop feedback is the topic of Chapter 10.

PROBLEMS

9-1. A amplifier in Fig. 9.2 is required to have significant loop feedback, 10-kΩ input impedance, and a 1-kΩ load impedance. The voltage and current gains with feedback are to be 10 and 100, respectively. (a) Does the TG-IR show that this is a possible design? (b) What voltage and current gains should the amplifier have before feedback? (c) Draw the circuit for your solution and calculate the loop gain.

9-2. Repeat Problem 9-1, using a shunt feedback amplifier.

9-3. Refer to Fig. 9.5 for the series feedback relations and assume that $\beta = 100$, $h = 1\Omega$, $R_L = 100\Omega$, $R_E = 10\Omega$, and $R_S = 100\Omega$. Make a new Fig. 9.5a to conform to these values. (a) Does the amplifier have adequate loop feedback? (b) Does the amplifier have significant loop feedback?

9-4. Consider the stage shown in Fig. P9.4 as a feedback amplifier. The loop gain is zero (0) with switch S closed. (a) What is the open loop voltage gain A? (b) What is the loop gain G_l? (c) What is the amplifier gain with feedback when S is open?

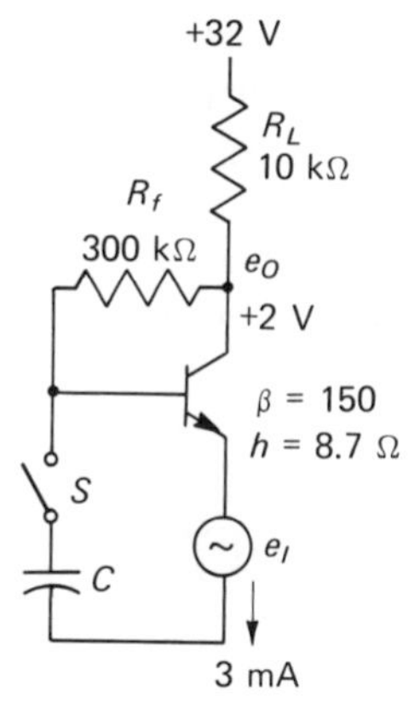

Figure P 9.4

Figure P 9.6

9-5. Refer to Fig. 9.6, and assume that $R_S = R_L = 100\ \Omega$, $R_f = 5\ \text{k}\Omega$, $e_c' = 1$ V. and $e_b' = 1$ mV. Find the loop gain for (a) $G_v = 100$ and for (b) $G_v = 10{,}000$.

9-6. For the amplifier shown in Fig. P9.6, assume that $R_S \gg R_E, R_L \gg R_E$, and call the gains with feedback $G_v' = R_L/R_E$ and $G_i' = R_S/R_E$. Call the gains without feedback $G_v = n_v G_v'$ and $G_i = n_s' G_i'$, respectively. (a) Show that the effect of the amplifier input impedance R_I is to make the loop gain,

$$G_l = \frac{n_v n_s}{n_v + n_s}, \qquad \text{where } R_I = \frac{n_s}{n_v} R_S$$

(b) Show that by making $R_S = R_I$, the loop gain is given by

$$G_l = \frac{1}{2}\frac{G_v}{G_v'} = \frac{1}{2}\frac{G_i}{G_i'}$$

DESIGNS

9-1. Measure the loop gain in one of the feedback amplifiers shown in Chapter 6 and observe the effect on the loop gain of shorting and of removing the signal source.

9-2. Select one of the feedback amplifiers shown in Chapter 6 and redesign for a 10 times higher or lower load impedance. Find whether your design has significant feedback.

REFERENCES

9-1. Terman, F. E., *Electronic and Radio Engineering*, 4th ed., Chapter 11, "Negative Feedback Amplifiers." New York: McGraw-Hill Book Company, 1955.

10

FEEDBACK-AMPLIFIER STABILITY

Feedback tends to change from negative to positive when the reactances in the loop cause increased phase shift at frequencies outside the amplifier's normal range. When the feedback signal aids the input signal, an amplifier develops an undesirable transient response and may oscillate. The advantages of negative feedback are realized only when the feedback loop is so designed that the amplifier does not have a tendency to oscillate.

One purpose of this chapter is to explain how the loop gain and phase characteristics limit the amount of usable feedback. A second purpose is to show how a designer controls the gain and phase characteristics so that an amplifier may be given more feedback and improved stability. These techniques are illustrated by the design of an audio-feedback amplifier. A part of the chapter is concerned with the unique relation that exists between the gain and phase response of ordinary (minimum-phase) circuits.

10.1 NYQUIST DIAGRAMS

The gain of a feedback amplifier is commonly expressed in the form

$$A' = \frac{A}{1 + A\beta} \tag{10.1}$$

This equation is similar to equation (9.6) except that the loop gain $A\beta$ may be calculated as a voltage gain and a voltage loss, or as a current gain and a current loss. The loop gain is numerically the same by either computation. The stability problem exists whenever $A\beta = -1$ because, if A' is infinite or very large, the feedback sustains any transient signal and the amplifier is useless.

The stability of feedback systems is commonly evaluated by constructing a Nyquist diagram, which is a polar plot of the loop transfer characteristic. This plot of the loop gain $A\beta$, shown in Fig. 10.1, simply depicts the gain and phase changes of a signal in making one round trip through the loop. The $A\beta$ curve represents the $A\beta$ term of equation (10.1) as the frequency is varied from zero to infinity. The midband gain of the amplifier is represented by the point where the curve crosses the positive real axis. Unless the loop transmits direct current, the curve begins at the origin, moves clockwise, and returns to the origin at a high, infinite frequency.

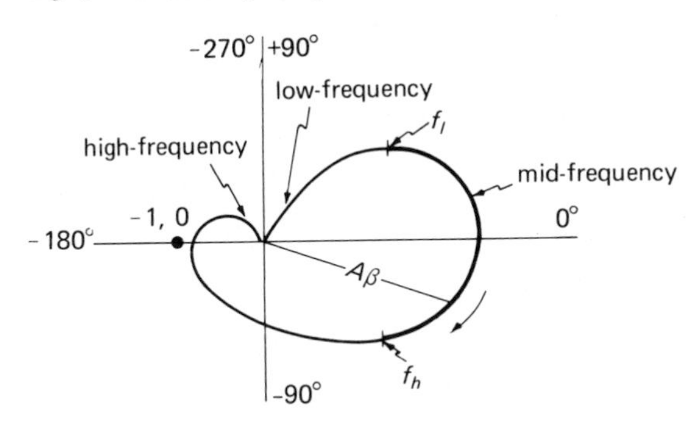

Figure 10.1. A Nyquist diagram.

The point -1.0 on the negative real axis is called the *critical point*. If the curve passes through the critical point $A\beta = -1$, the denominator in equation (10.1) is zero, and the gain with feedback becomes infinite. Infinite gain means that the amplifier oscillates. An amplifier is stable if the $A\beta$ curve does not pass through or enclose the critical point. An amplifier is said to have a phase margin of 30° if the phase shift through the loop is 30° less than 180° when the magnitude of $A\beta$ is 1. An amplifier is said to have a gain margin of 10 dB if the loop has a 10-dB loss when the loop phase shift is 180°. With these gain and phase margins an amplifier is generally considered stable, but for a well-damped response both margins should be increased.

A typical Nyquist diagram is shown in Fig. 10.2. The diagram for a stable amplifier shows that the phase shift at a very low frequency is -180°, as is produced by two coupling or bypass capacitors. As the frequency is increased, the loop phase rapidly approaches the phase of negative feedback and remains near the 0° angle throughout the normal frequency range of the amplifier. Near the high-frequency cutoff and above, the high-frequency reactances shift the phase to and beyond the inphase condition. As the example shows, the total high-frequency phase shift usually exceeds the low-frequency phase shift, because there are more places in the circuit where a spurious capacitance to ground adds another 90° of phase shift. A troublesome source of phase shift comes from the high-frequency cutoff in a transistor. Furthermore, a single transformer may produce a 90° phase shift in the leakage reactances caused by imperfect coupling and a phase shift of 90° or more from the winding capacitance, making a total exceeding 180°.

Diagram B in Fig. 10.2 represents an unstable amplifier that would be expected to oscillate because the $A\beta$ curve encloses the critical point -1.0. Diagram B is the same as diagram A except that the loop gain is increased at all frequencies by the same factor. This difference illustrates the fact that almost any amplifier becomes unstable if the loop gain is increased. Nearly any imaginable amplifier has enough spurious or excess phase shift to total 180° and make the amplifier oscillate if the loop gain is increased to unity at the in-phase frequency.

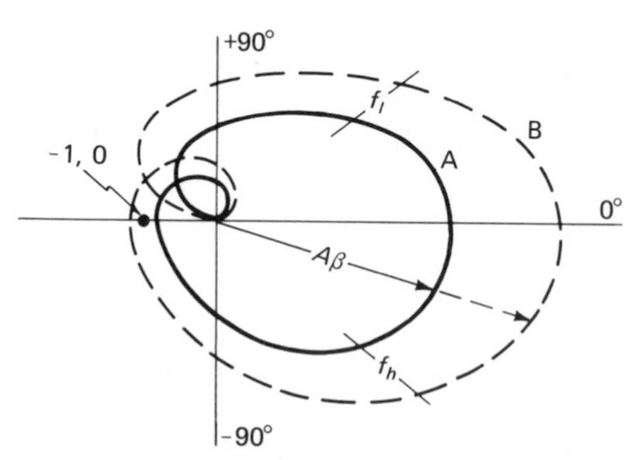

Figure 10.2. Nyquist diagram: A, low gain, B, high gain.

A direct-coupled amplifier is presumed to have no phase shift at zero frequency and is expected to be stable at low frequencies. However, the reactances in an unregulated power supply and certain thermal instabilities may cause a dc amplifier to oscillate or exhibit a peculiar low-frequency transient response. These conditions may be represented as an enclosure of the critical point, but it may be very difficult to obtain points on the curve experimentally.

10.2 NYQUIST-DIAGRAM EXAMPLES

Problems that are encountered in feedback-amplifier design are illustrated by the Nyquist diagrams in Fig. 10.3. Curve A shows the $A\beta$ curve of a single-stage, resistance-coupled amplifier. Curve A begins with zero frequency at the origin, moves along the circle clockwise, and returns to the origin at infinite frequency. The 3-dB cutoff frequencies, indicated as f_l and f_h, are in the low- and high-cutoff regions. The midband portion of the diagram may represent a very wide band where the amplitude is relatively flat. Curve B represents an amplifier having two identical R-C stages. Each stage contributes 45° at cutoff, so the total phase shift is 90° at cutoff. At zero and infinite frequency, curve B shows 180° of lead and lag, respectively.

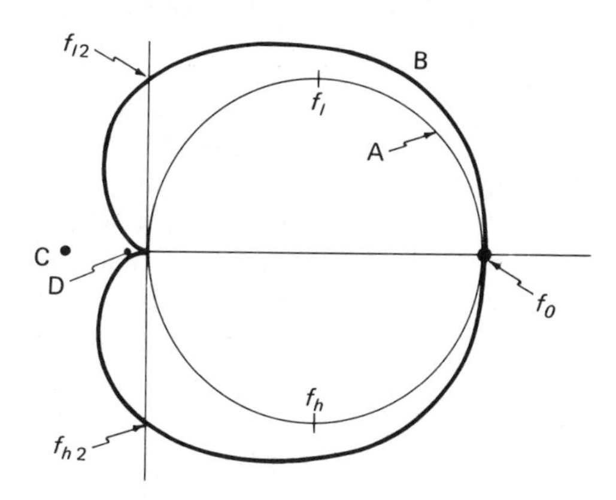

Figure 10.3. Nyquist diagram for R-C amplifier: A, single-stage; B, two-stage.

The stability of an amplifier is determined by observing the position of the $A\beta$ curves relative to the critical point -1.0. If the midband gain is $A = 4$, the critical point is at the left of the origin, as at C, by a distance that is one fourth the amplitude A at the midband frequency f_0. If the midband gain is 20, the critical point is at one twentieth of the amplitude at f_0, as at D in the figure. Observe that the higher gain causes the $A\beta$ curve to pass closer to the critical point and to decrease the phase margin. The one- and two-stage amplifiers represented by the curves are stable for any gain, regardless of the loop gain, because the curves cannot enclose the critical point. However, when the loop gain exceeds 20, a two-stage amplifier may exhibit a ringing transient because of the relatively small phase margin. Furthermore, an amplifier always has additional sources of phase shift that are neglected in the foregoing discussion.

The area of the Nyquist diagram that is of principal interest is within the unit circle at the left of the origin. Figure 10.4 represents a unit circle having the critical point as center and the origin at 3 o'clock. The circle is the locus

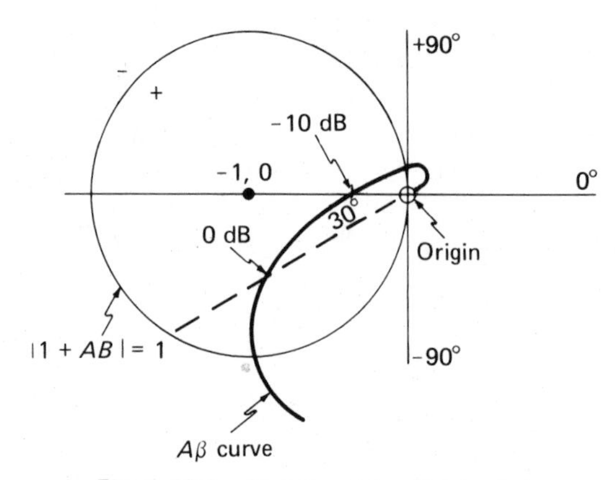

Figure 10.4. *Aβ* curve and unit-gain circle.

of all points for which $|1 + A\beta| = 1$. The $A\beta$ curve in the figure represents a loop-gain characteristic that passes the critical point, with a 10-dB gain margin and a 30° phase margin. Closing this loop should produce a stable amplifier. However, at the frequencies for which the curve is inside the unit circle, the $|1 + A\beta|$ is less than 1 so the gain with feedback is greater than the gain without feedback. When the curve is outside the unit circle, the gain with feedback is less than the gain without feedback. We say that the feedback is *negative* when the curve is outside the unit circle and *positive* when the curve is inside the circle

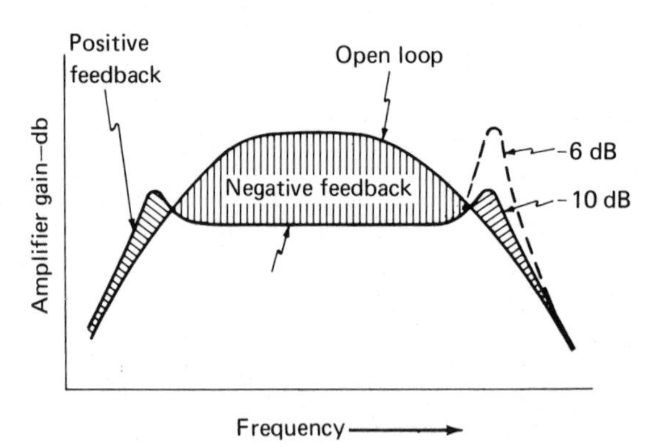

Figure 10.5. Feedback gain-frequency curves.

The curves in Fig. 10.5 illustrate the effects of positive feedback on the gain-frequency curves of a stable amplifier. With a 10-dB gain margin, the gain with feedback exceeds the gain without feedback by about 3 dB. When the gain margin is 6 dB, or less, the amplifier has a slowly decaying transient response and may break into oscillation if the loop gain increases, as by a change of temperature, load, or supply voltage.

10.3 RETURN RATIO AND RETURN DIFFERENCE

Nyquist diagrams are sometimes shown rotated, so that the critical point is at +1.0, as in Fig. 10.6. This form of the diagram is obtained by reversing the sign of the $A\beta$ term in equation (9.6). If the feedback equation is derived by writing the amplifier input 1 as the sum of the signal e and $A\beta$, then

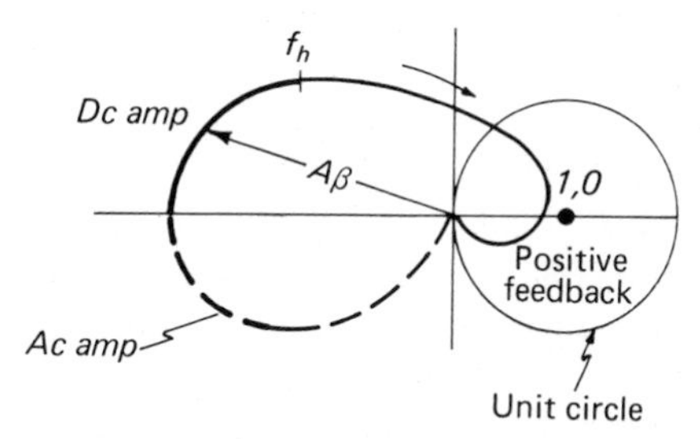

Figure 10.6. Nyquist diagram, second form.

$$1 = e + A\beta \tag{10.2}$$

and

$$e_O = A \cdot 1 \tag{10.3}$$

The feedback equation then takes the form

$$\frac{e_O}{e} = G_v' = \frac{A}{1 - A\beta} \tag{10.4}$$

In the latter form $-A\beta$ is called the *return ratio* and $(1 - A\beta)$ is called the *return difference.*

With the second form of the Nyquist diagram the feedback is negative when the $A\beta$ curve crosses the negative real axis, and the amplifier is unstable when the $A\beta$ curve encloses the critical point at +1.0. As is often the case, the scientific terminology describing a given subject is changed when a different point of view seems to aid the understanding. These changes are confusing at first, but it is easy to recognize and use both diagrams. Facility should be acquired in using either form of the feedback equation and the related terms.

10.4 GAIN-PHASE RELATIONS

Fortunately for the designer, in the networks ordinarily used in amplifiers the phase shift is simply and uniquely related to the amplitude-frequency response of the network. Therefore, the phase shift produced by a network can be found whenever the frequency response is given. Because the low- and high-frequency cutoffs in an amplifier are caused by capacitors and resistors, the designer can easily estimate the phase shift in the feedback loop because he knows the phase characteristics of simple combinations of resistors and capacitors. Networks excluded from this analysis include transmission lines, reactance bridges, and all-pass networks, all generally classed as nonminimum phase networks.

Examples of the relation between the phase shift and the variation of the amplitude-frequency response are shown for a number of simple cases in Fig. 10.7. Curve A is for a single *R-C* cutoff, where the amplitude is constant at low frequencies and decreases 6 dB/octave at high frequencies. When the amplitude is constant, the phase shift is small. At cutoff the phase shift increases most rapidly, and it becomes a 90° lag at high frequencies. If the *R-C* network produces a low-frequency cutoff, the phase shift is a 90° lead at low frequencies and is 0° at high frequencies.

Curve *B* represents the response of a parallel resonant *L-C* circuit. The corresponding phase curve has a 90° lead at low frequencies, which changes to a 90° lag at high frequencies. Curve C depicts an amplitude step where the

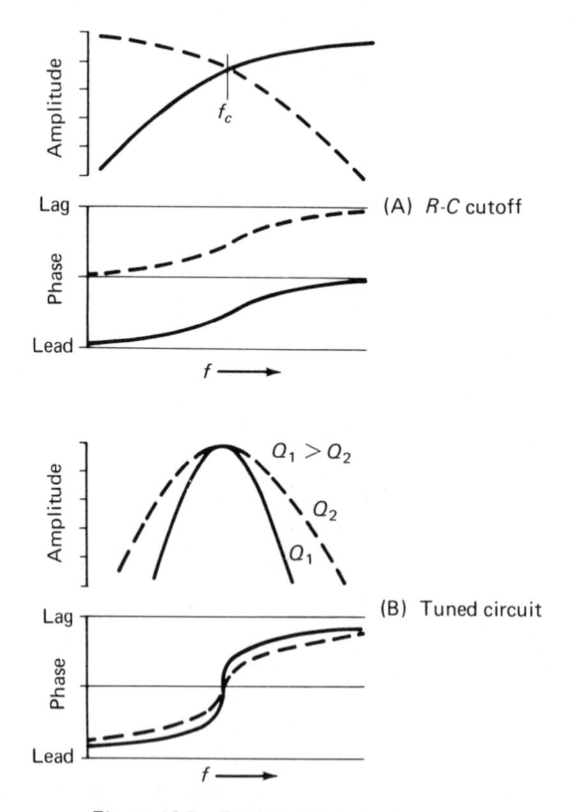

Figure 10.7. Typical gain and phase curves.

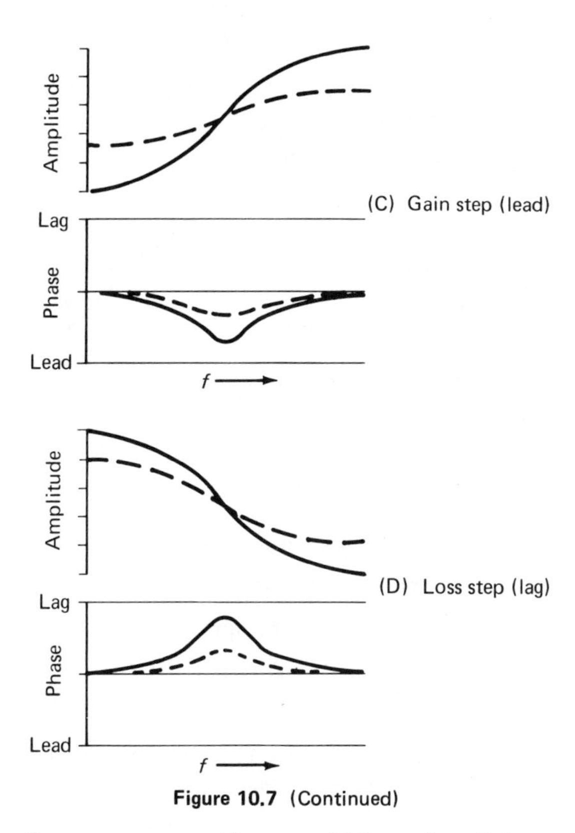

Figure 10.7 (Continued)

gain increases from a constant value to a higher value. Gain steps are used to produce a phase lead in a limited frequency range, as shown by the phase curve. Curve D is similar to C, except that the step is a loss and the phase shift is a lag in a limited frequency range.

Observe that in each case the amount of phase shift is determined by the slope of the amplitude characteristic. A rising characteristic produces a phase lead, and a falling characteristic produces a phase lag. In each case, when the amplitude characteristic is constant, the phase shift is zero.

The amplitude-frequency characteristic of a single *R-C* cutoff is shown in Fig. 10.8, with the related phase angles indicated along the curve: The phase angles are given for frequencies that increase by factors of $\sqrt{2}$. If the broken line with slopes of 0 and -6 dB/octave is used to represent the *R-C* curve, we assign the phase angles on the curve to the straight line at the same frequency. We are using the lines to represent both the amplitude and the phase angles of the curve.

The amplitude and phase characteristics in Fig. 10.8 represent the decade just above the half-power cutoff point of a resistor and shunt capacitor. Observe that over the two-decade frequency range the phase angle changes

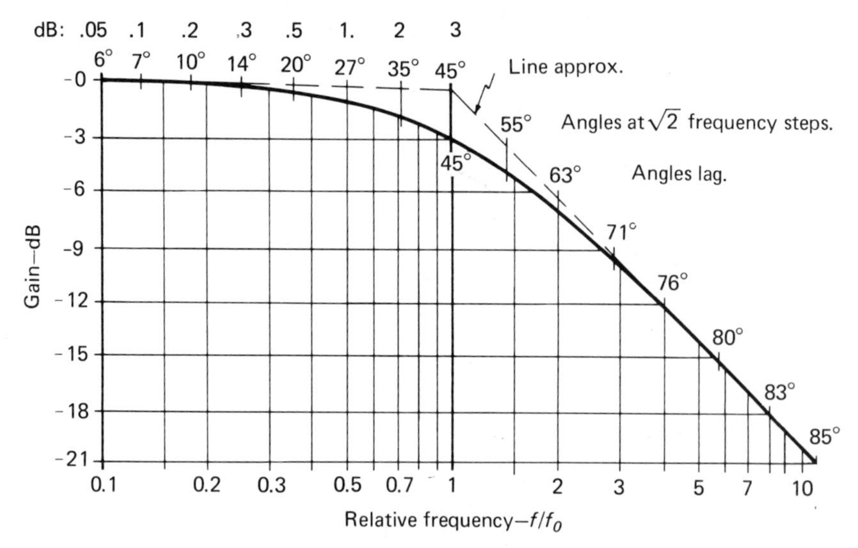

Figure 10.8. *R-C* phase-shift curve.

from 6 to 84°, and the phase shift may be approximated as 0 and 90° outside the two-decade frequency range. Observe also that the phase changes most rapidly near the cutoff frequency and is 45° at the breakpoint of the straight-line approximation.

The line approximations in Fig. 10.9 represent the amplitude and phase characteristics obtained with several combinations of *R-C* cutoff networks. Line A represents a single high-frequency cutoff having an asymptotic slope of -6 dB/octave of frequency. Line B is for two cutoffs with a -12 dB/octave slope from the same corner (cutoff) frequency, and line C represents two cutoffs with corner frequencies separated by approximately one decade.

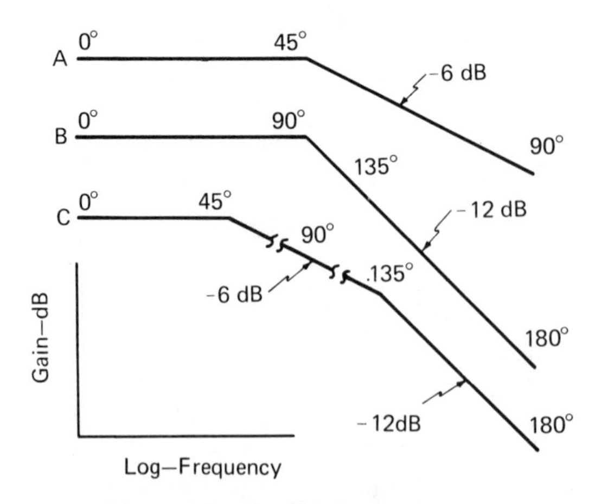

Figure 10.9. Cutoff gain-phase curves.

Observe that with two cutoffs at the same corner frequency, line B, the phase shift is 90° at the edge of the passband, and (referring to Fig. 10.8) the phase shift is 135° a little outside the passband. As line C indicates, when the corner frequencies are quite different, the phase shift does not become 135° until the second cutoff appears. Line C illustrates the advantage of using staggered cutoffs as a means of reducing the phase shift near the passband of an amplifier.

10.5 FINDING AN UNKNOWN PHASE CHARACTERISTIC

The lines in Fig. 10.10 show how two *R-C* cutoffs may be combined to find the phase shift produced by a gain step. Line D shows the phase angles produced by terminating a -6 dB/octave slope. The phase angles are given for even multiples of the cutoff frequency and are obtained from Fig. 10.8. The phase angles on the terminated +6 dB/octave slope are similarly obtained and indicated on the line. Line E, the step, is the result of combining the gain and phase characteristics of lines D and F. Observe that the phase lead at the first break on line E is the sum 45° - 63° = -18°, and that at the second break it is 27° - 45° = -18°. The step produces a phase lead of 18° at both breakpoints and a 20° maximum in between. The calculated phase is accurate to the extent that the gain step is the equivalent of the curves represented by lines D and E.

The phase characteristics of a given $A\beta$ amplitude-frequency curve may be found by suitably combining the phase characteristics of *R-C* cutoff networks. The procedure begins by constructing a straight-line approximation of the

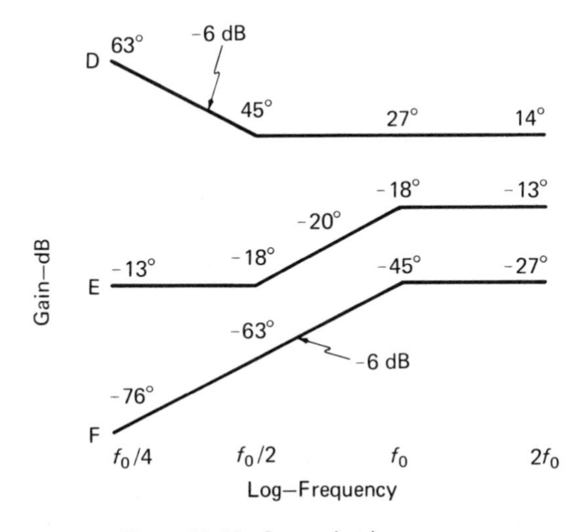

Figure 10.10. Step gain-phase curve.

given amplitude-frequency curve, from which we find the cutoff frequencies and slopes of the curve. We then duplicate the given frequency response by superimposing the response characteristics of networks for which we know the phase characteristics. The unknown phase-frequency response is obtained by summing the phase characteristics of the duplicating networks.

The construction begins by plotting the response-frequency characteristics on log-log scales. This is the only correct way to plot frequency characteristics, because the breakpoints show the cutoff frequencies, and the slopes of the curves show the number of equivalent *R-C* networks that produce the cutoff. To obtain the cutoff characteristics, the line approximations are drawn tangent to the curve, using only integral valued slopes, as shown in Fig. 10.11. A single stage *R-C* coupled amplifier with only one series input capacitor and one internal shunt emitter capacitance has a frequency response with a slope of +1 at low frequencies, a slope of 0 in the midrange, and a slope of −1 above the high-frequency cutoff. If the amplifier has two identical stages, the slopes will all be doubled, or +2, 0, and −2, respectively. The line approximations show that the amplifier, represented in Fig. 10.11, has a low-frequency cutoff of 6 dB/octave, and a high-frequency cutoff at f_2 of 12 dB/octave, which is terminated at f_3.

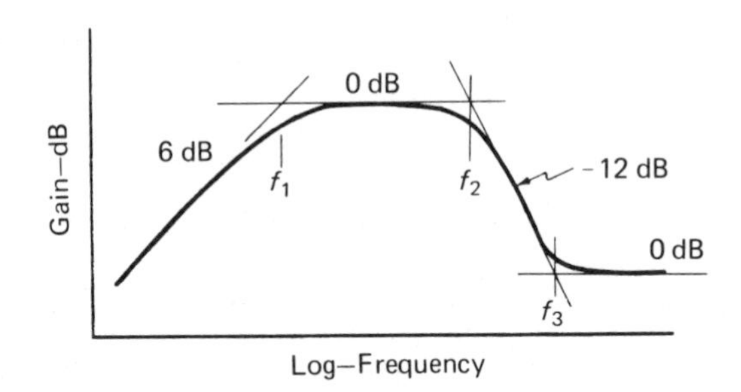

Figure 10.11. Line approximations of a curve.

10.6 AMPLIFIER STABILITY PROBLEMS

The tendency of an amplifier to oscillate increases rapidly as the number of stages is increased above two. When feedback is applied to amplifiers having more than two stages, stability problems are always expected, even with a few dB feedback. For adequate feedback special design procedures are required to keep the phase shift below 180° until the frequency is either so low or so high that the loop gain $A\beta$ has dropped below unity.

The frequency range over which the characteristics of the feedback loop must be controlled to avoid oscillations is unexpectedly large. Control must exist over approximately 1 octave for each 10-dB feedback, plus 1 or 2 octaves as a margin of safety. An amplifier which has a useful frequency range of 30 Hz to 30 kHz and 20-dB feedback must have the $A\beta$ curve controlled from 4 Hz to 500 kHz, which is 7 octaves beyond the useful range. Thus, an amplifier for service in a 3-decade frequency range requires very careful design of the feedback loop over an additional 2-decade frequency range. Our present purpose is mainly to explain the techniques that are used when an amplifier can be stabilized by minor changes in the circuit.

The stability of a given amplifier is studied by examining the asymptotic slopes of the open-loop, gain-frequency characteristic, particularly in the frequency range where the loop phase shift is 180°. The asymptotic slopes of an actual amplifier usually can be estimated by inspecting the circuit. Each resistance coupled stage and each transformer will contribute 6 dB/octave of rising slope at low frequencies. Similarly, each shunt capacitor will contribute 6 dB/octave of falling slope at high frequencies. Transformers are particularly troublesome sources of phase shift. The leakage reactance of a transformer contributes 6 dB/octave at the high-frequency cutoff, and the distributed capacitance may increase the slope by an additional 6 dB.

The capacitance effects within a transistor contribute one 6 dB/octave slope, which appears in a CE stage as the β cutoff. At frequencies near the alpha cutoff a transistor contributes a small additional phase shift (12 to 30°) known as the *excess phase*. As the bandwidth of an amplifier is extended to high frequencies, the control of the phase characteristics becomes increasingly difficult. Hence, feedback over two stages is generally the upper practical limit for a wide-band amplifier.

10.7 STABILITY ANALYSIS BY ASYMPTOTIC SLOPES

As an example of the stability analysis of a practical amplifier, consider a two-stage resistance-coupled amplifier that has two identical stages. As is shown in Fig. 10.12, the straight-line approximation of the frequency response of this amplifier has the line BC representing the midband frequency response and the line CD representing the cutoff response. Because the amplifier has

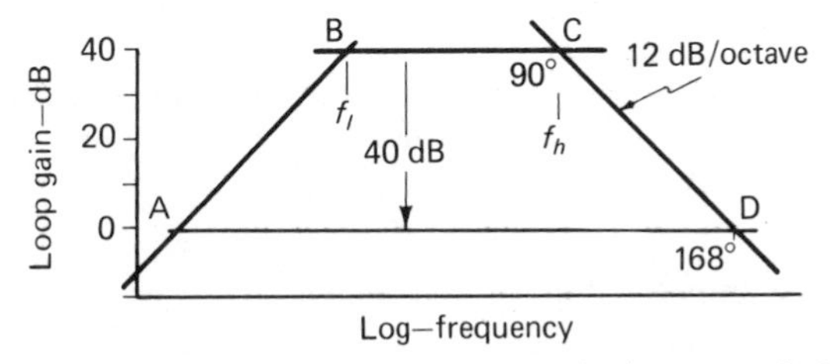

Figure 10.12. Loop gain and phase characteristics (two-stage R-C amplifier).

two identical stages, each with a single *R-C* cutoff, the line CD has a slope of −12 dB/octave, extending from f_h to infinity. Suppose that the amplifier is required to have the gain reduced by 40-dB feedback. The gain with feedback is represented by the line AD, which is 40 dB below the midfrequency open-loop response. The phase shift in the feedback loop is obtained by referring to Fig. 10.8 and is found to be 90° at f_h and 168° at the frequency at which the loop gain falls to 1 (0 dB). This figure shows that the phase margin with 40-dB feedback is only 12° and suggests that 40-dB feedback should be avoided when the stages are identical. (The reader should determine the phase margin that exists with 40-dB feedback when the frequencies are staggered to make one cutoff frequency 30 times the lower. The answer is 30°.)

Figure 10.13 shows a straight-line approximation representing the high-frequency cutoff of an *R-C* amplifier that has three identical stages. With three stages and a loop gain of 8, 18 dB, each stage contributes a gain of 2 and a phase shift of 60° when the loop gain falls to 1. We conclude that an amplifier with three identical stages will be unstable if the loop gain is 18 dB, and that a 10-dB gain margin means that the loop gain should be about 18 − 10, or 8 dB.

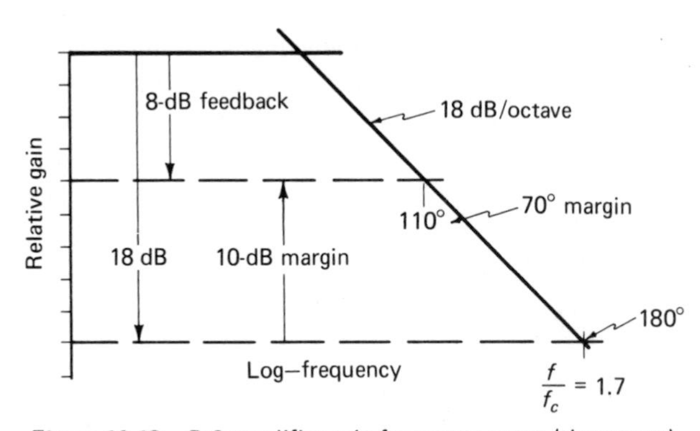

Figure 10.13. *R-C* amplifier gain-frequency curve (three-stage).

10.8 METHODS FOR INCREASING FEEDBACK

Two methods of stabilizing feedback amplifiers are in everyday use. One method is to stagger the cutoff frequencies; the other is to add a phase-reducing step. Both devices are easily described by straight-line approximations.

The three-stage amplifier represented in Fig. 10.14 has two identical stages with their corner frequencies at C and a third stage with a corner frequency at B. We assume f_c is over 10 times f_b. The $A\beta$ frequency response of the loop is

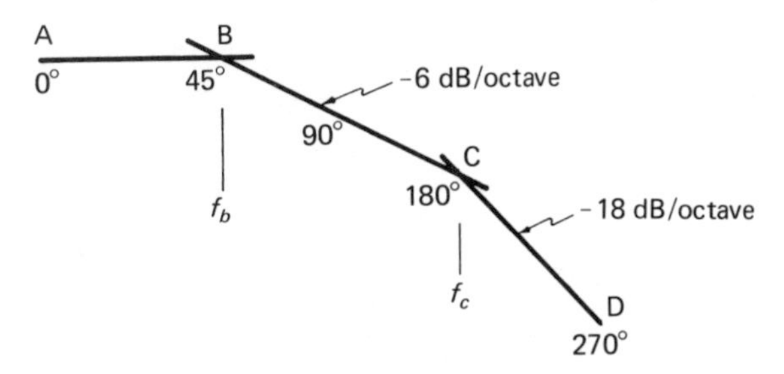

Figure 10.14. Gain and phase curve with staggered cutoffs.

represented by the lines ABCD. Phase angles indicated along the curve show that the phase shift is 45° at the first break and increases to 90° as the frequency approaches f_c. The second break is produced by two *R-C* cutoffs, each of which contributes 45° at the break. Adding 90° and 2(45°), we find that 180° is the phase angle at C. As the frequency approaches the high frequency at D, the phase angle is the total of three 90° phase shifts.

If the loop gain is adjusted to make the unity gain fall at the frequency of the second break, the phase margin is 0°, but the feedback in the midfrequency range can be made quite high. This amount of feedback is determined by the separation between the two breakpoints, and, theoretically, there is no limit to the feedback attainable in the midfrequency range below f_b. However, for a fixed f_c each 20-dB feedback reduces by a factor of 10 the frequency band over which the midband feedback is available, because less feedback is available at frequencies between f_b and f_c.

Staggered cutoffs are useful when an amplifier has two or more identical cutoffs at frequencies well beyond the frequency band in which feedback is required. For example, an amplifier with three stages, having identical β cutoffs at 300 kHz, will sing (oscillate) with 18-dB feedback. But a single *R-C* cutoff anywhere in the feedback loop that begins at 5 kHz will permit nearly 30-dB feedback in a frequency band extending up to 5 kHz. (Reducing the feedback to 20 dB gives a practical gain and phase margin.) Similarly, staggered cutoff frequencies on the low-frequency side of the band will permit more feedback than can be obtained with identical cutoffs. Generally, an amplifier can be direct-coupled to reduce the number of low-frequency cutoffs per stage, and staggered cutoffs are more acceptable at high frequencies where the control of phase shift is most difficult.

10.9 PHASE-SHIFT CONTROL BY GAIN STEPS

A second method of reducing the phase shift in a feedback loop is to introduce in the loop a gain step that reduces the phase shift at frequencies on the $A\beta$ curve near the critical point. As an example of a gain step, consider the amplifier response represented by the line approximations shown in Fig. 10.15. The line represents an amplifier having two high-frequency cutoffs that

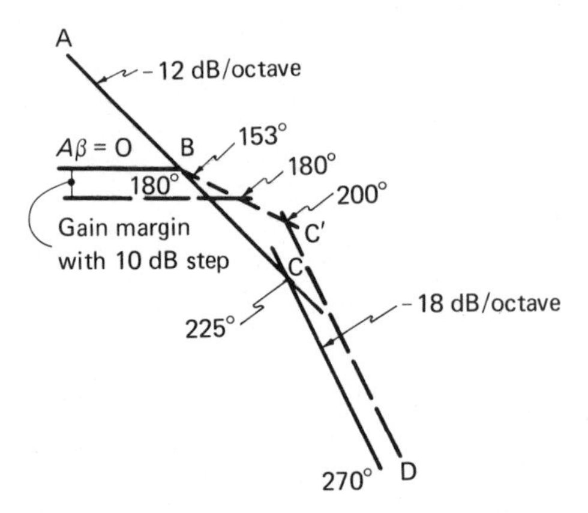

Figure 10.15. Gain and phase curve with gain step.

produce almost 180° of phase shift at B, and a third cutoff that increases the phase shift to 225° at the break point C. Suppose that we wish to use feedback that makes the loop gain 0 dB at B, where the phase shift is 180°. Introducing a 10-dB gain step in the loop increases the amplitude response between B and C′ and reduces the phase shift as shown. The effect of the step is to reduce the phase shift at B enough to provide a gain and phase margin. A Nyquist diagram representing the effect of a gain step is shown in Fig. 10.16.

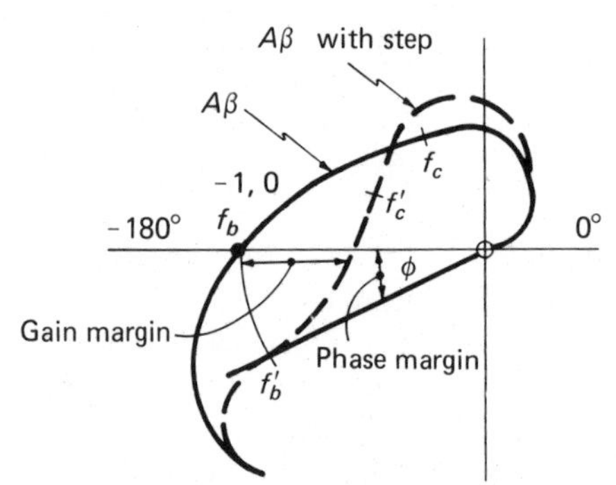

Figure 10.16. Nyquist diagram with gain step.

10.10 STABILIZING FEEDBACK IN A HIGH-FIDELITY AMPLIFIER

As a practical example of feedback system design, we consider a high-fidelity power amplifier that is to have 20-dB feedback. In a typical situation the problem is met when the amplifier is completed and is found to sing at a high frequency. We begin by removing all shunt capacitors that may cause phase shift in the loop, but the amplifier still sings. We know that the β

cutoff of the power transistors introduces one high-frequency cutoff that is probably between 10 and 40 kHz. By the rule in Section 10.6, 20-dB feedback with a margin of safety requires control of the loop-phase characteristics up to 4 octaves above the upper frequency limit of the amplifier. Four octaves above 20 kHz is $2^4 \times 20$ kHz = 320 kHz. The designer's problem is to control the gain characteristics between 20 and 320 kHz. Control of the gain-phase characteristics means that two cutoffs can be placed where required in the controlled frequency band, and that the control is effected without allowing the total number of cutoffs to exceed two.

The procedure usually given in textbooks on feedback design requires a measurement of the open-loop gain and phase characteristics of the amplifier in order to plot the $A\beta$ curve. In some amplifiers it is difficult even to identify the feedback loop, and in most it is a complicated procedure to open the loop without affecting the biasing and the impedance characteristics of the loop. To explain how an amplifier is stabilized, we assume that the $A\beta$ curve is known, but admit that meaningful gain and phase measurements are difficult and too time-consuming for practical work. An experienced engineer might be expected to measure and plot the $A\beta$ characteristic as a last resort, but in most cases a cut-and-try design, if accompanied by insight and understanding, will give a satisfactory result. The objective in the exposition below is to present a practical summary of the limitations and choices open to the designer.

Our procedure is to construct a table showing the phase angles introduced by the *R-C* cutoffs known to exist in the $A\beta$ loop. For this purpose we refer to Table III, which gives the phase shift produced by gain steps and *R-C* cutoffs at a series of frequencies each side of the reference or cutoff frequency. The phase angles are calculated for a series of frequencies that increase by factors of $\sqrt{2}$, so that the frequencies are spaced at equal intervals on a logarithmic frequency scale. Alternate frequencies are in the convenient 1:2 ratio. Table III tabulates the same kind of phase-frequency information that is shown in Figs. 10.8 and 10.10E.

TABLE III

Phase Angles for Gain Steps and *R-C* Cutoffs

f/f_c	.09	.13	.18	.25	.35	.50	.71	1.0	1.4	2.0	2.8	4.0	5.7	8.0	11	16
3 dB	1	2	3	4	5	7	9	10	10	9	7	5	4	3	2	1
6 dB	3	4	5	7	9	13	16	18	19	18	16	13	9	7	5	3
9 dB	3	5	6	9	12	17	21	26	28	28	26	21	17	12	9	6
12 dB	4	5	7	10	14	19	25	31	35	37	35	31	25	19	14	10
1 *R-C*	5	7	10	14	19	27	35	45	55	63	71	76	80	83	85	87
2 *R-C*	10	14	20	28	39	54	71	90	109	126	141	152	160	166	170	194

Note: For gain steps the lower corner frequency is at $f/f_c = 1$. The upper corner frequency is where the angle returns to the same value as at $f/f_c = 1$. The angles are degrees lag for an increase in gain and lead for a decrease.

The phase characteristic of the $A\beta$ loop is calculated by summing in Table IV the phase characteristics of the equivalent *R-C* networks in the loop. The phase angles in Table IV are obtained from Table III by matching the cutoff frequency in Table III with the frequency at which the cutoff occurs in Table IV. The phase angles are simply copied as a series into Table IV. Intermediate points may be calculated by referring to a larger table of phase angles given in Appendix A.5. However, the phase angles change so slowly with frequency that only a small error is introduced by adjusting a remote cutoff frequency to fit the chart. The error is negligible when the adjusted cutoffs are a factor of 10 removed from the critical frequency. The corner frequencies nearest the critical frequency should be carefully matched to the table values.

TABLE IV

Amplifier Phase-Frequency Calculations

1. *f* in kHz	20	40	80	160	320	640	dB at 180°,
2. 1 *R C* at 20 kHz	45	63	76	83	87	89	
3. 2 *R-C* at 320 kHz	7	14	28	54	90	126	
4. Initial phase in degrees	52	77	104	137	177	215	30
5. 1 *R-C* at 40 kHz	27	45	63	76	83	87	
6. Trial phase angles	97	122	167	213	260	302	24
7. 12-dB step at 160 kHz	−5	−10	−19	−31	−37	−31	
8. Net phase angles	92	112	148	182	223	271	30
9. Less 1 *R-C* at 320 kHz	89	105	134	155	178	208	45

The angles shown in Table IV represent the calculated phase-frequency characteristics of the high-fidelity amplifier. We assume there are two high-frequency cutoffs at 320 kHz at the upper edge of the controlled frequency band and that any cutoff above 320 kHz may be neglected. We calculate the amount of feedback that can be obtained with one and two R-C cutoffs inside the controlled frequency band.

Line 1 in the table is the series of frequencies in the cutoff region at which we are calculating the phase shift. Line 2 shows the phase angles produced by the β cutoff in the power-stage transistors. Taking 20 kHz as a reasonable value for the 3 dB cutoff frequency, we write 45° in Table IV under 20 kHz in line 2. The rest of line 2 is simply copied from Table III.

Line 3 in Table IV represents the phase characteristics of the two additional R-C cutoffs which contribute 90° phase shift at 320 kHz. After the 90° angle is located, the remaining angles are copied from Table III. Line 4 is the sum of the angles in lines 2 and 3, and we find that the loop phase shift may be assumed to total 180° at 320 kHz.

The loss associated with phase shift is calculated by referring to the curve in Fig. 10.8, since a straight line approximation represents the gain-frequency characteristic only at frequencies four or more times the corner frequency. At a corner frequency the loss is 3 dB, at twice a corner frequency is 7 dB, and at four times the frequency is 2 × 6 dB, etc. The loss with a gain-step is approximately one-fourth the step at the corner frequency and is one-half the step at the mid-frequency.

The amount of feedback which makes the amplifier unstable is calculated by summing the loss in the cutoff networks at 320 kHz. The loss in the power-stage cutoff is 4 × 6 dB and the loss in the R-C cutoffs at 320 kHz is 2 × 3 dB. Thus, the phase shift is 180° when the loss totals 30 dB, and the amplifier can be expected to sing with 30 dB feedback. A similar calculation shows that the loss totals 20 dB at 160 kHz, which means that with 20 dB feedback there is a 10 dB gain margin and a phase margin of 177-137 = 40°.

As a more typical example, suppose that the amplifier has an additional high-frequency cutoff in the driver stage with a corner frequency at 40 kHz. Line 6 in Table IV shows that the phase turn-over frequency is lowered to 100 kHz. The cutoff loss at 80 kHz is 19 dB and at 100 kHz is 24 dB when the phase shift is 180°. We conclude that 20 dB feedback cannot be used with a reasonable margin of stability. However, 20 to 40° of phase shift can be removed by adding a gain step, so we examine the effect of a step in improving the gain and phase margins.

Line 7 in Table IV represents a 12 dB step beginning at 160 kHz, and line 8 shows the phase-frequency characteristics of our completed design. The corresponding amplitude characteristic in Fig. 10.17 shows that the loop gain at the 160 kHz singing point is 30 dB. With 20-dB feedback the gain margin is 10 dB and the phase margin is 28°.

The foregoing example shows that the 12-dB gain step is of considerable help in the successful application of feedback. Moreover, we show by line 9 in Table IV that the removal of one of the *R-C* cutoffs at 320 kHz considerably simplifies the design problem. A comparison of lines 3 and 7 shows that the gain step approximately compensates for one of the 320-kHz cutoffs and illustrates the importance of beginning a feedback design with as few high-frequency cutoffs as possible.

The gain step is the most easily recognized stabilizing technique. These steps are usually created by connecting a small capacitor across the main feedback resistor. Most hi-fi audio amplifiers require a gain step beginning above 100 kHz. A resistor is sometimes used in series with the capacitor to limit the amount of the gain step and, thereby, to avoid increasing gain and instability at still higher frequencies. A gain step may also be produced by inserting a resistor in series with a shunt capacitor that is used to increase the loop cutoff rate. The step may be placed anywhere in the feedback loop.

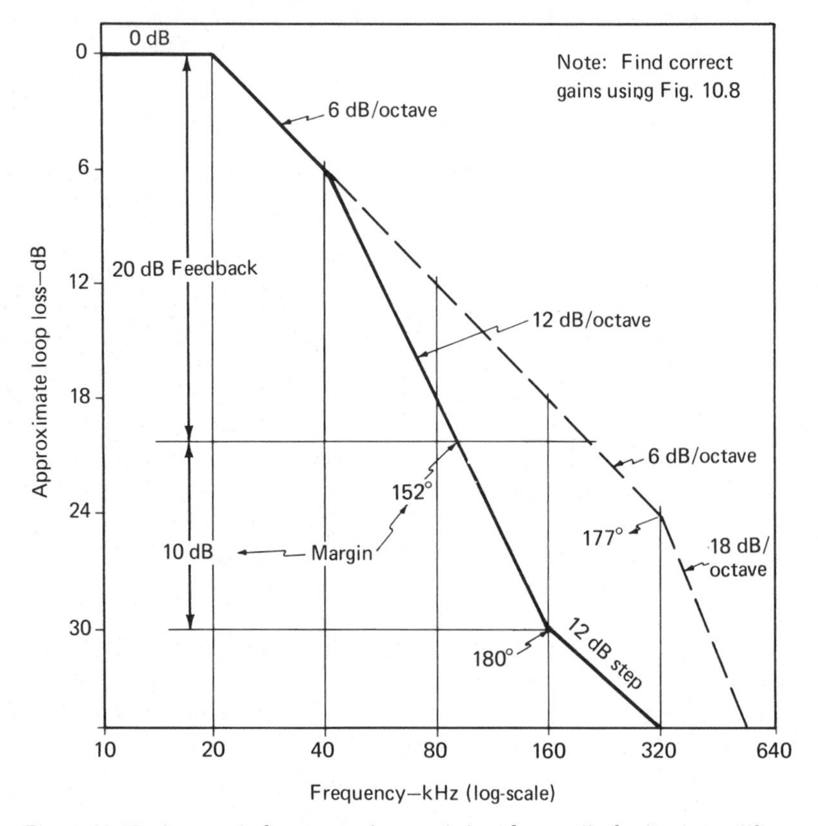

Figure 10.17. Loop gain-frequency characteristics of an audio feedback amplifier.

10.11 AMPLIFIER STABILIZING TECHNIQUES

The results of the preceding section may be summarized as a list of techniques that can be used to achieve the successful application of feedback in multistage amplifiers. We have seen that a number of conditions must be simultaneously satisfied.

1. The amplifier must be designed with as few high-frequency cutoffs as possible. The β cutoff of the power stage must be accepted, and the β cutoff of the remaining stages must be placed as high as practical. Perhaps the simplest way of reducing the high-frequency phase shift is to reduce the number of stages in the feedback loop or to stagger the high-frequency cutoffs.
2. The amount of attainable feedback depends on the frequency range over which the cutoff can be continued at a rate of 12 dB/octave or less.

Two cutoffs may be allowed to dominate the cutoff region, providing one of these may be terminated, as with a series resistor, or offset by a gain step.

3. A gain step beginning near the singing frequency is a considerable help in improving the phase margin and in increasing the feedback. A gain step at too low a frequency may reduce the gain margin. Sometimes a resistor is used in series with the gain-step capacitor to prevent an undue increase of the high-frequency gain
4. When the loop phase is 180°, the gain with feedback is higher than the midband gain and reaches a maximum near the 180° phase-shift frequency. At this frequency the positive feedback produces a peak in the gain characteristic and ringing transients. These gain peaks are reduced by reducing the feedback or by increasing the gain and phase margins.
5. The stabilization of feedback on the low-frequency side of the amplifier response is essentially the same as for the high-frequency side. The phase shift at low frequencies is caused by the series-coupling capacitors and by shunt emitter and filter capacitors. Transistor amplifiers are easily direct-coupled and the accumulated low-frequency phase shift is usually insufficient to produce low-frequency instability. When instability is observed, the designer may eliminate emitter capacitors, stagger the coupling cutoffs, and use a regulated power supply or Zener diodes for decoupling low-power stages. Table III and the characteristic curves in this chapter are symmetrical with respect to frequency and may be used for the low-frequency region by replacing f/f_c by its reciprocal f_c/f.

10.12 OPERATIONAL-AMPLIFIER STABILITY

Operational amplifiers (op amps) have voltage gains of 10^5 or more and are always used with overall feedback. To simplify their use in feedback systems, the amplifiers are stabilized, with a dominate capacitor furnished as a part of the amplifier. The capacitor is usually sized to make the high-frequency cutoff begin at about 10 Hz, as shown in Fig. 10.18.

The long 6-dB/octave slope makes an operational amplifier stable with any resistance-feedback network; and the 90° phase margin and very high β-cutoff frequencies give the amplifier a well-damped response that is free of gain peaks outside the band. When operated as a unit gain inverter, the feedback network has only a 6-dB loss, so a small capacitor is sometimes connected across the feedback resistor to provide a gain step.

The low-frequency gain of an operational amplifier is sometimes increased by a chopper and ac amplifier. This additional gain is used as feedback at very

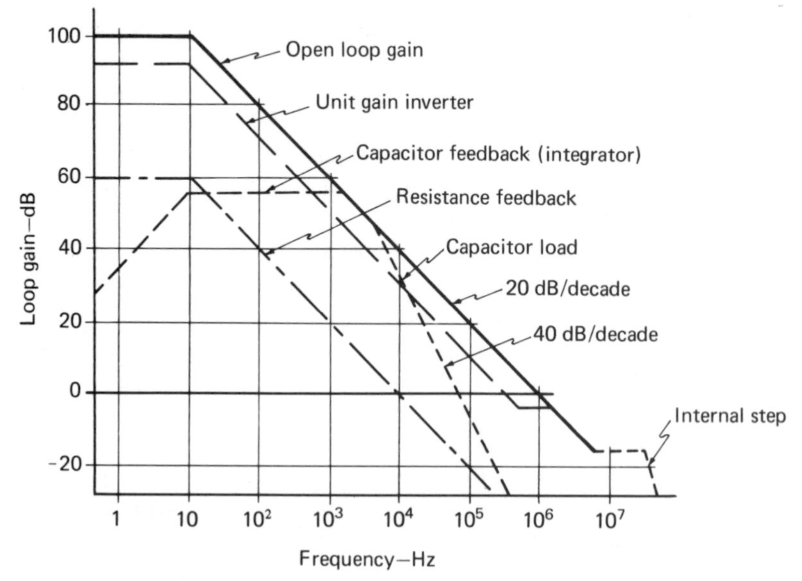

Figure 10.18. Loop gain of an operational amplifier.

low frequencies as a means of reducing the Q-point drift away from a desired setting.

When the feedback is partly reduced so that the amplifier has an overall gain of about 1000, the system is stable enough to permit considerable flexibility in the choice of the feedback circuit. With a feedback capacitor, as in an integrator, the amplifier is stable, provided the internal high-frequency cutoff is properly designed. With a capacitor load the loop phase shift approaches 180°, and a small resistor may be needed in series with the load to prevent stability problems. The resistor has the effect of increasing the loop gain, as with a gain step. In general, the stability problems that arise in the application of operational amplifiers are solved by the same techniques that we have described for the audio-feedback amplifier.

10.13 SUMMARY

The stability and transient characteristics of a feedback amplifier are evaluated by plotting a Nyquist diagram of the loop transfer function. An amplifier is unstable if the $A\beta$ curve includes or encloses the critical point -1.0. An amplifier is generally considered stable if the $A\beta$ curve passes inside the critical point with a 30° phase margin and a 10-dB gain margin.

An amplifier has negative feedback when the $A\beta$ curve is outside the unit circle enclosing the critical point. When the curve is inside the unit circle, the gain with feedback exceeds the gain without feedback, and the amplifier may exhibit ringing transients and gain peaks at frequencies just beyond the negative-feedback region.

Because of the practical difficulties of measuring the open-loop amplitude and phase characteristics, the $A\beta$ curve is usually obtained from the cutoff frequencies that are known to exist in the loop. From the cutoff frequencies we plot the loop amplitude-frequency response as a broken-line approximation. From the breakpoints and slopes of the line we calculate the loop phase shift.

A feedback amplifier is stabilized by modifying the amplitude and phase characteristics or reducing the loop gain until suitable gain and phase margins are provided. Techniques that are used for increasing the amount of feedback or for improving the stability are given in Section 10.11. The terminology and the components of a system change with the field of application, but all feedback systems are stabilized by amplitude and phase control of the feedback loop.

PROBLEMS

10-1. (a) Using Table III in the text, plot the high-frequency gain and phase response of a three-stage *R-C* amplifier in which the cutoff frequency of one stage is at 10 kHz, a second is at 14 kHz, and the third is at 20 kHz. (b) At what frequency is the phase shift 180°? (c) How much may the gain be reduced by feedback without oscillations?

10-2. (a) Plot the gain and phase response of a three-stage *R-C* amplifier in which the cutoff frequencies are at 10, 14, and 100 kHz. (b) What is the permissible amount of feedback if the phase margin is 45°? (c) What is the permissible amount of feedback with a 10-dB gain margin?

10-3. An amplifier has one *R-C* cutoff at 20 kHz, a second and third at 320 kHz, and a 12-dB gain step at 160 kHz. Assume that there are no additional cutoffs at the high frequencies. (a) What is the minimum amount of feedback that will cause the amplifier to oscillate? (b) What is the permissible amount of feedback with a 10-dB gain margin and a 30° phase margin?

10-4. A feedback amplifier has one *R-C* cutoff at 20 kHz, a second at 40 kHz, and a third at 320 kHz. (a) Where should a 12-dB gain step be placed to obtain the maximum amount of feedback, and (b) how much feedback can we expect without singing?

10-5. An *R-C* amplifier has one low-frequency cutoff at 20 Hz, a second at 10 Hz, and an emitter-bypass capacitor that places an 18-dB gain step centered at 2.5 Hz. (a) With how much feedback will the amplifier oscillate? (b) What is the permissible amount of feedback with 10-dB gain and 30° phase margins?

10-6. An amplifier with three identical high-frequency *R-C* cutoffs has feedback with a 9-dB gain margin. (a) What is the phase margin? (b) If the loop gain is increased to 18 dB and a 6-dB step is placed in the loop, what is the maximum obtainable phase margin? (c) What is the gain margin with the maximum phase margin?

REFERENCES

10-1. *RCA Transistor Manual*, SC-13, p. 492. Harrison, N.J.: Radio Corporation of America, 1967. See also available vacuum-tube and transistor manuals for descriptions of techniques used to stabilize audio and servo amplifiers.

10-2. Giles, J. N., *Frequency Compensation Techniques for an Integrated Operational Amplifier*, App-117/2. Mountain View, Cal.: Fairchild Semiconductor, 1969.

10-3. Terman, F E., *Electronic and Radio Engineering*, 4th ed., Chapter 11, "Feedback Amplifiers." New York: McGraw-Hill Book Company, 1955.

DESIGNS

10-1. Observe the square-wave response of an audio-feedback amplifier and design a feedback gain step for the best transient response. Similarly, see if the low-frequency transient response can be improved by a low-frequency gain step.

10-2. Design frequency-shaping networks that will give an op amp the response characteristic shown in Fig. 10.17 with the frequency scale reduced by a factor of 100. Close the feedback network and study the transient response.

11

VIDEO AMPLIFIERS

An amplifier that has a frequency response extending from a low audio frequency to tens of megahertz is called a *wide-band* or *video* amplifier. These amplifiers are generally required when the input signal has abrupt changes of wave shape that must be faithfully reproduced. Wide-band amplifiers are found in TV systems, radars, satellite relays, and oscilloscopes.

For the faithful reproduction of pulse and TV signals an amplifier must have a flat frequency response and a minimum of phase shift. To make the midband gain independent of frequency and to extend the frequency response, a wide-band amplifier usually has overall feedback. However, the remaining low- and high-frequency cutoffs that produce wave-form distortion can be offset by adding equalizing networks. The designer's problems stem from the facts that (1) the gain-bandwidth product of a transistor is fixed so that a wide bandwidth implies low gain and (2) high gain-bandwidth products are obtained only from small devices that have a limited power capability. This chapter shows that the high-frequency cutoff of a video amplifier may be easily calculated and describes techniques used in the design of wide-band video amplifiers.

11.1 HIGH-FREQUENCY CUTOFF IN WIDE-BAND AMPLIFIERS

The high-frequency cutoff of a video amplifier is many orders of magnitude above the low-frequency cutoff, so the bandwidth is determined mainly by the upper cutoff frequency. With such a wide band the parameters that control the low-frequency response may be considered separately from those that determine the high-frequency response. However, the designer is somewhat constrained by the fact that large components, which are required for a low cutoff frequency, are bulky and, therefore, add shunt capacitance that may adversely affect the high-frequency response.

The low-frequency response of a video amplifier is usually determined by the coupling capacitors that are used at the input, the output, and between stages. Emitter capacitors are generally avoided because a large capacitor is required to prevent low-frequency cutoff, and the impedance of a large capacitor tends to increase at video frequencies and thus to reduce the gain at high frequencies. For these and other reasons, the emitter capacitor may only complicate the design at both the low- and high-frequency limits. The coupling capacitors needed for a given low-frequency response may be found by the methods outlined in Chapter 4.

The high-frequency cutoff in a video amplifier is generally produced by capacitance in the transistor and by the capacitance between the leads and components of the circuit. Although very small, the capacitors that limit the high-frequency response are as easily considered in design as are the large components that affect the low-frequency cutoff.

The high-frequency cutoff in a *CE* stage is approximately equivalent to a single *R-C* cutoff that is the combined effect of two capacitors which shunt the input. One capacitor, shown as C_M in Fig. 11.1, is a result of the Miller

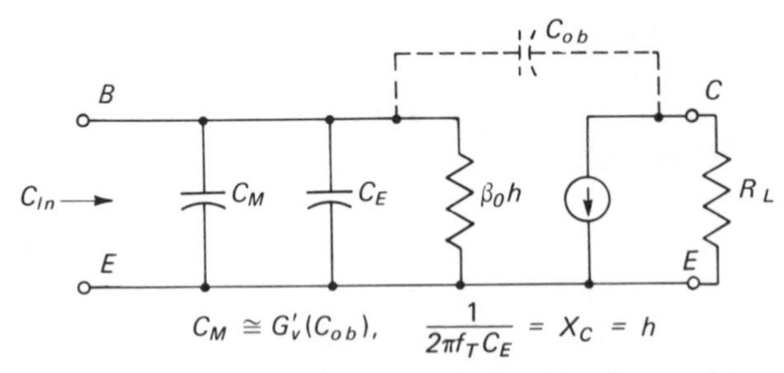

Figure 11.1. *CE* equivalent circuit for video frequencies.

effect. This input capacitance equals the collector-to-base feedback capacitance C_{ob} increased by the voltage gain from the base to the collector. The voltage gain is the gain that exists after the local feedback is taken into consideration.

The second part of the input capacitance is the intrinsic base-emitter capacitance C_E of the transistor. This capacitance causes the transistor current gain to decrease 6 dB/octave at high frequencies, and the capacitance can be calculated from the high-frequency gain-bandwidth product f_T, which is usually given for high-frequency transistors.

11.2 TRANSISTOR GAIN-BANDWIDTH PARAMETERS

For an explanation of high-frequency β falloff consider the short-circuit *CE* equivalent circuit that is shown in Fig. 11.2. In the high-frequency literature the symbol β designates the complex current gain, which varies with frequency, and the low-frequency current gain, which is independent of frequency, is designated by β_0. At low frequencies the transistor input impedance is $\beta_0 h$, and the SC collector current is $\beta_0 i_b$. At high frequencies the intrinsic base-emitter capacitance C_E shunts the emitter diode, so only part of the input current i is effective as base current i_b. When the signal frequency is $f = f_T$, the reactance of C_E is h, so the current in C_E is $\beta_0 i_b$ and the signal input current is $(\beta_0 + 1)i_b$. Thus, as shown in the Fig. 11.2, the transistor input-to-output current gain is reduced to 1 when the current in the shunt capacitor is β_0 times the base current i_b. Observe that the current gain falloff begins when the reactance of C_E is $\beta_0 h$. Hence the β cutoff frequency f_β is

$$f_\beta = \frac{f_T}{\beta_0} \tag{11.1}$$

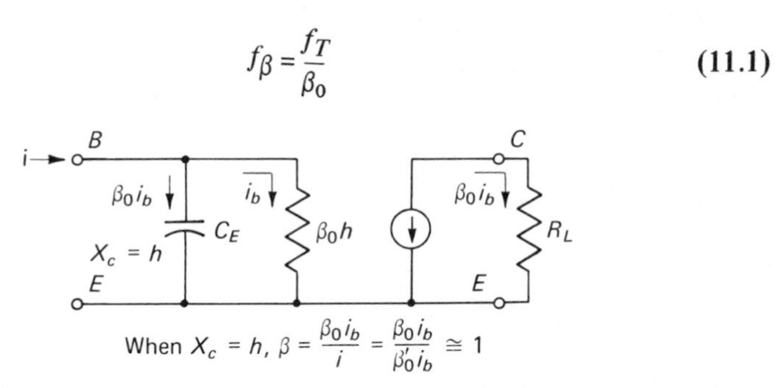

Figure 11.2. *CE* equivalent circuit when $f = f_T$.

and the current gain falls off inversely with the frequency (-6 dB/octave) until $\beta = 1$ at $f = f_T$. These characteristics of the high-frequency *CE* current gain are illustrated by the curve shown in Fig. 11.3. For circuit calculations C_E may be obtained from a reactance chart by using $f = f_T$ and the reactance $X_E = h$. The value of h is found by Shockley's relation, using the Q-point

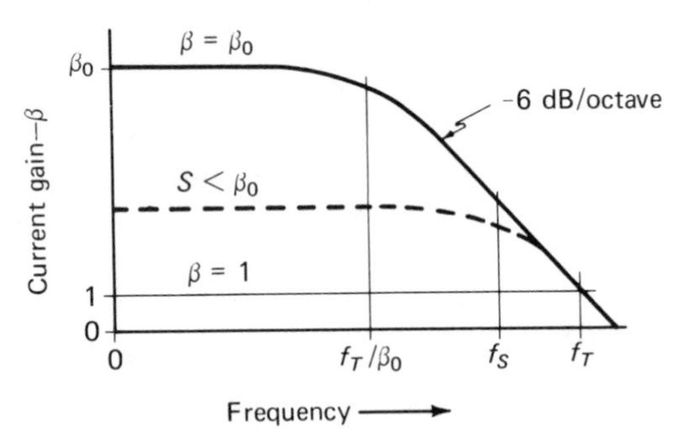

Figure 11.3. Beta-frequency curve.

emitter current. However, a calculation of C_E is usually avoided in design by using the transistor f_T parameter instead.

The transistor gain-bandwidth parameter f_T tends to vary with the collector current somewhat in the same way that the low-frequency β_0 varies with the collector current. For this reason the transistor f_T tends to increase with collector current until the low-frequency β_0 begins to fall off with increasing collector current. Generally, the manufacturer gives curves showing f_T as a function of the Q-point parameters.

Curves for a 2N3137 ultrahigh-frequency transistor, showing how f_T varies with the collector current and voltage, are shown in Fig. 11.4. Character-

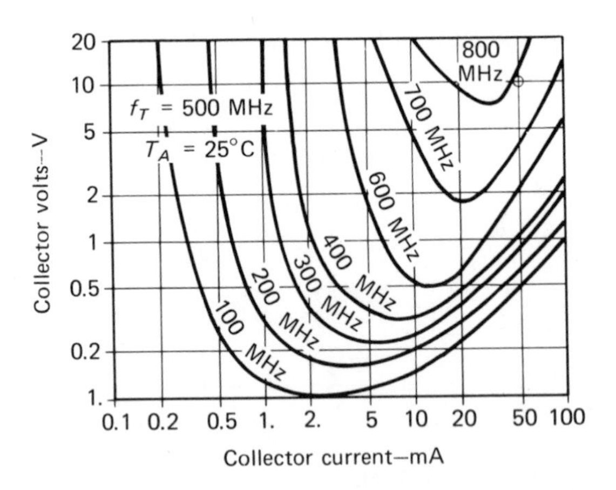

Figure 11.4. Gain-bandwidth curves for an UHF transistor.

istically, the gain-bandwidth product depends mainly on the collector current, and near the maximum is not overly sensitive to a 2-to-1 change of the current. The curves shown in Fig. 11.4 may be used for some high-current planar transistors (e.g., 2N870) by increasing the collector currents by a factor of 10 and decreasing the f_T values by a factor of 10. As this statement suggests, one way of manufacturing a transistor for higher frequencies is to decrease the dimensions of the junction. If the area is reduced by a factor of 10, the capacitance and the collector current are reduced by the same factor of 10. For this reason, high-frequency transistors tend to be low-current devices and must be operated near a current that gives the maximum gain-bandwidth product.

11.3 VIDEO BANDWIDTH CALCULATIONS

The equivalent circuit shown in Fig. 11.2 suggests correctly that the frequency limits imposed by the input capacitance C_E may be reduced by shunting the input by a low impedance or driving the stage with a low-impedance source. This method of increasing the cutoff frequency lacks the advantages of feedback and is impractical, because the required values of source impedance, usually less than 10 Ω, are too low for a high-frequency design.

If the stage current gain is reduced by emitter feedback that makes the ac current gain S, equation (11.1) which gives the amplifier cutoff frequency becomes

$$f_S = \frac{f_T}{S} \tag{11.2}$$

As shown by equation (11.2), the product of the gain S and the bandwidth f_S is constant. Thus, the bandwidth may be increased by accepting a lower current gain. In other words, the cutoff frequency produced by input capacitance is moved to a higher frequency, because emitter feedback reduces the transistor input capacitance relative to the external source impedance.

The high-frequency cutoff f_h in a stage is caused by the combined effect of the Miller capacitance and the transistor input capacitance acting as high-frequency shunts across the equivalent base-to-ground impedance R'_S. Since the Miller capacitance is proportional to the stage voltage gain and the transistor input capacitance depends on the stage current gain, a formula for calculating the stage cutoff frequency f_h tends to be complicated and without meaning for the designer.

However, the calculation of C_E may be avoided altogether by calculating separately the cutoff frequency f_M produced by C_M alone and combining f_M with f_S to calculate f_h. This procedure is outlined as follows:

1. Find the base-to-ground ac impedance R'_S, which is the parallel equivalent of the source impedance and the stage input impedance. With significant feedback the transistor input impedance may be neglected.
2. Find the equivalent ac current gain $S = R'_S/R_E$ of the stage.

3. Find the current-gain cutoff frequency f_S by dividing the transistor f_T by the equivalent ac current gain found in step 2.
4. Find the Miller-effect input capacitance C_M by multiplying the collector-to-base capacitance C_{cb} by the base-to-collector voltage gain G'_v.
5. Find the Miller-equivalent cutoff frequency f_M from the impedance R'_S of step 2 and the Miller input capacitance C_M.
6. Find the video cutoff frequency f_h by combining f_S and f_M, as if parallel resistances. Thus

$$f_h = \frac{f_S f_M}{f_S + f_M} \tag{11.3}$$

The technical literature outlining the design of broad-band amplifiers is usually quite complicated and requires the use of the hybrid-pi equivalent circuit. The method outlined above, by recognizing that feedback produces an exchange of current gain for increased bandwidth, gives the designer the same result much more easily. The separate calculation of f_S and f_M provides a means of comparing the two sources of cutoff and leads to a clearer understanding of design problems.

11.4 EXAMPLE OF HIGH-FREQUENCY β CUTOFF

As an example of high-frequency β cutoff, we show that the effect may be observed at audio frequencies and that the cutoff frequency may be easily confirmed by a simple calculation. The β cuttoff of a transistor is demonstrated by the circuit shown in Fig. 11.5. The transistor is connected

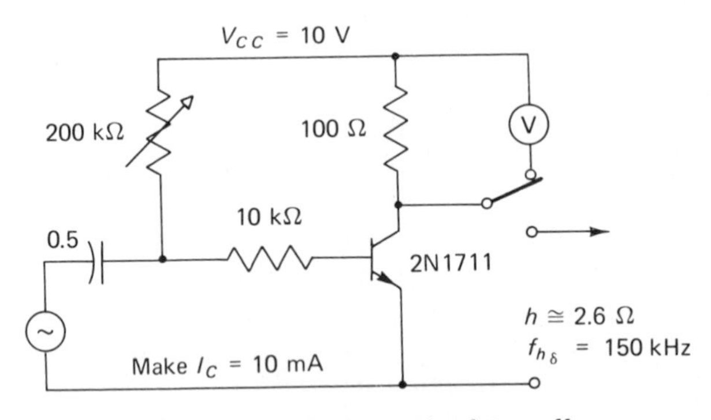

Figure 11.5. *CE*-stage example of β cutoff.

with a 100-Ω load resistor. A 10-kΩ series base resistor is placed near the base to make the signal a current source and to minimize feedback through the external collector-to-base circuit capacitance. Almost any general-purpose planar transistor may be used, but high-frequency data needed for the calculations are given for the 2N1613, the equivalent of a 2N1711, in reference 11-5. The circuit is arranged to protect the transistor from excess collector current regardless of the value used for R_A, and a low value of the load resistor is used to limit the base-to-collector voltage gain so that the Miller effect may be neglected.

With a 10-V collector supply the bias is adjusted to make the collector current approximately 10 mA. If preferred, the bias may be adjusted to make the dc drop approximately 1.0 V in the collector resistor. However, the dc voltmeter must be removed when making ac measurements.

From the collector current we know that $h = 2.6$ Ω. From the base current I_B, which is 0.05 mA, we find $\beta_0 = 10/0.05 = 200$. The transistor data sheet shows that β is 3 at 20 MHz when the collector current is 50 mA. The current gain-bandwidth product is the product of the high-frequency β and the frequency of measurement, so the data-sheet values imply that $f_T = 3(20) = 60$ MHz. Since f_T is given for a 50-mA collector current and the circuit in Fig. 11.4 uses a 10-mA current, a corrected f_T can be estimated by using the fact that β falls off approximately as the square root of the collector current. Thus, f_T is about 30 MHz, and from equation (11.1) we find that the β-cutoff frequency is $f_\beta = 30/200 = 0.15$ MHz.

The collector-to-base feedback capacitance is given as 25 pF and, with a 5-pF allowance for the stray collector to-base capacitance, the total feedback capacitance is about 30 pF. For convenience the transistor parameters that we now have are tabulated in Table V.

TABLE V

Parameters for the 2N1711 used in Fig. 11.5

Given Data	Calculated
V_C = 5V	β_0 = 200
I_C = 10 mA	$\beta_0 h$ = 520 Ω
I_B = 0.05 mA	C_{cb} = 30 pF
f_T = 30 MHz	f_β = 150 kHz

A measurement of the response characteristic shows that the amplifier high-frequency cutoff is at 140 kHz. If the cutoff is caused by the Miller effect, the cutoff frequency should increase substantially when the stage voltage gain is reduced by a factor of 10. Shunting the collector resistor by 10 Ω increases f_h to only 200 kHz; hence, the observed cutoff confirms the calculated β cutoff shown in Table V.

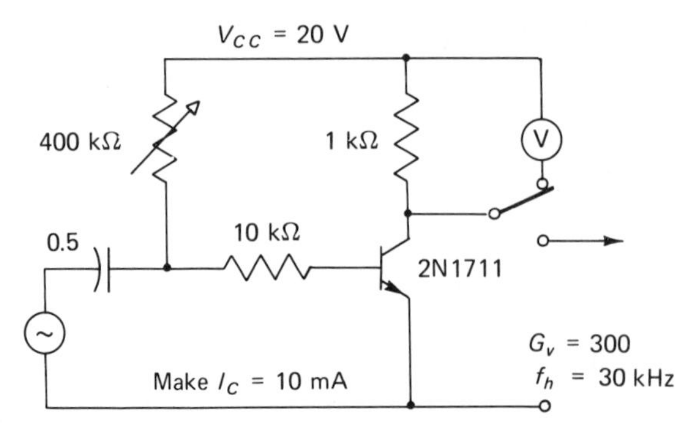

Figure 11.6. *CE*-stage example of Miller cutoff.

11.5 EXAMPLE OF MILLER CUTOFF

Cutoff produced by the Miller effect is observed when a CE stage has a high voltage gain. The voltage gain of the stage used in the previous example is increased by increasing the collector load resistor to 1 kΩ. Because the dc voltage drop in the load resistor is 10 V with 10 mA, the supply voltage is increased to 20 V, as shown in Fig. 11.6. To maintain the same bias current, the resistor R_A is doubled and the collector voltage is checked to confirm that the collector current is 10 mA. The base-to-collector voltage gain measures 300, and the amplifier cutoff frequency is now observed at approximately 30 kHz. The circuit changes have not changed the β-cutoff frequency, so we calculate the cutoff frequency produced by the Miller effect.

The Miller-effect input capacitance C_M is the collector-to-base feedback capacitance multiplied by the stage voltage gain. The measured voltage gain is 300, so the equivalent input capacitance is 9000 pF. The transistor input impedance is 520 Ω and 9000 pF has a reactance of 520 Ω at f_M = 34 kHz, which confirms the observed cutoff frequency.

Suppose that the designer decides to reduce the voltage gain by a factor of 6 so that f_M is increased from f_M = 34 to 200 kHz. Combining f_M with f_β (200 kHz), equation (11.3) gives f_h = 100 kHz. By this calculation we see that the high-frequency cutoff is limited equally by the Miller effect and the transistor input capacitance.

A careful reevaluation of the foregoing example reveals that the Miller effect cannot be entirely neglected when the collector load resistor is 10 Ω, and, similarly, the β cutoff cannot be entirely neglected when the voltage gain is 300.

In a different application the calculated f_h may be much more approximate than in the present example. Nevertheless, the calculated f_M and the f_S show the designer the main limitations of a given circuit and show where changes must be made to meet a given objective.

The cutoff frequencies observed in the circuits of Figs. 11.5 and 11.6 may be increased by a factor of 10 by driving the amplifiers from a 50 Ω source. A further increase of the bandwidth probably requires the use of a transistor intended for UHF applications. Transistors designed for UHF service may have $f_T = 400$ MHz and a collector-to-base capacitance of about 3 pF. These values suggest the possibility of another factor-of-10 increase of the amplifier gain-bandwidth values. The design of amplifiers having cutoff frequencies more than 100 times the values observed for the 2N1711 examples presents a substantial challenge.

11.6 HIGH-FREQUENCY CUTOFF AND LOAD CAPACITANCE

The transistor collector-to-ground capacitance C_{ob}, which acts as a collector-load capacitance in a CE stage or as an input capacitance in a CC stage, may contribute a significant reduction of the amplifier bandwidth. Within a multistage amplifier the collector-to-ground capacitance usually may be neglected, as compared with the following stage input capacitance. In an output stage that drives a high-impedance load the collector capacitance and the load capacitance together may be the principal limitation on the amplifier bandwidth. In most cases, however, the designer is able to offset the load capacitance by compensation, as discussed in a following section. The values of C_{ob} and the load capacitance are usually given, and their importance in the design may be determined easily by a designer. For these reasons additional details and an example are omitted here.

11.7 EXCHANGE OF GAIN FOR BANDWIDTH

The 2N1711 transistor used to demonstrate the β cutoff and the Miller effect is a general-purpose transistor not usually recommended for broad-band applications. Nevertheless, the device provides an interesting study to determine whether it can be used in a video amplifier. Suppose that a signal obtained from a 50 Ω source is to be amplified in a broad-band 6-MHz amplifier that uses a 2N1711 transistor. Two problems must be solved to obtain the 6-MHz bandwidth. The stage current gain must be reduced so that the gain-bandwidth capability of the transistor is not exceeded, and the voltage gain must be reduced to keep the Miller effect from limiting the bandwidth. For a trial design the designer decided to use a current gain that results in a 3-dB loss at 10 MHz. Hopefully, the total cutoff loss will be about 3 dB at 6 MHz.

The transistor, a 2N1711, has a gain-bandwidth product $f_T = 60$ MHz. If the current gain is reduced to 6 by emitter feedback, the bandwidth is $f_S = f_T/S = 60/6 = 10$ MHz. On consulting a reactance chart in Appendix A.10, we find that a Miller input capacitance $C_M = 300$ pF across a 50 Ω source impedance produces a 3-dB cutoff at 10 MHz. Thus, with $C_{cb} = 30$ pF, the voltage gain cannot exceed 10. These estimates give the designer trial values for the video-stage current and voltage gains.

Assuming a 50-Ω source impedance, the designer chose a 9.1-Ω emitter resistor and a 100-Ω load resistor, which make the current gain $S = 5.5$ and the voltage gain $G_v = 11$. The 2N1711 is biased for a 50-mA collector current to make $f_T = 60$ MHz, so the dc voltage drop in the load resistor is 5 V. If a 15-V supply voltage is used, the collector dissipation is 0.5 W, which is nearly the maximum recommended for 25°C ambient. With a high junction temperature the dc S-factor should be 20 or less, so R_B is made approximately 20 R_E. The amplifier has the circuit shown in Fig. 11.7, and R_A is set by trial to make the dc emitter voltage $V_E = 0.05(10) = 0.5$ V.

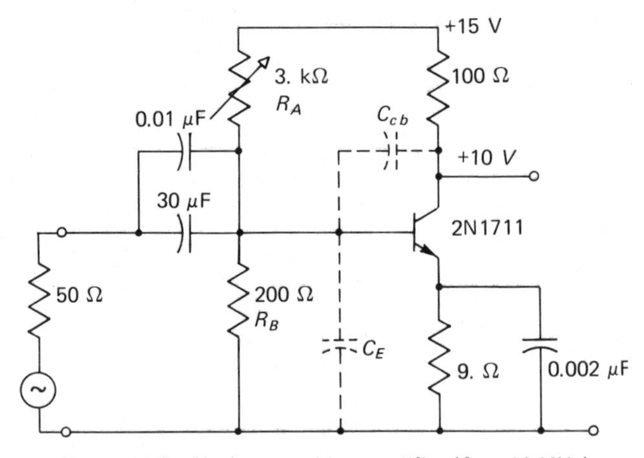

Figure 11.7. Single-stage video amplifier (f_h = 10 MHz).

The video amplifier was carefully checked without compensation and found to give a voltage gain of 10 from 20 Hz to 6 MHz with a 3-dB loss at these frequency limits. The loss at 10 MHz is 6 dB. Because of the difficulty of making accurate voltage measurements at 10 MHz, the gain may be measured by constructing the loss network shown in Fig. 11.8. When driven by a 50-Ω source, the network presents a 50-Ω source impedance and a 10:1 voltage loss. Ordinary composition resistors are satisfactory in the required values up to about 100 MHz. With the loss network in place, the signal input and the amplifier output are nearly the same value, so the loss can be measured beyond the calibration range of an ac voltmeter. If the signal generator does not present a 50-Ω impedance, the impedance should be shunted to 50 Ω.

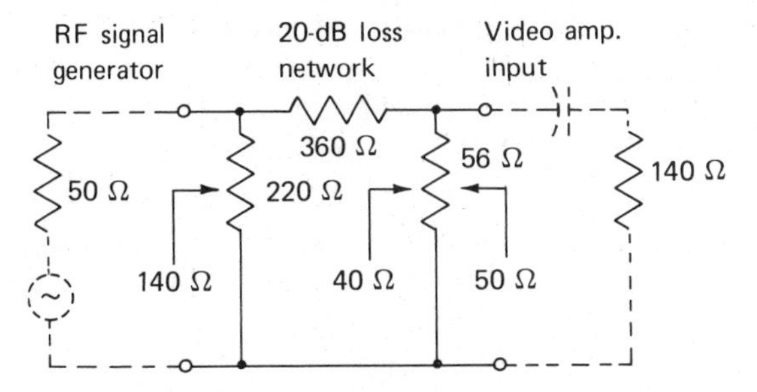

Figure 11.8. A 20-dB RF loss network.

Although the calculations neglected the effect of R_B in reducing R'_B, the effect of C_{ob} as a reactance load, and other effects the circuit was found to perform as expected. The designer's first objective is to construct a circuit and to evaluate the performance. The calculations are a means to this end and may always be repeated and improved.

For higher overall gain a series of video stages may be cascaded, provided each is shielded from the others and the power-supply connections are individually filtered and bypassed to ground. Measurements indicate that the stage shown in Fig. 11.7 gives an iterated gain of 10 over nearly the same frequency range as measured for the single stage. However, a small loss in each stage at the extreme frequency limits adds up in a series of stages and reduces the bandwidth. Thus, the stages in a multistage amplifier should have a slightly wider band to avoid degrading the transient response.

As in the present example, a video amplifier should generally use a transistor with an f_T value no higher than is actually required by the application. In using a transistor with a higher f_T rating the designer may unknowingly sacrifice reliability by operating the high-frequency transistor nearer the maximum current, voltage, or power rating. Moreover, high-frequency transistors usually cost more and have higher noise levels than lower frequency types.

11.8 AMPLIFIER BANDWIDTH WITH SHUNT FEEDBACK

Shunt feedback is useful in video amplifiers when a stage can be driven from a current source and a reduction of the input impedance can be tolerated, as in low-level stages. Because feedback biasing and a grounded emitter produce a simpler circuit, shunt feedback is sometimes preferred in high-frequency applications. In some high-frequency circuits alternate stages of shunt and series feedback may be used with significant advantages.

A stage that has shunt feedback and is driven by a current source is illustrated in Fig. 11.9. To examine the effect of feedback on the transistor f_β

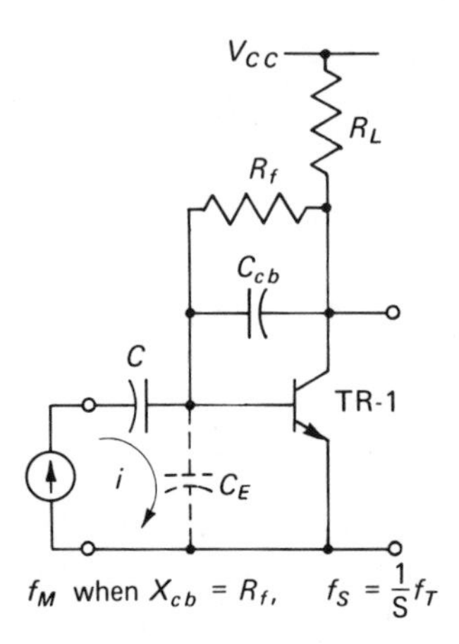

Figure 11.9. Video stage with shunt feedback.

cutoff, we neglect the feedback capacitance C_{cb}. The transistor input capacitance C_E is not changed by shunt feedback, but the input impedance seen by the input signal is reduced from $\beta_0 h$ to Sh. Thus, the effect of the shunt feedback is to reduce the stage current gain to S and to increase the upper cutoff frequency from f_β to f_S, as given by equation (11.2). The amplifier gain at the high-frequency cutoff may be increased by inserting a small inductance in series with the feedback resistor. This form of compensation increases the stage current gain with the advantage that the input impedance is increased also.

The effect of the collector-to-base feed back capacitance C_{cb} is clearly to increase the feedback when the reactance of C_{cb} becomes less than the resistance R_f. In other words, the Miller cutoff frequency may be found on a reactance chart by using the C_{cb} and R_f values that characterize the stage.

Because the emitter capacitance and the Miller-equivalent capacitance are both in shunt with the input impedance R_f/G_v', the high cutoff frequency f_h is found by substituting f_M and f_S in equation (11.3). Thus, the procedure for finding the upper cutoff frequency with shunt feedback is somewhat simpler than with emitter feedback, and an example is left as a problem for the reader.

11.9 PULSE SIGNALS IN VIDEO AMPLIFIERS

In most applications of video amplifiers the output signal must faithfully reproduce the shape of the input signal. Therefore, the transient-response characteristics of a video amplifier are more important than the amplitude-frequency response. The form of a signal pulse is determined mainly by the cutoff characteristic at the ends of the amplifier passband, and a simple *R-C* cutoff does not produce an optimum transient response. In most video amplifiers additional reactive elements are introduced to shape the response in the cutoff region and thus improve the transient response. This technique for improving the transient response is referred to as *compensation.*

11.10 LOW-FREQUENCY COMPENSATION

If an output signal is required to have a flat top without a sag, as in Fig. 11.10, the amplifier either must transmit exceedingly low frequencies, or the low-frequency response must fall off relatively slowly. Thus, for the reproduction of TV signals a video amplifier must have a flat amplitude response from 30 Hz to about 4 MHz, and the cutoff rate must be carefully controlled at both low and high frequencies outside the band.

In an amplifier that has one low-frequency *R-C* cutoff, the sag at a time t_1 following a step-like pulse is approximately 10 per cent per degree of phase shift at the frequency $f_1 = 1/t_1$. For example, if an oscilloscope or hi-fi amplifier has only 5° of phase shift at 60 Hz, the step response will sag nearly 50 per cent after $\frac{1}{60}$ of a second. With a single *R-C* cutoff the phase shift exceeds 5° at 60 Hz unless the 3-dB cutoff is below 5 Hz. These values show that a flat-topped transient specification may require a surprisingly low 3-dB cutoff frequency. For a square-wave signal, as in Fig. 11.11, the slope of the flat top is less by a factor of 2, since the droop of a square wave is only 3 per cent per degree of phase shift at the frequency $f = 1/T$.

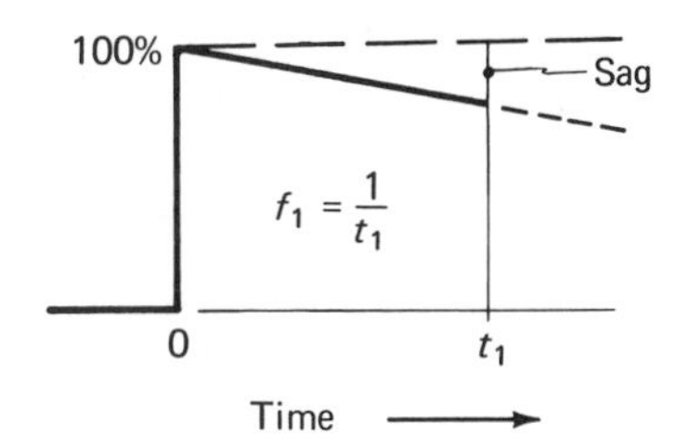

Figure 11.10. Step response with sag.

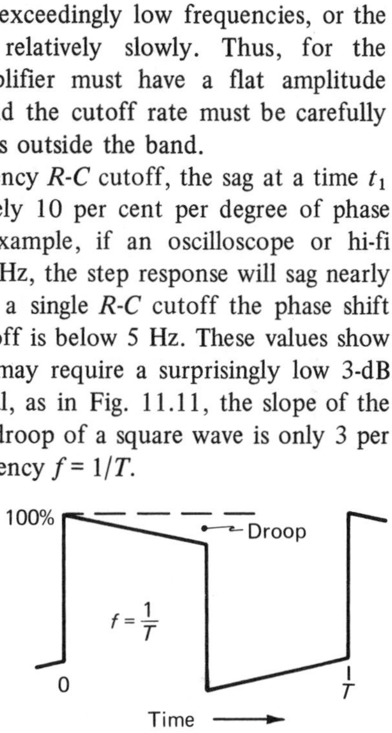

Figure 11.11. Square-wave response with droop.

The low-frequency response of a wide-band amplifier is degraded by the loss in the coupling capacitors, by transformers, and by emitter-bypass capacitors. For practical reasons a designer often uses compensation instead of using a larger capacitor. For example, a small coupling capacitor may be used to reduce a capacitance to ground, which produces high-frequency cutoff. Hence, compensation may be used to offset low-frequency cutoff both in and external to the amplifier.

For an explanation of low-frequency compensation, consider the two-stage amplifier shown in Fig. 11.12 and assume that the input capacitor introduces a loss and phase shift that are to be offset by compensation. The network for compensation is the parallel-connected R_C and C_C, which are shown in series with the base resistor R_2. Assume that the collector load resistor R_L can be very large compared with the reactance of the coupling capacitor C_2 and the

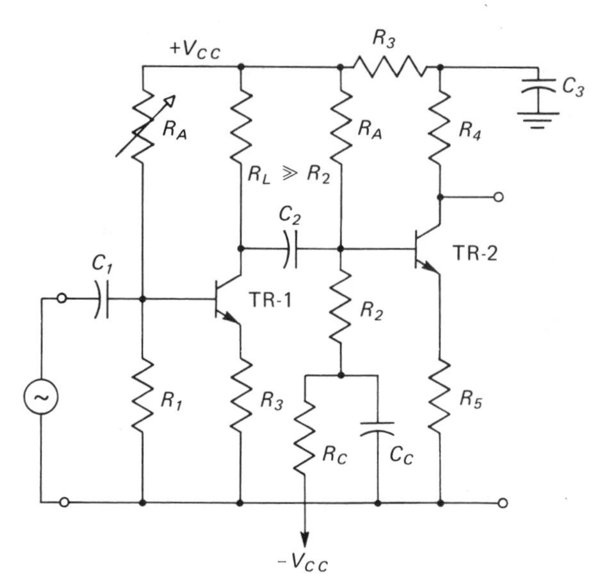

Figure 11.12. Amplifier with low-frequency compensation.

resistance R_2. If the resistance R_C is large compared with R_2, the load seen by TR-1 is R_2 in series with C_C, and compensation for the low-frequency loss in C_1 is achieved by making the product R_2C_C equal to R_1C_1. With this adjustment the low-frequency loss and phase shift in C_1 are exactly offset by an increased gain in the first stage. If R_C can be made very large compared with R_2, the compensation removes both the loss and the phase shift produced by the input capacitor

In a practical situation R_C cannot be made very large compared with R_2, and, for example, R_C may equal R_2. In this case the capacitor C_C is selected by trial to achieve at least an approximate compensation, as shown in Fig. 11.13. In a similar manner the compensation network R_3 and C_3 may be used in series with the load resistor R_4 of the transistor TR-2. Compensation by adjusting the product R_4C_3 may be used to offset a loss and phase shift that are external to the amplifier or in the load on TR-2.

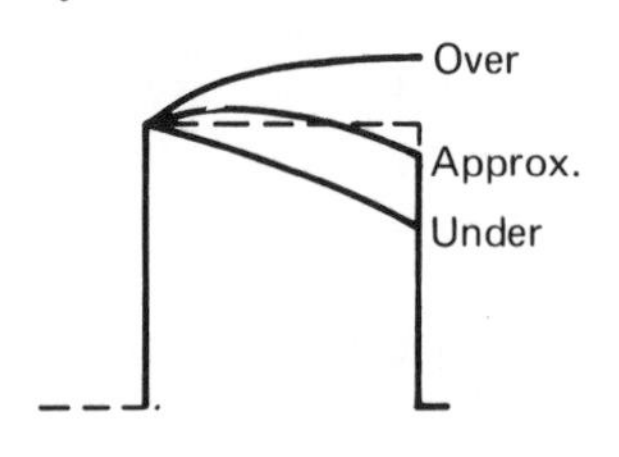

Figure 11.13. Wave form with low-frequency compensation.

Generally the series resistors R_C and R_3 cannot be made very large, and the compensation is at best only an approximation. In some cases compensation is used to offset sag introduced by several sources present simultaneously. Because a single approximately compensating network cannot exactly compensate for a number of sources, the design of a compensating network is normally a compromise in which a simple trial-and-error adjustment is an accepted procedure.

11.11 HIGH-FREQUENCY COMPENSATION

If an output pulse is required to rise very rapidly and without overshoot, an amplifier must transmit a very wide band and the response must not fall off too rapidly outside the band. The rise time of a signal leaving a wide-band amplifier with a single *R-C* cutoff is approximately one third the reciprocal of the high cutoff frequency. Thus

$$t_r \cong \frac{1}{3f_h} \cong \frac{0.33}{BW} \qquad (11.4)$$

Similarly, the minimum pulse length t_p that can be transmitted is usually given as

$$t_p = \frac{0.5}{BW} \qquad (11.5)$$

An input pulse shorter than t_p will appear at the output reduced in amplitude and lengthened to a duration of t_p. A compensating network that increases the gain of a wide-band amplifier by 3 to 4 dB at the 3-dB cutoff frequency may increase the bandwidth enough to nearly halve the pulse rise time or the pulse length.

The simplest means of increasing the high-frequency response or of increasing the bandwidth of an *R-C* amplifier is by the addition of a small inductance in series with the base resistor or the collector load resistor, as shown in Fig. 11.14. This arrangement, called *shunt peaking*, decreases the transient rise time but may cause an overshoot, as indicated in Fig. 11.15. The fastest transient response without overshoot is obtained by increasing the high-frequency sine-wave gain about 2 dB at the 3-dB cutoff. By keeping the response at high frequencies, always slightly below the response at the lower frequencies, there is no overshoot. If the frequency response is only 1 dB greater at the high frequencies than at lower frequencies, as shown in Fig. 11.16, the rise time is slightly faster, but the resulting overshoot and the transient ringing are generally undesirable.

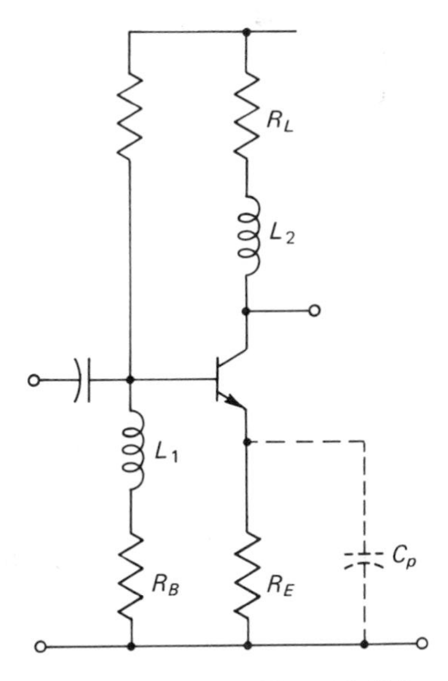

Figure 11.14. Amplifier with high-frequency compensation.

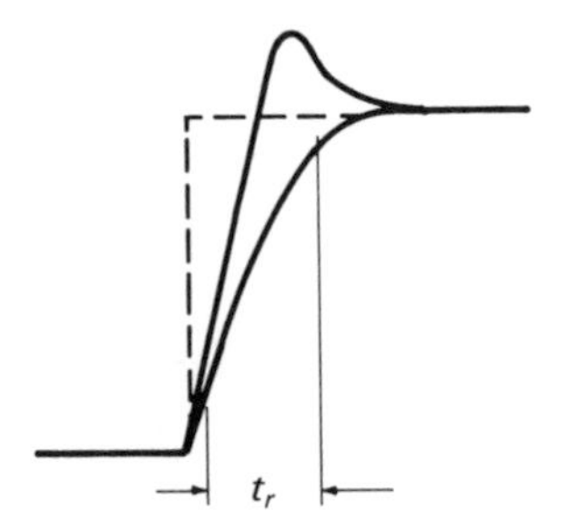

Figure 11.15. Pulse rise time and overshoot.

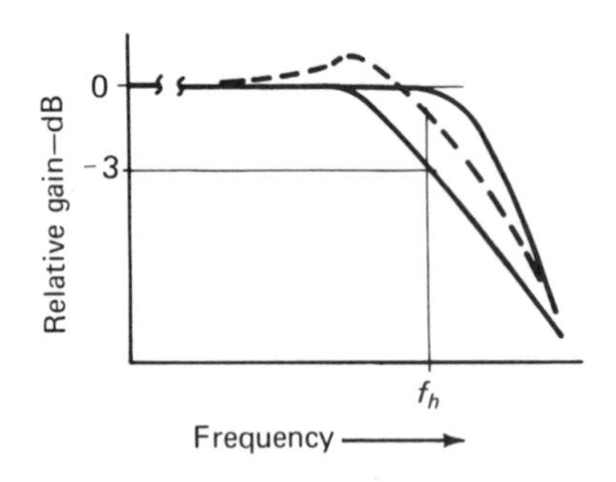

Figure 11.16. Video high-frequency response curves.

The high-frequency performance of a video amplifier can be improved further when two capacitors can be separated by a series inductor, as shown in Fig. 11.17. With series peaking the capacitors act separately and the transient rise time may be reduced to about two thirds the time with shunt peaking alone. Design values for similar circuits may be found in the vacuum-tube literature, but a trial value of inductance may be obtained by making the inductive reactance equal to the reactance of the average value of the two capacitors at the desired cutoff frequency. Series peaking is most effective when the two capacitors are approximately equal.

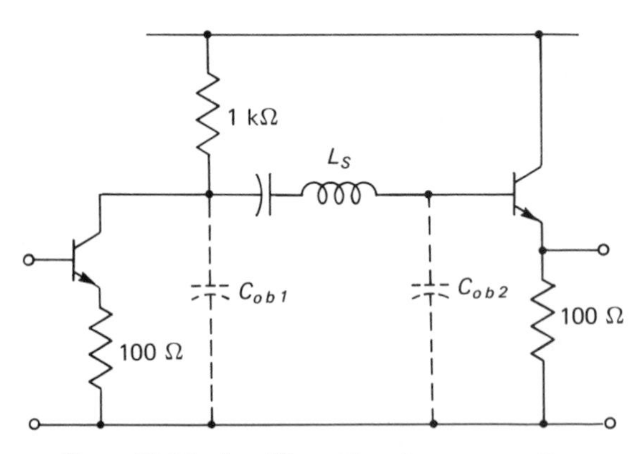

Figure 11.17. Amplifier with series compensation.

The response characteristics of a video amplifier are influenced greatly by temperature, by the supply voltage, and by changes of stray capacitance caused by the relocation of lead wires. The high-frequency response and compensation are particularly sensitive to the change of a critical parameter. The effect of measurement probes and tools should be considered during measurements and adjustments.

11.12 HIGH-FREQUENCY COMPENSATION: AN EXAMPLE

The video amplifier shown in Fig. 11.7 has a 6-MHz bandwidth when a 2N1711 transistor is used without compensation. With high-frequency compensation the bandwidth may be increased to 10 MHz. The simplest way to increase the high-frequency gain is to shunt the emitter resistor by a capacitor that is just large enough to raise the gain at 20 MHz. (A large capacitor that increases the midband gain tends to reduce the bandwidth and rise time.) The simplest way to select the capacitor is to try one that has a reactance equal to twice the resistance that is in series with the emitter, and to observe the effect on the frequency response in the decade from 1 to 10 MHz when the capacitor is connected across the emitter resistor. For the amplifier shown in Fig. 11.7, a 0.002-μF disc capacitor moves the 3-dB cutoff frequency up to 12 MHz and essentially doubles the bandwidth.

A second way to compensate the amplifier is to insert an inductance in series with the collector resistor so that the effective load impedance is increased 3 dB at 10 MHz. An inductance of 2 μH has 100-Ω reactance at 10 MHz. Reference to the inductance design chart in Appendix A.6 shows that 2 μH may be constructed by winding 10 turns of wire in a helix having an inside diameter (ID) of 10 mm and a length of 20 mm. The inductance may be decreased by collapsing turns, and a fine adjustment is made by spacing the turns closer or farther apart. Compensation with the inductance alone should produce the same response as with the emitter capacitor. If the 2-μH inductance and a smaller value of the emitter capacitor are used together, the amplifier bandwidth may be increased to at least 15 MHz, but the transient response has an overshoot that may be undesirable. Because all circuit components and connecting leads have series inductance and stray capacitance, a calculation of the compensating reactances becomes more and more difficult as the operating frequency is increased above 10 MHz.

Unfortunately, compensation is rarely without undesirable effects. Both methods of compensation tend to decrease the amplifier input impedance, and an inductance in the collector circuit increases the output impedance. The effect on the input impedance is easily observed by connecting an ac voltmeter to the input and changing the compensation. However, the amplifier is designed to minimize the Miller effect at 10 MHz for voltage gains less than 10. For that reason, decreasing the load impedance, as with tuning, does not reflect a large change in the input impedance. This result may be easily observed also. Compensation is always expensive, although not usually requiring as much increase of the input power as is exhibited by the foregoing examples.

The dependence of a high-frequency design on the source and load impedances is always in evidence. Both the low- and high-frequency cutoffs of the broad-band amplifier are determined as much by those impedances as by the other parameters. This characteristic of high-frequency circuits makes a particular design somewhat inflexible, because a change of the source or load impedance either degrades the frequency response, or resistors added to offset an impedance change usually reduce the power gain. Transmission lines, cables, and high-frequency amplifiers must be correctly terminated or matched to obtain the intended transient-response and gain characteristics.

11.13 A TV VIDEO AMPLIFIER

The video amplifier in a TV receiver usually drives a high impedance load, and the output transistor, which must be physically small, is required to operate near its maximum voltage and power capability. An amplifier for this application usually needs two closely coupled stages, and the design problems that we now consider have interesting differences from those explained earlier.

The circuit of a video amplifier typical of the present-day designs found in an all-transistor receiver is shown in Fig. 11.18. This amplifier uses a high-voltage silicon power transistor in the output stage, and the input stage has a low-power transistor operated CC for impedance step-down. The amplifier has an overall voltage gain of 50, a current gain of approximately 10, and a bandwidth of 4 MHz. The amplifier is used to couple a video IF detector to the cathode of a picture tube. The intermediate gain control adjusts the picture contrast by changing the magnitude of the output signal. Adjusted for a maximum contrast, the peak-to-peak output signal is 90 V, which is adequate for a 19-in picture tube.

The CC input stage couples the demodulated video signal at a 4-kΩ impedance level to the gain control at a 300-Ω level. This stage needs a current gain of 13 up to 4 MHz, so the transistor gain-bandwidth product should exceed 50 MHz. The 2N699 transistor has a minimum $f_T = 50$ MHz. However, the collector-to-base capacitance C_{ob} adds a 20-pF input capacitance, which is easily compensated by the 0.2-mH shunt peaking coil.

The source impedance of the output stage varies with the setting of the gain control but is usually less than 50 Ω. The compensation capacitor in the emitter circuit reduces the emitter impedance to 25 Ω at 4.5 MHz, which makes the stage current gain about 2 at 4.5 MHz. the 2N3440 power transistor has an $f_T = 15$ MHz. Therefore, the transistor input capacitance limits the bandwidth to $15/2 = 7.5$ MHz. The collector-to-base feedback capacitance is given as a maximum $C_{ob} = 10$ pF with 10 V on the collector. The collector-to-base capacitance decreases with increased collector voltage. Hence, with 70 V on the collector and with careful shielding, we may assume

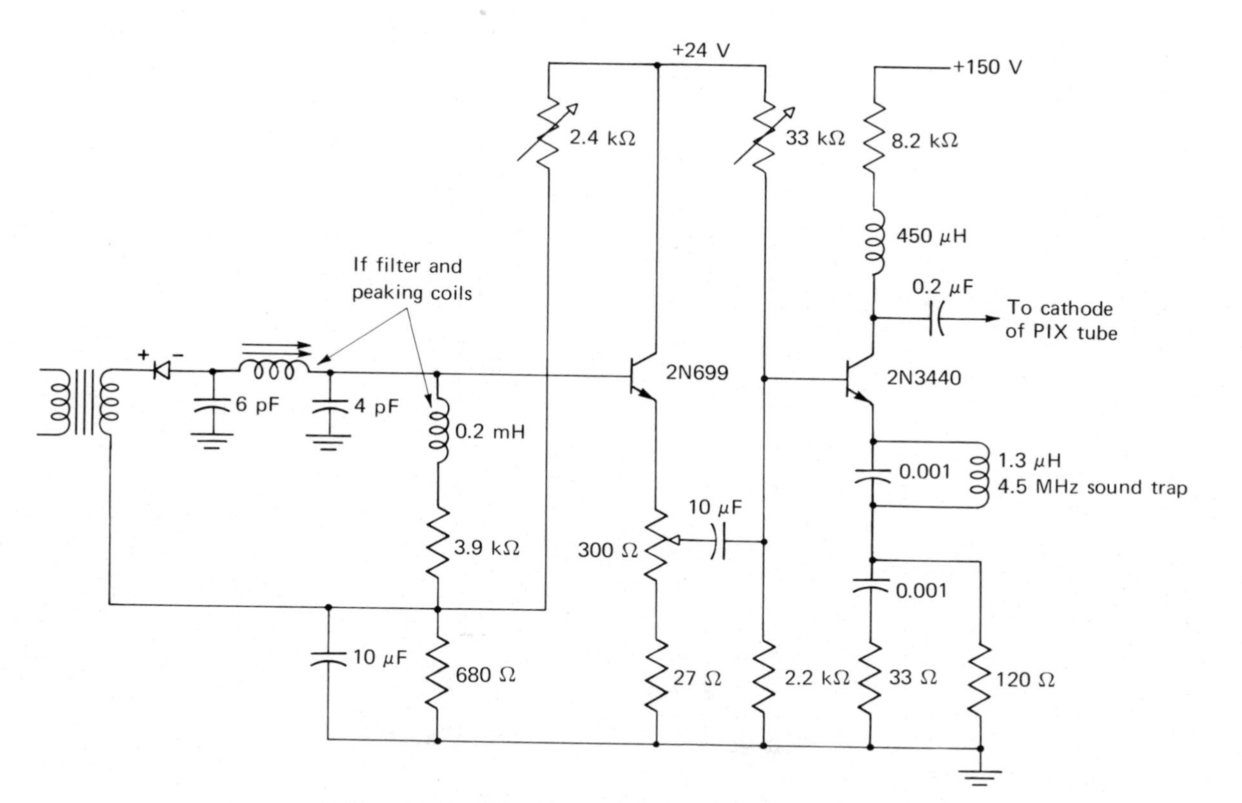

Figure 11.18. TV video amplifier (10 Hz to 4 MHz).

that the total-feedback capacitance is less than 10 pF. The R_L/R_E ratio indicates a stage voltage gain of 70, but with a TV tube as load, the base-to-collector voltage gain measures about 50. For this voltage gain the equivalent Miller input capacitance is less than 500 pF. The 50-Ω source impedance with an input capacitance of 500 pF indicates a 6-MHz cutoff. The Miller-effect cutoff f_M and the transistor cutoff f_S together, by equation (11.2), indicate a 3.3-MHz high-frequency cutoff, whereas a TV signal needs 4 MHz.

The foregoing calculation has used "worst case" estimates, and the effect of the TV tube as a capacitance load has been tacitly neglected. Because the reactance elements dominate the circuit in the high-frequency cutoff region, a more accurate calculation of the gain and the bandwidth is both impractical and too time-consuming for a practical design. However, the approximate calculations show the designer that a video bandwidth is probably attainable with a two-stage voltage gain of about 50. With this information as a guide, the designer usually proceeds more rapidly and more accurately by experimental procedures.

The designer of the TV amplifier knows that the input signal from the IF amplifier is relatively large, since a rectified signal must be at least 1 V peak-to-peak. Because of the large input signal, the amplifier gain may be reduced to obtain increased bandwidth. The bandwidth of the video amplifier may be increased by reducing the 3.9-kΩ base resistor and by reducing the 8.2-kΩ output load resistor. Thus, the values shown in Fig. 11.18 are the designer's final compromise between gain and bandwidth.

A designer usually sketches a proposed circuit, sets up a prototype for study, adds compensation coils, and experimentally evaluates and adjusts his design. For each assembly of the amplifier the compensation is adjusted by spacing the inductance winding until a desired transient response is obtained. In the input stage the reactance of the 0.2-mH inductance is approximately 3.9 kΩ at 4 MHz. Hence, by adjusting the coil, the high-frequency gain may be increased approximately 3 dB. Similarly, in the output stage the reactance of the 0.45-mH peaking coil at 4 MHz is approximately 8.2 kΩ. Thus, the shunt peaking coil has approximately correct inductance for adjustment at assembly.

The final design of a TV circuit is usually a compromise of costs and performance. Compromises are the essence of design, while costs and customer acceptance are important and ever-present components of a final design. Because a video amplifier is sensitive to the physical layout, to lead capacitance, and to the compensation adjustments, the success of a design can only be evaluated by experimental tests and performance studies. For these reasons RF and video amplifiers may be more a product of the designer's practical skill than of his calculations and circuit theory.

11.14 INTEGRATED-CIRCUIT VIDEO AMPLIFIERS

A high-gain, multistage video amplifier may be designed by cascading low-gain stages that use discrete components and local feedback. Direct-coupled pairs require fewer components, particularly coupling capacitors, and have a higher loop gain, which makes the feedback more effective. For bandwidths up to 10 MHz the discrete construction offers high power output and the high signal voltage required for driving a TV or radar picture tube. Above 10 MHz a designer is forced to use low-power transistors, and at these frequencies the capacitance and inductance of leads make it difficult to use feedback over more than a single stage.

Above 10 MHz an integrated circuit offers the advantages of a circuit with exceedingly small lead lengths and with high f_T values in the transistors. One such monolithic wide-band video amplifier (Sylvania SA 20) has a voltage gain of 10 with a 3-dB cutoff at 100 MHz. This amplifier uses three direct-coupled stages in a *CE-CE-CC* connection with overall feedback from the output to the first emitter. The amplifier is intended to use nearly 40-dB feedback, which improves the linearity and maintains a high input impedance. The circuit of the video amplifier, with minor omissions for simplification, is shown in Fig. 11.19.

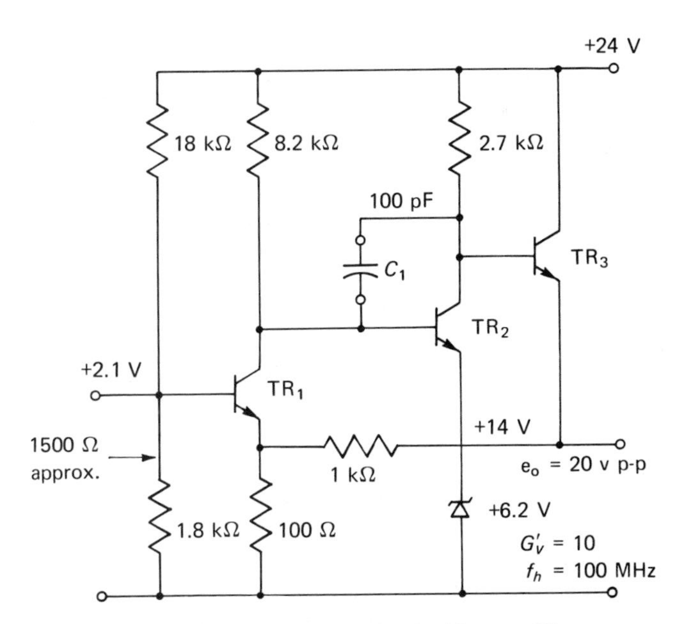

Figure 11.19. Integrated-circuit video amplifier.

With the first emitter bypassed to ground, the amplifier has an open-loop gain $G_v = 1000$, which may be shaped in the high-frequency cutoff region by local feedback in the second stage. The collector-to-base feedback capacitor C_1 may be adjusted to shape either the closed-loop gain or the transient response. In this way the frequency response may be maximally flat or caused to peak, as desired. With a larger value of the capacitor to shape the transient response, the amplifier has pulse rise and fall times of less than 10 nsec without overshoot.

11.15 SUMMARY

The bandwidth attainable in a video amplifier depends mainly on the circuit and transistor capacitances, which limit the upper cutoff frequency of the amplifier. The capacitance of the emitter junction determines the transistor gain-bandwidth product f_T, and the cutoff frequency f_S produced by the emitter capacitance is f_T divided by S, the ac current gain of the stage. The cutoff frequency of a stage with collector-to-base feedback is the frequency f_M at which the Miller-equivalent input capacitance has a reactance equal to the ac impedance from the base to ground. Generally, a stage has both Miller effect and gain-bandwidth limiting. With both effects acting together, the cutoff frequency is calculated by combining f_S and f_M, as if adding resistors in parallel.

The upper cutoff frequency of a video amplifier is increased by selecting transistors with high f_T values, by operating the transistors at relatively high collector currents and voltages, by reducing the stage current and voltage gains with feedback, and by using low-impedance circuits. Thus, a wide-band, high-gain video amplifier is usually a series of low-gain, low-power stages, each with local feedback. A stage designed for a high voltage gain or a high current gain may exhibit bandwidth limiting at frequencies more than 3 decades below the transistor f_T rating.

For the faithful reproduction of pulse and TV signals a video amplifier must have a wide-band response and a gradual cutoff at the low-frequency and high-frequency limits of the frequency response. With a square-wave input signal, a flat-top response is maintained only if the amplifier has a 3-dB, low-frequency cutoff more than a decade below the fundamental signal frequency.

The rise time of an output pulse is approximately one third the reciprocal of the high cutoff frequency. The rise time of an amplifier may be reduced by high-frequency compensation, which is generally achieved by shunt peaking. Because the transient response of a video amplifier is usually the main concern of the designer, the peaking components and their adjustment are often decided by experimental methods.

The characteristics of a video amplifier depend on the capacitance between components and on the circuit layout. They are often too complicated for accurate paper design. Any design approaching the state of the art depends to a large degree on the skill and ingenuity of the designer. For low-power applications, integrated circuits offer predictable gain-bandwidth values in excess of values obtainable with discrete circuits.

PROBLEMS

11-1. Find the high cutoff frequencies for the shunt feedback video stage shown in Fig. 11.9 when driven by a 50-Ω line. Assume that TR-1 is the 2N1711 transistor described in the text, and use R_L = 100 Ω, R_f = 2kΩ, I_C = 50 mA, and β_0 = 100. (b) What is the value required for capacitor C if f_1 = 5 Hz? How does this design compare with the video stage shown inFig. 11.7?

11-2. What is the expected high cutoff frequency for the single-stage video amplifier shown in Fig. 11.7 if the transistor is changed to a 2N708 transistor biased to operate with I_C = 10 mA when f_T = 300 MHz, β_0 = 50, and C_{ob} = 6 pF?

11-3. Carefully examine the values for all components shown in Fig. 11.18 and evaluate the design. (a) Give the Q-point for both transistors. (b) Find the high cutoff frequency f_h for both stages. (c) Show that the peaking components have approximately the correct values.

11-4. Show how low-frequency compensation can be added to the collector circuit of the video amplifier shown in Fig. 11.7. Give component values that approximately compensate for the input coupling capacitor.

DESIGNS

11-1. Design and experimentally test a single-stage video amplifier that is suitable for low-power application as a 500-Ω line-amplifier. The amplifier is expected to have a maximum practical power gain with a frequency range from 100 Hz to 1 MHz.

11-2. Construct and verify operation of the single-stage amplifier shown in Fig. 11 7. Design a similar amplifier that uses shunt feedback.

11-3. Construct and make performance tests of the TV amplifier shown in Fig. 11.18.

REFERENCES

11-1. Terman, F. E., *Electronic and Radio Engineering,* 4th ed., "Voltage Amplifiers for Video Frequencies." Chapter 9, New York: McGraw-Hill Book Company, 1955.

11-2. "Solid-State Television Video Amplifiers," Application Note AN-165. Phoenix, Ariz.: Motorola Semiconductor Products, Inc., 1967.

11-3. *RCA Power Circuits Manual,* SP-51, pp. 328-352. Harrison, N.J.: Radio Corporation of America, 1969.

11-4. *RCA Transistor Manual,* SC-13, pp. 52-56. Harrison, N.J.: Radio Corporation of America, 1967.

11-5. Cowles L. G., *Analysis and Design of Transistor Circuits.* New York: Van Nostrand Reinhold Company, 1966.

12

RADIO - FREQUENCY TUNED AMPLIFIERS

In many applications the bandwidth of frequencies to be amplified is only a small percentage of the center frequency, so a tuned amplifier may be used to select the desired band of frequencies and reject all others. Tuned circuits are commonly used at radio frequencies to eliminate interfering signals or the harmonic distortion of oscillators and class-C amplifiers.

Tuned circuits offer several advantages in high-frequency circuits. The transistor and circuit capacities are made a part of the tuned circuit instead of a low-impedance problem for the designer. When the design frequency approaches the frequency of unit current gain f_T, a stage has power gain, provided the load impedance exceeds the input impedance. Tuned circuits provide a means for coupling a high-impedance collector to a low-impedance base without a power loss. In this way the relatively low-power gain of a high-frequency amplifier is conserved.

12.1 TUNED AMPLIFIERS

A typical tuned amplifier, as shown in Fig. 12.1, has tuned coupled circuits in both the input and the collector circuit. The resonant impedances are selected to provide suitable input and load impedances, and the inductances are so selected that the resulting Qs provide the desired selectivity. The capacitors may be found from the resonant frequency and the given inductance, except that the transistor input capacitance usually supplies an appreciable part of the capacitance.

Amplifiers at IF, RF, and UHF frequencies generally require tuned coupled circuits because adequate gain is obtained in the β-cutoff region only by making the transistor capacitance part of a tuned circuit. By this means the amplifier input impedance may be increased by almost the Q of the tuned circuit. If the transistor has more capacitance than can be permitted in the design, the base may be connected to a tap on the input circuit, and the tuned circuit will operate as a step-down transformer.

RF transformers are tuned either by capacitors or magnetic cores, and one winding usually has so many turns that the collector or the base connects to a tap on the winding. The excess winding is used either to accommodate the tuning slug better or to increase the impedance so that the capacitor is physically small. The collector and base connections are tapped down on the windings to limit the effect of the internal collector resistance on the Q or to reduce the stage gain and avoid the need for neutralization.

12.2 TUNED-AMPLIFIER DESIGN

Tuned amplifiers are generally used in an iterated series of stages that are designed to give a maximum power gain. In a low-frequency tuned amplifier, the ac current gain S may be selected somewhat arbitrarily between a high value to increase the input impedance or a low value to reduce the variation of the gain with the transistor β. The collector load impedance R_L is chosen to provide a desired voltage gain. The actual load shunts the tuned output circuit, or is connected to a tap on the inductor, or coupled, so that the collector is loaded by the required impedance.

The power gain that is attained in a stage is simply the product of the current and voltage gains; thus

$$G_p = SG_v \tag{12.1}$$

With tuned circuits the load may be inductively coupled or connected to a tap with a high efficiency of power transfer. For this reason the collector may be

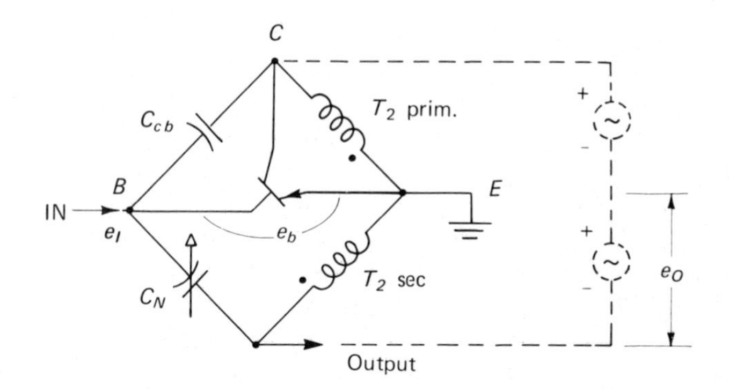

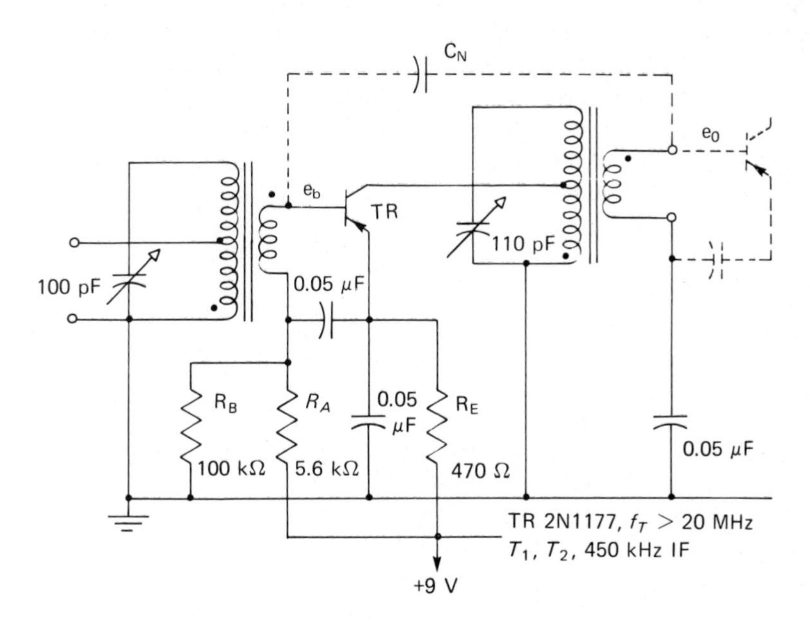

Figure 12.1. IF amplifier—450 kHz.

terminated in any desired load impedance, while the load power is transferred to the following base without a power loss. Hence, high gain in a tuned amplifier is obtained by designing each stage with a maximum ac current gain S and a maximum base-to-collector voltage gain G_v.

For small-signal applications a tuned stage is biased class A the same way a stage is biased in an audio-frequency amplifier. As shown in Fig. 12.1, a tuned stage usually has an emitter resistor and a base-circuit voltage divider so that the dc collector current is controlled by the bias resistor and the dc S-factor. In RF applications the emitter resistor is usually bypassed to make the ac current gain as high as possible.

The small physical dimensions of RF transistors and the effects of capacitance make these devices significantly different from audio transistors. RF transistors have lower breakdown ratings because the junctions are relatively thin. Therefore, the operating Q-points are usually nearer the maximum voltage and power ratings of the transistors. The capacitance effects profoundly change circuit design at radio frequencies.

At high frequencies the transistor current gain β decreases 6 dB/octave of frequency, and the collector current is not in phase with the base current. The transistor data sheets generally give the current gain of a typical transistor and a frequency at which the current gain is between 5 and 10. The product of the given frequency and the current gain is the transistor gain-bandwidth product f_T. Because the internal phase shift and feedback tend to make a high-frequency stage unstable, each stage of a tuned amplifier must have a relatively low voltage gain.

12.3 THE HIGH-FREQUENCY *CE* EQUIVALENT CIRCUIT

As an aid in understanding the high-frequency problems in transistor circuit design, consider the transistor equivalent circuit shown in Fig. 12.2. This

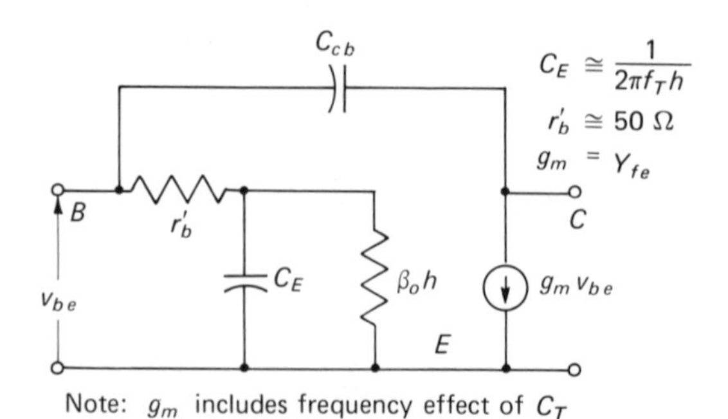

Figure 12.2. Transistor high-frequency equivalent circuit.

circuit is similar to the one shown in Fig. 11.2 except for a resistance r_b' in the base lead and the collector-to-base feedback capacitance C_{cb}, which is connected to the base input terminal B. The ohmic resistance r_b', called the *base-spreading resistance*, is a consequence of the exceedingly small dimensions used in high-frequency transistors. A low-power RF transistor usually has a value of r_b' between 20 and 100 Ω.

As explained in Chapter 11, the base-emitter junction capacitance* C_E is a relatively large capacitance that causes the β cutoff at f_β and the reduction of β to $\beta = 1$ at the frequency $f = f_T$. The resistance r_b' is relatively low in audio-frequency transistors and may be neglected in low-frequency circuits. At high frequencies r_b' is important because it tends to separate the capacitance C_E from the input terminal, making it difficult to offset the capacitance by tuning or by shunting with a low source resistance. When the dimensions of a transistor are decreased in order to reduce capacitance effects and to obtain a better high-frequency device, r_b' may be 50 to 500 Ω and is, therefore, an important component in the equivalent circuit. Except for smaller capacitance values and the addition of r_b', the equivalent circuit shown in Fig. 12.2 has exactly the same component values found in a low-frequency device.

The equivalent circuit shown in Fig. 12.2 is useful to about 100 MHz. At this frequency the inductance of a very short length of lead becomes important, particularly in the emitter, and cannot be neglected in circuit calculations or applications. The important point for present purposes is that a circuit must be carefully laid out in order to minimize the inductance and stray capacitance.

The problems of high-frequency design stem from the effects of capacitance in the transistors and circuit components and from the tendency of high-gain amplifiers to oscillate when feedback through the collector-to-base capacitance is in phase with the input signal. The voltage gain of a stable amplifier is limited by the positive feedback that is both internal and external to the transistors.

The feedback internal to the transistor is partly in the base-spreading resistance r_b'. In present-day transistors this base resistance may cause positive feedback at frequencies above 100 MHz, but can be neglected generally in low-frequency designs. A second cause of feedback is in the collector-to-base capacitance C_{cb} internal to the transistor and in stray capacitance between the connecting leads. The total collector-to-base capacitance is designated by C_{cb}.

Feedback through the capacitance C_{cb} cannot be neglected, even at audio frequencies. For the present we shall neglect feedback in r_b' and consider only the collector-to-base feedback problem.

*The emitter transition capacitance C_T that is sometimes given in the transistor data is the emitter capacitance with $I_E = 0$. The emitter capacitance C_E is usually much greater than C_T and the latter is useful when I_E is small.

12.4 TUNED-AMPLIFIER GAIN LIMITATIONS

As in any high-gain system, an amplifier becomes an oscillator and is useless when the forward gain exceeds the loss from the output back to the input. At frequencies just below resonance, capacitance feedback returns a signal in phase with the input signal and effectively shunts the transistor input with a negative component of resistance. If the magnitude of the negative resistance is low enough to offset the positive resistance of the input circuit, the stage is forced to oscillate. As in any feedback system, oscillations begin when the loop phase shift is 180° if the base-to-collector voltage gain exceeds the feedback loss.

If R_b is the equivalent base-circuit resistance at resonance, and X_{cb} is the relatively high collector-to-base reactance, the feedback signal loss is R_b/X_{cb}, and the phase shift is 90°. However, there is an additional 45° phase shift in each of the tuned circuits at the 3-dB low cutoff frequency, so the total loop phase shift is 180°. Thus, if

$$G_v \frac{R_b}{X_{cb}} = 2 \tag{12.2}$$

the loop gain is 1 when the phase shift is 180°, and the amplifier is unstable. Therefore, a tuned amplifier oscillates at a frequency near the 3-dB low-frequency cutoff when the stage voltage gain at resonance is

$$G_v = 2\frac{X_{cb}}{R_b} \tag{12.3}$$

Equation (12.3) gives the minimum gain that produces oscillation, and thus gives the designer approximate upper limits of capacitance and gain that can be tolerated in a design. In a practical problem the designer must use a lower gain or a smaller capacitor to avoid the effects of production variations and tuning difficulties.

12.5 NEUTRALIZATION

The effects produced by feedback through the collector-to-base capacitance and internal feedback through r_b' can be offset by providing additional feedback in the opposite sense If an external feedback circuit cancels the effects of all resistance and capacitance feedback, the amplifier is said to be *unilateralized*. If the circuit cancels the effect of only the reactance feedback, the amplifier is said to be *neutralized*.

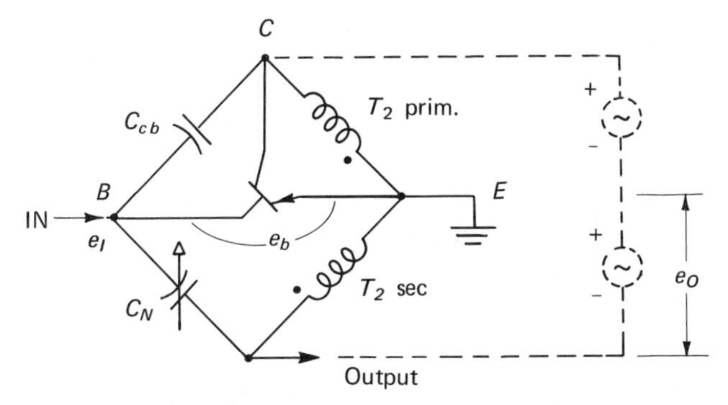

Figure 12.3. Neutralization bridge.

The neutralization of amplifiers operating well above the β-cutoff frequency is a useful technique for increasing the gain and stability and may be effected by connecting a neutralizing capacitor C_N to a winding on the output transformer, which provides a phase reversal. The capacitors C_N and C_{cb} form two sides of a bridge, as shown in Fig. 12.3. By applying a collector-to-ground signal, C_N is adjusted to minimize the signal observed at the base. Sometimes the capacitor is set merely by observing its effect on the symmetry of the response characteristic. When adjusted for balance, the base-to-ground signal e_b is independent of the output signal e_O so the input and output circuits may be designed and tuned independently.

The unilateralization of tuned stages is generally considered impractical because the circuit requires a delicate balance of several feedback effects which makes the adjustments difficult and the amplifier is unstable for small changes away from the balance conditions.

For similar reasons the neutralization of cascaded high-gain amplifiers is generally costly, time-consuming, and is optimized for only one frequency. Neutralization and tuning problems are particularly troublesome in cascaded stages, because the tuning of a collector circuit affects the tuning of the previous base circuit and may even interact from stage to stage. However, these difficulties may be removed by deliberately reducing the gain per stage enough to ensure a desired degree of stability and ease of tuning. The price paid for the improved performance is lower stage gain, but we obtain ease of design, ease of alignment, and a simpler circuit. When the cost of an additional stage is important, the designer may use a combination of reduced gain and neutralization, a technique generally used in the IF amplifiers of transistor radios.

A typical tuned amplifier that uses neutralization is shown in Fig. 12.1. The neutralization capacitor C_N is connected between the base and the output winding, which is phased 180° out of phase with the collector voltage. The neutralized condition is determined by adjusting C_N until the signal voltage at the base does not change when the collector circuit is tuned. With low available circuit gains, where neutralization is most needed, this adjustment is easy and effective. Residual effects that may be observed either require a resistive component of feedback (unilateralization) or indicate additional stray circuit coupling.

12.6 TUNED-AMPLIFIER STABILITY–AN EXAMPLE

As a practical example of the instability problem caused by the collector-to-base feedback, consider the amplifier shown in Fig. 12.4. The amplifier operates at the 450-kHz IF frequency often used in broadcast-band radios, and is useful for demonstrating the stability problems met in designing an IF stage. The input circuit is a series-resonant circuit that can be connected to a signal source. Regardless of the source impedance, the Q of the tuned circuit at 450 kHz is approximately 3.4, so the equivalent parallel resistance is approximately $50(Q^2)$ Ω. When corrected for the transistor input resistance, the equivalent base impedance R_b is 450 Ω at resonance. The collector load impedance may be adjusted by the variable resistor R_L.

The transistor-feedback capacitance is given in the data sheets at 15 pF, which at the oscillating frequency, 380 kHz, is 23 kΩ. Substituting these values for X_{cb} and R_b in equation (12 3), we get $G_v = 2(23{,}000)/450 = 100$. Hence, the 2N4355 transistor should permit a voltage gain of nearly 100 at 450 kHz.

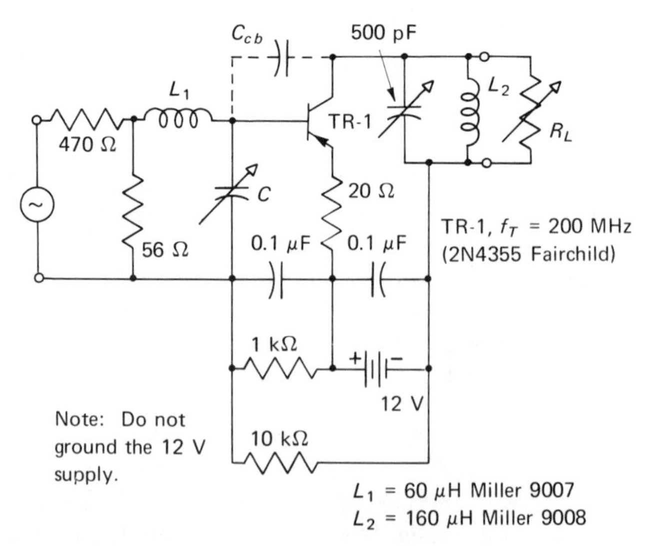

Figure 12.4 Experimental 450-kHz tuned *CE* amplifier.

The amplifier shown in Fig. 12.4 was found to oscillate when the voltage gain exceeds about 150. This gain is in satisfactory agreement with the calculated value since the feedback capacitance is not known accurately.

The amplifier shown in Fig. 12.4 illustrates several characteristics of tuned amplifiers that are of interest to the designer. The frequency of oscillation is below the resonant frequency. When the inductance and capacitance are both increased by a factor of 10 so that the resonant frequency is 50 kHz, the observed maximum gain is 2000, approximately as expected. This observation shows that the gain of a tuned amplifier must be reduced as the operating frequency is increased.

The amplifier illustrates the fact that, although the stage is stable for a voltage gain of 100 without neutralization, a higher stability, as that required for manufacturing, may be obtained by reducing the collector load. Moreover, unless the stage gain is reduced to about 20, which is a severe reduction, the stage shows interaction between the output and input that makes tuning difficult and unstable. The stage illustrates the fact that neutralization, which reduces the interaction, may be needed, even though not required for stability.

In summary, amplifiers designed to operate near the knee of the power gain versus the frequency curve require neutralization or a relatively low gain to ensure stability and ease of tuning. Any stage with over 30-dB voltage gain is approaching an unstable condition, and fixed neutralization is usually not practical. Mismatching the load to reduce the gain at least 10 dB below the critical gain is generally considered good design and a practical compromise of cost and reliability.

12.7 HIGH-GAIN TUNED-AMPLIFIER DESIGN

The prodedure for the design of a high-gain tuned amplifier is as follows:

1. Select a transistor that can be operated near to or below the β-cutoff frequency f_T/β_0.
2. Select a trial circuit and provide biasing, as for a low-frequency class-A amplifier. Use emitter feedback to improve gain stability and transistor interchangeability (unless overall AGC is planned) or to reduce the input capacitance and increase the input impedance.
3. Decide which resistances, including the transistor input impedance, determine the tuned-circuit load. Calculate the equivalent input load R_b. Both primary and secondary loads must be included. Select a loaded Q that gives the required selectivity. The loaded Q of transistor IF transformers is generally about 50.

4. Find the required inductance $L = R_b/\omega Q$ and calculate the capacitance needed for resonance at the midband frequency.
5. Calculate the critical voltage gain G_v, using equation (12.3). Select a practical voltage gain that is 3 to 10 times below the critical gain, depending on the required stability. (If the estimated gain is unacceptably low, neutralization may permit a 6- to 10-dB increase of power gain.) Find the collector load resistor that gives the desired voltage gain,
6. Design the collector tuned circuit, using the load resistor selected in step 5 and the Q of step 3. Find the inductance and tuning capacitor, as in step 4 above. Confirm by experiment that the stage will operate with the desired gain and stability.
7. Decide whether the output load is to be connected to a tap for low cost or to a tuned secondary for higher selectivity. For impedance ratios less than 10 to 1, the tap on a tuned circuit may be found by using the square of the turns ratio, as in transformer calculations; that is,

$$\left(\frac{n_1}{n_2}\right)^2 = \frac{R_1}{R_2} \tag{12.4}$$

12.8 COUPLING CIRCUITS

The coupling circuits used in RF amplifiers have different forms, depending on the center frequency, the amplifier bandwidth, and the source and load impedances that are coupled to the transistor.

The amplifier shown in Fig. 12.4 uses a *single-tuned L-network* for coupling a 50-Ω source to the relatively high-impedance base. The collector is coupled to a *parallel-tuned* circuit with the load connected directly across the tuned circuit. If the load impedance is relatively low, say 50 Ω the load may be connected to a tap on the inductor or may be inductively coupled, using a closely coupled untuned secondary with the same number of turns. For very low impedance loads a large capacitor may be connected in series with the tuning capacitor, and the load is connected across the large capacitor. The series capacitor, called a *capacitance divider*, provides efficient coupling to a low impedance, provided the reactance of the large capacitor is less than the resistance of the load.

Double-tuned input and output networks of the forms shown in Fig. 12-5 are often used in preference to single-tuned circuits because a second variable capacitor permits independent control of the bandwidth and the impedance ratio. The double-tuned circuits are particularly useful when experimental performance studies are made or when the source and load impedances are changed. The transformation ratio of the L-network is proportional to Q^2.

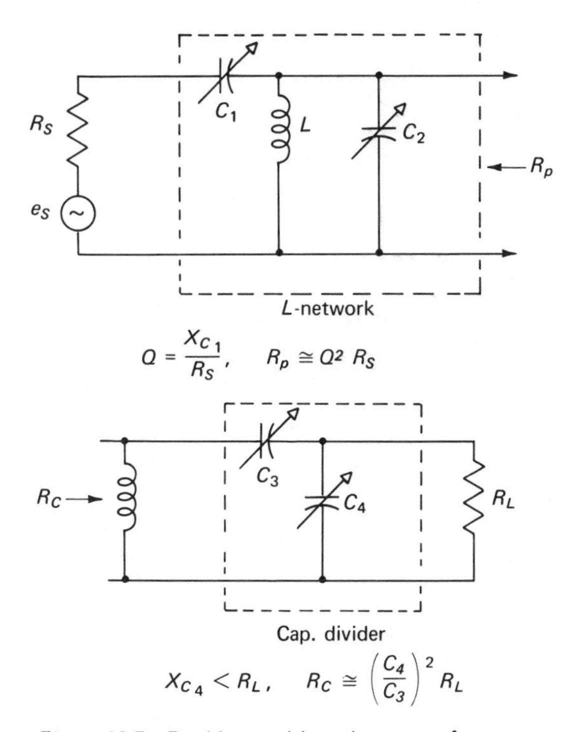

Figure 12.5. Double-tuned impedance transformers.

The impedance ratio of the capacitance divider is proportional to the square of the capacitance ratio. If the input is connected across the capacitor C_3, the capacitance divider is a CLC pi.

Tuned, inductively coupled transformers are used in IF amplifiers when isolated dc circuits are needed to supply bias currents. With separate primary and secondary coils the collector and base currents are supplied through the inductors, as in Fig. 12.1, so that a choke and coupling capacitor are eliminated. If only one winding is tuned, the coils should be closely coupled, and the impedance ratio is given by equation (12.4).

When the primary and secondary are both tuned, the circuits may be designed with high Qs and with a high efficiency of power transfer. When the primary and secondary are tuned to the same frequency (synchronous tuning) and are coupled for efficient power transfer at the center frequency, the response falls off rapidly at the ends of the passband and has a high degree of attenuation far removed from resonance. With close coupling, tuned coupled circuits may have a flat response over most of the passband. When the circuits in separate stages are stagger tuned, the flat passband may be extended even farther by accepting slightly reduced gain. The design of tuned coupled circuits is described in Appendix A.4.

Above the transistor β-cutoff frequency the attainable power gain of a tuned amplifier varies inversely with the center frequency. The power gain is independent of the amplifier bandwidth up to bandwidths of one fifth the center frequency. The bandwidth is determined by the Q of the tuned circuits. Bandwidths of nearly the transistor f_T rating may be obtained in tuned coupled stages by accepting a sufficient gain reduction. Thus, a wide-band amplifier may have as little as 6-dB power gain per stage and use six or more identical stages.

In distributed amplifiers a series of identical low-gain stages is connected in parallel by two transmission lines. An input line feeds power to each base, and an output line directs the output of each stage toward the load. Distributed amplifiers are used above 500 MHz where lumped components are displaced by transmission lines, thin-film circuits, and microwave techniques.

12.9 UHF AMPLIFIERS

In the UHF frequency region, 100 MHz to 1 GHz, the problems and methods of tuned-amplifier design are very different from those at low frequencies near the transistor β cutoff, and power gains are more significant than voltage gains. The reactances of the transistor become increasingly important, and the base-spreading resitance r_b' separates the junction capacitance from the external base terminal. Thus, the transistor transconductance i_c/v_{be} has a reactive component that may contribute 90° of phase shift when the signal frequency approaches the frequency f_T.

At frequencies a decade below $f = f_T$ the load in a stable amplifier usually equals the input impedance, and the power gain is a result of the transistor current gain. In the UHF region when the frequency f is f_T, the current gain is 1, but power gains of 10 to 15 dB are obtained by using load impedances that are 3 to 5 times the input impedance. As the following paragraphs show, the latter type of amplifier is stable if the load impedance is reduced, or mismatched, until the power gain is about 3 dB below the maximum, or matched, power gain.

The transistors used at UHF frequencies have internal feedback, which reduces the internal output impedance 2 or 3 orders of magnitude, and the load impedance is perhaps only one fifth the internal output impedance of the transistor. The reader may easily verify by Fig. 12.6 that, if the load impedance is 0.2 times the internal impedance, the output power is only 3 dB below the maximum obtainable power output. If the load impedance is lowered by another factor of 2, making it 0.1 times the internal impedance, the power output is only 5 dB less than the maximum output power, while the voltage across the load is reduced 15 dB.

Our interest in these changes stems from the fact that the voltage across

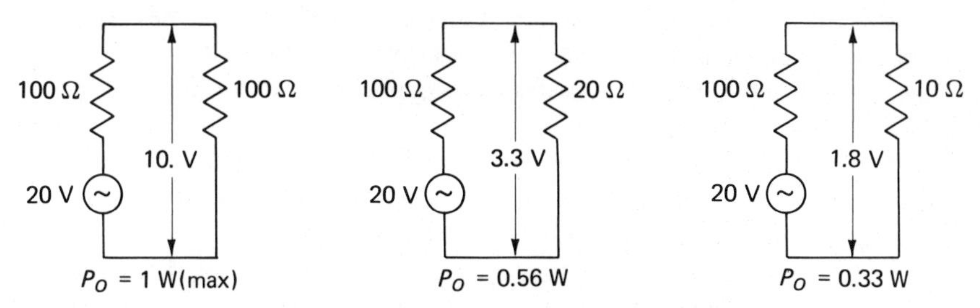

Figure 12.6. Load power and voltage with mismatch.

the load is the feedback signal, which determines the stability of a tuned amplifier. Since reducing the feedback signal by a factor of 5 is accompanied by only a 3-dB loss of output power, mismatching is an attractive method for increasing the stability of a tuned amplifier. Thus, a tuned amplifier may be stabilized *at any frequency* by reducing the collector load impedance until the required degree of stability is obtained. Neutralization may be used to reduce tuning interaction, but, when used to obtain both increased power gain and stability, the design and adjustment of neutralization become difficult and sensitive to any change of the transistor parameters.

12.10 UHF TRANSISTOR PARAMETERS

The transistor manufacturers often publish curves for UHF transistors, as in Fig. 12.7, showing the maximum available gain (MAG) at frequencies near f_T. The power gain implied by these curves cannot be realized in a practical

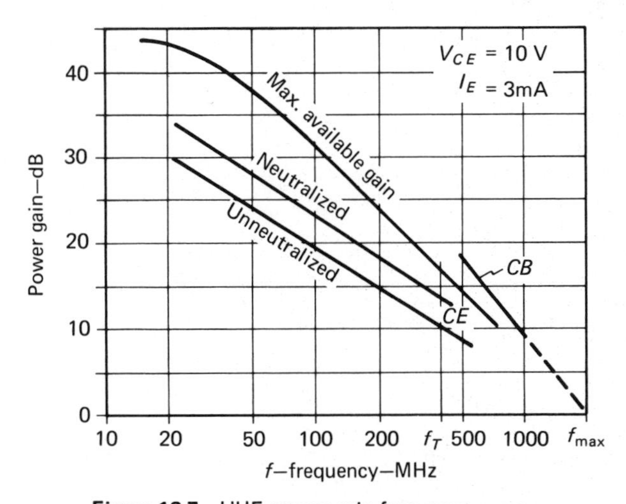

Figure 12.7. UHF power gain-frequency curves.

situation. The curves imply perfect unilateralization, which may be approached by experimental trial and error, but even to find a satisfactory form for the unilateralization circuit is generally considered impossible. The MAG curves are useful in giving the designer an upper limit of power gain and, by extrapolation, in showing the maximum frequency at which the transistor can be made to oscillate. Presumably, the maximum frequency of oscillation, f_{max}, is the frequency at which the attainable power gain is 1.

The curve of maximum neutralized power gain gives a more realistic value of the attainable power gain at high frequencies. These curves decrease approximately 6 dB/octave and usually indicate 15-dB power gain at $f = f_T$. In a practical situation the actual power gain of a stage may be 3 to 6 dB lower, depending on the loss in components and on the loss introduced to obtain stability and ease of tuning.

At extremely high frequencies when a transistor must be operated above the f_T rating, the power gain of the *CE* amplifier is limited partly by the inductance of the emitter lead. Because the *CB* amplifier is not limited by the emitter lead inductance and has positive internal feedback, the power gain of a *CB* stage is about 3 dB greater at very high frequencies than for a *CE* stage. However, the *CB* stage tends to be unstable at high frequencies and, for this reason, is more commonly used in high-frequency oscillators than in amplifiers. At frequencies above 50 MHz, a *CB* stage with a tuned collector load can be made to oscillate by the addition of a small capacitor from the collector to the emitter. A high-frequency oscillator that uses the *CB* connection and an emitter inductance to increase the loop phase shift is illustrated in Fig. 12.8.

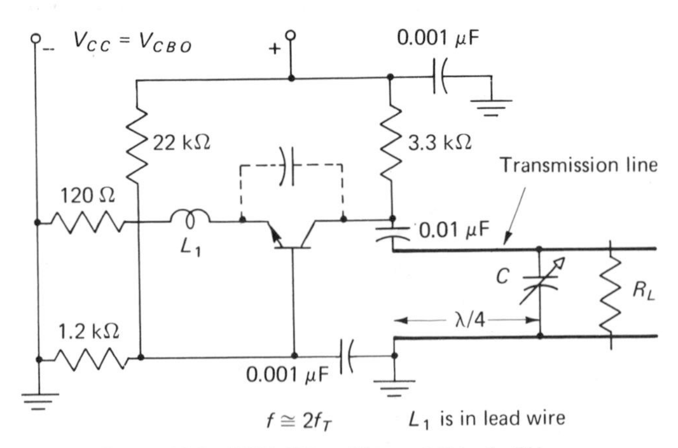

Figure 12.8. UHF *CB* oscillator, 0.5 to 1. GHz.

At frequencies approaching f_T, the transistor parameters change with frequency and may be more reactive than real. For example, the *CE* input impedance decreases to approximately r_b' with a small equivalent series reactance (a capacitance). The current gain approaches 1 with a 90° lagging phase angle, and the output impedance drops to a few kilohms with a 45° phase angle.

12.11 TRANSISTOR *y*-PARAMETERS

Above 100 MHz the electrode capacitances, lead inductance, and response times complicate the equivalent circuit and measurement of transistor-impedance parameters. For these reasons UHF transistor characteristics are expressed by curves depicting admittances, measured with either the input or the output short circuited.

The admittance parameters are *y*-coefficients of the transistor input and output current equations,

$$i_i = y_{ie}e_i + y_{re}e_o \tag{12.5}$$

$$i_o = y_{fe}e_i + y_{oe}e_o \tag{12.6}$$

The currents and voltages are the input and output variables shown in Fig. 12.9, and the *y*-parameters are measured by short circuiting e_i and e_o in turn. For example, y_{ie} is the admittance i_i/e_i measured with $e_o = 0$. The real part of y_{ie} is $\mathrm{RE}(y_{ie})$, or g_{ie}, and the imaginary part is $\mathrm{Im}(y_{ie})$, or b_{ie}. The real and imaginary components of all four *y*-parameters are measured with an admittance bridge and are presented as curves that are a function of frequency. These curves are given with UHF transistor data, and usually the designer uses only two of the curves.

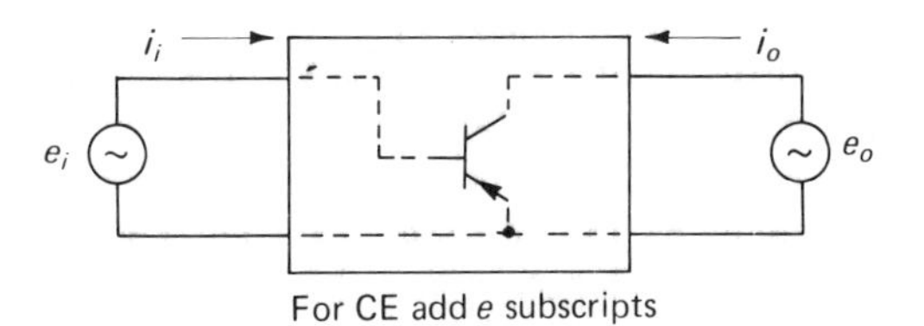

Figure 12.9. *y*-parameter voltages and currents.

For most designs in the UHF region the collector load impedance is low compared with the internal impedance, so the input admittance is reasonably approximated by the short-circuited admittance y_{ie}. Similarly, the transistor internal output admittance is approximately y_{oe}.

When a parallel-tuned circuit is connected across the transistor input, the shunt resistance at resonance is the reciprocal of the conductance g_{ie}, that is, $1/g_{ie}$. Similarly, with a parallel-tuned collector circuit the transistor internal output resistance is $1/g_{oe}$. Thus, the equivalent parallel input and output resistances of a transistor are easily estimated from the g_{ie} and g_{oe} conductance-frequency curves. For practical design purposes these parameters are useful mainly in selecting a transistor and in estimating the matching input and load impedances.

12.12 UHF TUNED-AMPLIFIER DESIGN

A procedure for the design of a tuned UHF amplifier is as follows:

1. Select a transistor with an f_T rating at least as high as the desired operating frequency, and select a Q-point that gives the required f_T. Design a low S-factor bias circuit that gives the desired collector current.
2. From the manufacturer's g_{ie} curve, a recommendation, or by experiment, obtain an estimate of the parallel input resistance for use in the design of the input coupling network.
3. Construct an input coupling circuit suitable for an experimental evaluation of the amplifier. Construct a tuned collector load circuit in which the load resistor can be varied from 2 to 20 times the transistor input resistance with corresponding Qs of about 5 to 50.
4. Measure the maximum stable load resistance and, depending on the required stability, select a 3 to 5 times lower value for the final load impedance. Use capacitance neutralization to reduce the tuning interaction, if necessary.
5. Design the collector coupling circuit so that the load may be coupled with a tuned circuit Q that gives the selectivity desired.
6. Evaluate the design by experimental performance studies. When possible, the design should be reevaluated by someone other than the designer.

At UHF frequencies one of the most important considerations is the layout of the circuit and the choice of components. All leads must be as short as possible, grounds must be carefully placed, and good shielding must be maintained between stages and between the input and output circuits of each stage. At high frequencies bypass capacitors have series inductance and chokes have shunt capacitance, so unexpected resonant effects may occur. Consequently, these elements must be carefully selected, preferably with the aid of a high-frequency bridge. Low-frequency blocking, often observed, is generally a symptom of high-frequency oscillations caused by inadequate bypassing of bias and power circuits. These circuits should be supplied through feed-through capacitors and bypassed by high-frequency ceramic capacitors connected close to the transistor.

Practical help in the design of high-frequency amplifiers may be obtained from the manufacturer's data sheets that describe RF transistors. The data sheets usually show the circuit used for power-gain measurements and often include detailed information concerning the construction of inductors, transmission lines, and microwave cavities. The designer usually can change these circuits to serve another purpose. Much additional information concerning circuits and construction techniques may be obtained from the references and by examining equipment in the marketplace.

Field-effect transistors are being used increasingly toward 100 MHz in circuits that are designed by the methods used with vacuum tubes. FETs and vacuum tubes have very similar characteristics, and in many circuits the devices are practically interchangeable.

At gigaherz frequencies, 10^9 Hz, the configuration of a circuit and the methods of design change considerably. The transistor parameters, called *S-parameters*, are measured by forward and reflected microwave power measurements, and both the transistor and the circuit are treated as components of a microwave transmission line.

Under some conditions RF amplifiers are not biased as class-A amplifiers. Amplifiers for high-level FM signals can be used without bias because only the zero crossings of the signal are important, and these stages are generally designed to limit the signal by peak clipping. Amplifiers for pulse signals are designed as wide-band amplifiers and are often operated without bias in order to shape the signal. Class-C amplifiers are usually biased by clipping the signal peaks and storing the charge in a capacitor so that the bias varies with the signal amplitude. By this means the bias tends to suppress most of the input signal, so the collector current is a series of high-energy pulses at the signal frequency. The tuned collector tank circuit converts these pulses to a sine wave at the signal frequency. If the tank is tuned to a harmonic frequency, the amplifier is called a *frequency multiplier.*

12.13 UHF AMPLIFIER—AN EXAMPLE

A typical low-noise, broad-band, 200-MHz amplifier circuit is shown in Fig. 12.10. The amplifier has tuned input and output circuits with neutralization to increase the stability and ease of tuning. The amplifier has 20-dB power gain between a 50-Ω source and a 50-Ω load, but a 300-Ω line may be capacitor-coupled to the base and/or a 300-Ω load may be coupled to the collector with similar performance characteristics but reduced gain.

The transistor is a high-frequency, low-noise, germanium device, which has an f_T exceeding 1 GHz. A germanium device generally offers a higher f_T than is available with silicon, although for applications at 200 MHz a silicon 2N4135 gives equivalent characteristics and performance. The transistor is operated with a 3-mA emitter current that is adjusted by the variable resistor in the +6-V emitter supply. At 200 MHz the transistor current gain is about 5, so 20-dB power gain requires a collector load impedance that is at least 20

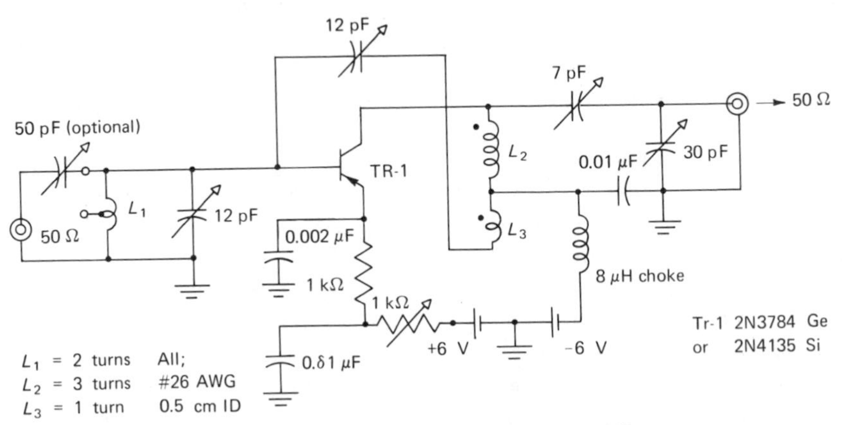

Figure 12.10. Broad-band UHF 200-MHz amplifier.

times the input impedance. Since the g_{ie} curve shows an input resistance of 50 Ω the collector load resistance should be about 20(50) = 1000 Ω.

The reactance of the two-turn input coil is about 30 Ω at 200 MHz, so the loaded Q with the 50-Ω source and the 50-Ω input resistance is about 1. However, the low-reactance coil makes tuning easier with a high shunt capacitance. By connecting the source to a center tap on L_1 and accepting a power mismatch, the Q and selectivity may be improved.

The collector coil L_2 has a reactance of 70 Ω, so the loaded Q is about 10. The 50-Ω load is coupled through a capacitance divider, which allows some adjustment of the Q and bandwidth. The bandwidth is 10 to 20 MHz, depending on the tuning and the effects of internal positive feedback. Neutralization is required because of the relatively high power gain, but may be omitted by using a transistor that has $f_T = 200$ MHz and by accepting about 15-dB gain.

Since the amplifier is designed to connect between equal source and load impedances, a second stage may be connected for additional gain. If we assume a 1000-Ω collector load impedance and a 6-V supply, the class-A power output is, by equation (12.7), about 18 mW. With a 10-V supply the power output may be increased to 50 mW. Power outputs of several watts may be obtained by operating silicon transistors in class C.

12.14 RADIO-FREQUENCY POWER AMPLIFIERS

When an amplifier is required to deliver a large amount of power, the collector load impedance may be considerably lower than the impedance values used to obtain a high power gain. In a class-A power design the collector load impedance is determined by the same relation as that used at audio frequencies, equation (7.11):

$$R_L = \frac{V_{CC}^2}{2P_O} \tag{12.7}$$

The use of equation (12.7) implies that the peak collector voltage may be driven by the available input signal from 0 to $2V_{CC}$ and that the peak current $I_{Cp} = V_{CC}/R_L$ is within the capabilities of the transistor and the input signal. When loading the transistor for a maximum power output, the output network must be designed to resonate with the circuit and transistor reactances and to provide the required collector loading.

When an amplifier is required to supply several watts of power, the transistors are generally operated class C because the higher efficiency permits more output power from a given transistor. As with class-A power amplifiers, the collector load resistor for a class-C amplifier is determined by the supply voltage and the power output, but the optimum design relations are not easily represented by a formula. Moreover, the characteristics of transistors operated class C are quite different from the low-power characteristics.

The circuits used for class-C operation are similar to those described in this chapter except that there is usually no way to forward bias the transistors. In class-C operation a transistor is operated with zero bias or a small reverse bias so that the transistor is driven ON for less than a half-cycle of the input signal. The reverse bias is usually developed by rectification of the signal peaks, so a large capacitor in the base circuit supplies a reverse bias that changes with the input signal. The amplifier shown in Fig. 12.10 may be operated class C by removing the emitter bias circuit and either connecting the emitter directly to ground or through an RF choke shunted by a 45-pF variable capacitor.

The manufacturers of RF power transistors usually supply circuits and component values for the construction of class-C amplifiers. A multistage class-C amplifier is designed by selecting a power transistor and circuit that give the required power output. This circuit may be adapted to operate at a lower frequency, provided all inductors and all capacitors are increased by the inverse frequency ratio. When changed in this way, the reactances and circuit impedances are substantially unchanged, and the amplifier should be operable at the lower frequency.

From the power-gain or the drive-power requirements of the output stage, the designer finds the output power requirements of the driver stage. Repeating the procedure outlined for the output stage, the designer can select and adapt a low-power class-C amplifier for use as a driver stage. At still lower power levels, when the input signal peaks are less than 1 or 2 V, the amplifier must operate class A. Similarly, an FM amplifier may use zero bias at high-signal levels but must have class-A stages when the signal level is less than 1 or 2 V.

The power output attainable from a single transistor is approximately 50 W at 100 MHz, and the power decreases by the first power of frequency to about 1 W at 2 GHz.

12.15 SUMMARY

Amplifiers designed for high-frequency applications usually have tuned input and output circuits because adequate gain is obtained only by including the transistor and circuit capacitance as a part of the tuned circuit. Above the transistor β-cutoff frequency, the transistor current gain drops off 6 dB/octave and introduces approximately 90° of phase shift. With tuned circuits an amplifier tends to be unstable with more than 30-dB gain and is difficult to adjust and tune.

The critical gain at which a tuned amplifier oscillates may be calculated by equation (12.3), which is a useful design guide. The equation shows that, for a given input circuit resistance, the critical gain is increased by reducing the transistor feedback capacitance. Tuned amplifiers may be stabilized by reducing the power gain 10 dB below the critical gain and by capacitance neutralization. Neutralization is a means to offset the capacitance feedback, but the internal resistance components of feedback make neutralization impractical as a means of eliminating feedback.

Transistors operated near the frequency f_T have unit current gain, but will produce approximately 10-dB power gain by making the collector load resistance higher than the input resistance. At these frequencies the parallel equivalent input resistance is approximately r_b', the transistor base-spreading resistance. Since r_b' in low-power transistors is generally about 40 Ω, the input impedance is low and the impedance of the collector load for a maximum stable power gain is approximately 10 times higher.

The transistor input resistance and internal output resistance are the reciprocal of the real components of the y_{ie} and y_{oe} admittance parameters. Class-C amplifier design requires the use of class-C transistor parameters, which are found in power-transistor data sheets. Typical circuits for high-frequency class-C applications are generally given there also.

PROBLEMS

12.1. Calculate the voltage and power relations shown in Fig. 12.5.

12.2. An 2N3478 transistor has a collector-to-base feedback capacitance of 0.7 pF. The transistor is to be operated with a tuned input circuit, which makes the resonant base-circuit resistance 100 Ω, and the tuned collector load resistance is to be 400 Ω. If a power gain of 16 dB is required, what is the maximum frequency of operation without neutralization? What maximum frequency would you recommend for stable operation without neutralization?

12-3. A transistor has a current-gain-bandwidth product of 50 MHz and a low-frequency current gain of 50. (a) What is the equivalent short-circuit input capacitance when $I_E = 1\,\text{mA}$? (b) At what frequency is the transconductance 71 per cent of the low-frequency transconductance if the base-spreading resistance is 50 Ω? (c) Draw an equivalent circuit showing the reactance and resistance values at the frequency f_T. What is the low-frequency value of g_m in your equivalent circuit?

12-4. A 2N3784 Motorola transistor is to be operated at 1000 MHz with a 3-mA emitter current. (a) Draw a circuit showing the equivalent parallel input resistance and reactance. (b) Draw a circuit showing the equivalent parallel output resistance and reactance. (c) What value of inductance resonates with the output capacitance? (d) What would you do to match the input circuit to a 50-Ω line?

12-5. An FET with a feedback capacitance of 0.2 pF is to be operated with tuned gate and tuned drain circuits. (a) If the gate circuit has a resonant resistance of 20,000 Ω and the power gain is 40 dB with a 4-kΩ load resistance, what is the maximum frequency of operation without neutralization? (b) Specify L and C values for input and output tuned circuits that resonate at $\frac{1}{10}$ the frequency of oscillation. The circuits should have $Q = 20$ with the load resistors connected in parallel.

REFERENCES

12-1. *RCA Transistor Handbook,* SC-13, pp. 45-65, 508. Harrison, N.J.: Radio Corporation of America, 1967.

12-2. *RCA Silicon Power Circuits Manual,* SP-51, pp. 262-352. Harrison, N.J.: 1967.

12-3. Terman, F. E., *Electronic and Radio Engineering,* 4th ed., Chapter 12, "Tuned Voltage Amplifiers," pp. 400-447. New York: McGraw-Hill Book Company, 1955.

DESIGNS

12-1. Construct the experimental amplifier shown in Fig. 12.4. Find the load impedance that makes the amplifier unstable and decide what power gain can be obtained with a reasonable stability. Observe the tuning interaction with several load impedances.

12-2. Construct and evaluate the performance of an impedance transformer. Use air-core coils that may be hand wound or the windings removed from a filament transformer, and use a frequency of 10 kHz or higher, which can be measured easily.

12-3. Design and construct a UHF broad-band amplifier and use the amplifier as a TV preamplifier with a transmission line between the amplifier and a TV receiver.

13

TRANSISTOR LIMITATIONS—NOISE, BREAKDOWN, AND TEMPERATURE

This chapter is concerned with the three most common transistor limitations that affect the performance and reliability of a circuit. Noise originating in the resistors and transistors limits the useful gain and places a lower limit on the input signals, which can be observed in high-gain systems. Transistor noise may be reduced by using an optimum collector current and source resistance.

Breakdown limits the ability of a transistor to produce a large signal output, or makes a circuit unreliable when subjected to an unusual stress. Circuit reliability is obtained by operating transistors within their capabilities and with an adequate margin of safety. Similarly, high power output is obtained by careful observance of transistor current, voltage, and dissipation ratings.

All transistors are damaged by a permanent change of the junction semiconductors when the junction temperature exceeds a critical temperature. The temperature of a transistor is controlled by reducing the thermal resistance so that the heat developed in the junction is transferred to the surrounding environment.

By an understanding of transistor limitations and their control, a designer is able to design circuits with high reliability, better performance characteristics, and at a reasonable cost.

13.1 NOISE

The resistors and transistors of an amplifier generate a steady, random electrical noise that fixes a lower limit of the useful input signal, limiting the maximum useful gain of an amplifier. The noise in an amplifier is caused by the particle nature of all electrical conduction and originates from several different sources. For example, a video amplifier with a 5-MHz bandwidth and a 1 kΩ input resistance will have a 9 microvolt (μV) input noise signal, which is produced mainly by the input resistance.

All resistances produce a fluctuating voltage called *resistance noise*, *thermal noise*, or *Johnson noise*. Because of the fundamental nature of resistance noise, the long-time average of this noise may be used as a precise signal standard or as a reference for comparing other noise signals.

The current in a vacuum or in a semiconductor is a random flow of electrons or holes that produces a noise called *shot noise*. Semiconductors and composition resistors also exhibit a noise at subaudio frequencies called $1/f$ or *flicker noise*. High-frequency semiconductor amplifiers produce a noise, just below the high-frequency limit of amplification, that is called *cutoff noise*.

The frequency distribution of noise can be expressed by a curve that shows the noise power of a system as a function of frequency, as in Fig. 13.1.

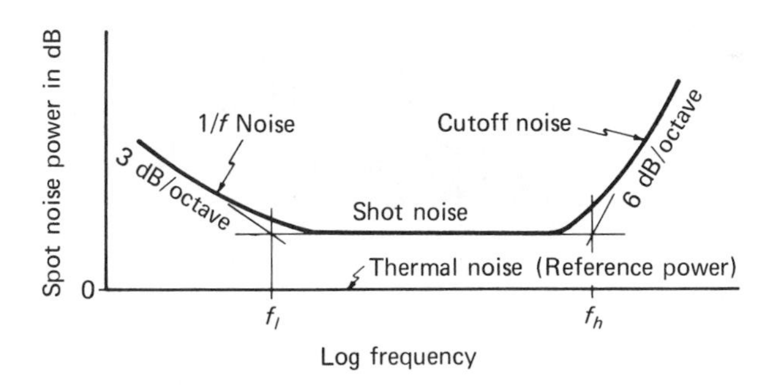

Figure 13.1. Transistor spot-noise-power curves.

Because the noise produced by a transistor departs from a uniform (white) noise-frequency distribution, in both the $1/f$ and cutoff regions, the noise must be measured using narrow-band equipment. Such noise, measured in a reasonably narrow band and referred to an equivalent 1-Hz bandwidth, is called *spot noise*. A spot *noise factor* is the factor by which the total noise exceeds the resistance noise. Noise factors are expressed as power ratios, or are given in dB units and called a *noise figure*.

13.2 RESISTANCE NOISE

The thermal noise power in a resistor is proportional to the absolute temperature of the resistor and to the frequency band used to observe the noise. For quality resistors at room temperature the rms thermal noise voltage is approximately

$$e_n = \sqrt{1.6R(f_h - f_l)} \times 10^{-4} \qquad \textbf{(13.1)}$$

where the units in equation (13.1) are rms microvolts, ohms, and hertz. A noise chart in Appendix A.9 gives the resistance-noise voltage in terms of the circuit resistance and the bandwidth. Equation (13.1) assumes a flat frequency response within and no response outside the frequency band. Measurements are often made with a simple *R-C* cutoff at both low and high frequencies, for which the equivalent bandwidth is given by the 5.4-dB cutoff frequecies, close to half value. Thermal noise should be measured by a power measurement, but, when observed on an oscilloscope, the peak-to-peak noise voltage is from four to six times the rms value, and only rarely is more than eight times.

The simplest way of measuring a noise voltage is to sweep an oscilloscope alternately with a low-frequency square wave superimposed on the noise signal. When the square wave is reduced so that the two noise traces just merge, as illustrated in Fig. 13.2, the rms value of the noise is just one half the square-wave peak-to-peak amplitude.

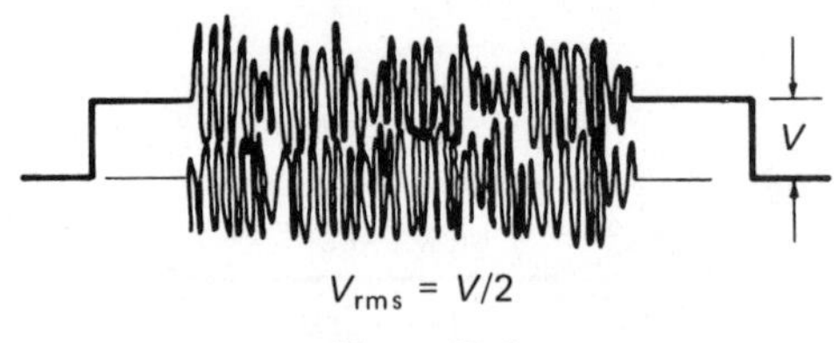

$V_{rms} = V/2$

Figure 13.2.

The measurement of a noise factor usually requires a device capable of measuring the noise power output of an amplifier, both with and without a calibrated noise input. The use of a calibrated noise source instead of a sinsusoidal source offers the advantage that an average noise factor can be obtained without measuring or calculating the equivalent noise bandwidth of the system.

The chart in Appendix A.9, or a calculation by equation (13.1), shows that the noise voltage measured across a 1-MΩ resistor with a frequency band of 0.5 MHz is approximately 100 μV. The noise observed at full gain of an oscilloscope is normally resistance noise. An oscilloscope having a sensitivity of 1 mV/cm with a 500-kHz bandwidth should exhibit a noise that is approximately the trace width. If the observed noise is appreciably higher than the value given by equation (13.1), the noise probably can be reduced.

The noise observed in an amplifier is evaluated by comparing the observed noise with the theoretical noise of equation (13.1), where R is the source resistance used in the measurement. The square of the ratio of the measured voltage to the theoretical noise voltage is the amplifier noise factor F_N. The noise factor expressed in decibels is called the *noise figure*. The square of the ratio is used because the noise factor is a comparison of noise powers. If the noise is measured at the output of an amplifier, the equivalent input noise is calculated by dividing the output noise by the amplifier gain.

The thermal noise of an amplifier is reduced only by eliminating all unnecessary resistors at the input, by reducing the circuit bandwidth, and by having enough gain in the input stage to make the signal exceed the noise produced in the second stage. Shot noise, flicker noise, or cutoff noise may sometimes be reduced or alleviated by changing the circuit or the semiconductors.

The thermal noise of a signal source or a transducer is usually beyond the control of a designer of amplifiers. Generally, the thermal noise of a signal source may be reduced only by reducing the absolute temperature of the source. High-gain antennas directed into free space are low-temperature signal sources, which sometimes require the use of refrigerated preamplifiers.

13.3 SHOT NOISE

Shot noise is caused by random changes in the number of current carriers found in electron emission and in semiconductor conduction. Shot noise has a frequency dependence that is the same as for resistance noise, and both are said to have a "white-noise" frequency dependence. The shot noise in transistors makes the total noise 1 to 4 dB above the theoretical resistance noise. The noise in the center-frequency portion of the transistor noise-figure curve, shown in Fig. 13.1, is called shot noise, although it is partly thermal noise produced in the transistor resistances. High-frequency transistors have a higher shot noise and a high ohmic resistance in the base, which makes the noise in the center-frequency region a few decibels higher than in audio transistors.

13.4 FLICKER ($1/f$) NOISE

Flicker noise is observed at subaudio frequencies (Fig. 13.1) and is characterized by a 3-dB rise of noise power for each octave decrease of the center frequency. Also called *excess noise*, or *pink noise*, this noise is found in transistors, FETs, vacuum tubes, and carbon resistors. Because $1/f$ noise depends on the frequency, the noise voltage is generally specified by a spot-noise figure at a given frequency on the $1/f$ low-frequency slope.

The spot-noise figure of a low-noise planar transistor is about 6 dB at 100 Hz, which is in the $1/f$ noise region. This means that the total noise power measured in a 1-Hz frequency band is 4 times the resistance noise power. At 10 Hz the $1/f$ spot noise is 10 times larger, or 31 times the resistance noise. Thus, the 10-Hz spot noise factor is 31, or about 15 dB. By comparison, a low-noise FET with the same input source resistance has at 10 Hz a spot-noise figure of only 3 dB. However, the low-frequency noise characteristics of the field-effect transistor are similar to those of the vacuum tube, so a noise advantage is realized only with source resistances exceeding 10 kΩ.

The $1/f$ noise of a quality transistor is more or less proportional to the emitter current and may be minimized by operating at low emitter current. The excess noise of high-frequency transistors may be significant at frequencies as low as 1 MHz in a wide-band RF amplifier.

Flicker noise may usually be neglected when the upper frequency limit of the amplifier is more than 10 times the $1/f$ noise corner frequency, where the spot-noise figure is 3 dB. The low-frequency drift of a dc amplifier is partly $1/f$ noise, but again the contribution of the $1/f$ noise to the total of a wide-band amplifier may be neglected.

13.5 HIGH-FREQUENCY NOISE

The amount of noise in high-frequency amplifiers is always an important consideration, because high-frequency losses make the available signal levels small, and high-frequency systems are generally free from man-made and natural electrical noise. However, all transistors are noisy in the frequency decade below the transistor f_T. In this frequency range, the spot-noise figure of a high-frequency transistor increases 6 dB/octave, and the noise is cutoff noise. The frequency at which the noise figure is 6 dB may be considered the upper useful frequency limit of the transistor in applications needing high signal sensitivity.

13.6 TRANSISTOR NOISE

The noise that originates in a wide-band transistor amplifier is the summation of the noise power from the resistors, the current in semiconductors, and the $1/f$ and cutoff noise. When the total noise is measured for various source impedances and transistor Q-points, the noise is called *transistor noise*.

The transistor noise of devices intended for low-noise applications is usually described by curves of constant, narrow-band noise figures that are plotted on charts of source resistance versus collector current. An example of these charts

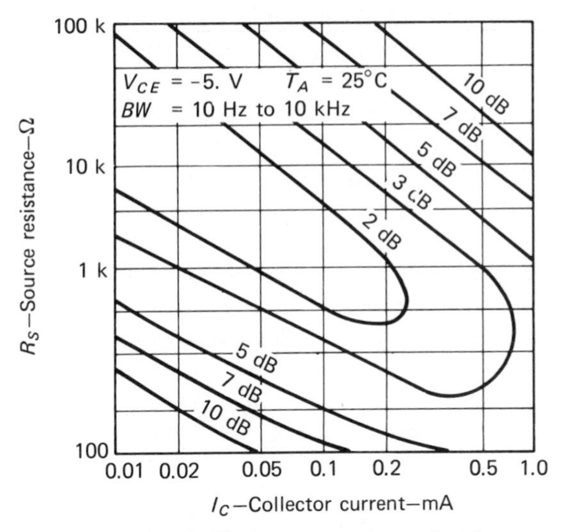

Figure 13.3. Wide-band noise-figure contours, low-noise planar *pnp* transistor.

is shown in Fig. 13.3. Low-noise planar transistors generally have narrow-band noise figures of less than 4 dB when the source resistance is 3 to 10 kΩ and the collector current is 0.1 mA. For low-noise applications the best general rule is to use a low-noise high-β transistor or an FET, a low source resistance, and low collector current. The collector current should be between 1 and 100 μA, with the higher current level used with low source impedances. The transistor β should exceed 100 at the Q-point current. Resistors and resistance losses should be avoided in the input circuit, and the emitter resistor should be bypassed.

The frequency distribution of transistor noise is shown by noise-figure versus frequency curves similar to the curves shown in Fig. 13.4. Because the

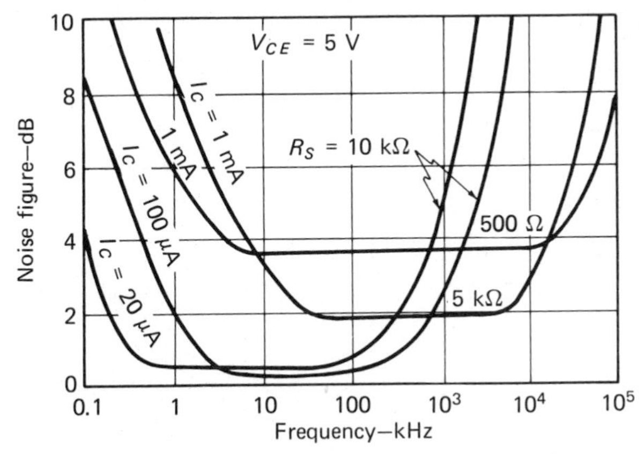

Figure 13.4. Noise figure versus frequency, low-noise npn planar transistor.

noise in a wide-band amplifier is mainly contributed by the noise at high frequencies, where the bandwidth is high, a wide-band amplifier should use a higher emitter current and a lower source resistance than a low-frequency amplifier. As Fig. 13.4 shows, an audio preamplifier that emphasizes the low-frequency noise should use a lower emitter current than a wide-band audio amplifier. Similarly, a wide-band video amplifier with a 5-MHz high-frequency cutoff, 1-mA emitter current, and 5-kΩ source resistance should have a 2-dB wide-band noise figure. The same video amplifier with a 0.1-mA emitter current would have a wide-band noise figure of more than 6 dB, because the amplifier frequency band is mainly in the cutoff-noise region.

At radio frequencies the minimum noise figures are obtained with collector currents of approximately 1 mA and a source resistance of 1 kΩ. The stated figures are useful guides for noncritical applications, but noise curves for the transistor should be consulted when a low noise figure is important.

13.7 LOW-NOISE AUDIO STAGE

The circuit of a low-noise audio stage is illustrated in Fig. 13.5. The amplifier is designed to use any low-noise transistor that is intended to be used at 0.1-mA collector current with a 5-kΩ source. The stage is intended to be biased for a 4-V collector-to-emitter voltage, at which point the collector current is 0.1 mA. Because the collector supply provides a voltage that is three times the desired collector voltage, it is convenient to use equal collector and emitter resistors and a low *S*-factor. The low *S*-factor ensures that the *Q*-point is independent of temperature.

An emitter-bypass capacitor is used to eliminate the emitter resistor as a noise source and to reduce noise feedback from the collector circuit. A good quality composition resistor carrying the dc current of this stage has a noise level nearly as low as the thermal noise of a wire-wound resistor. However,

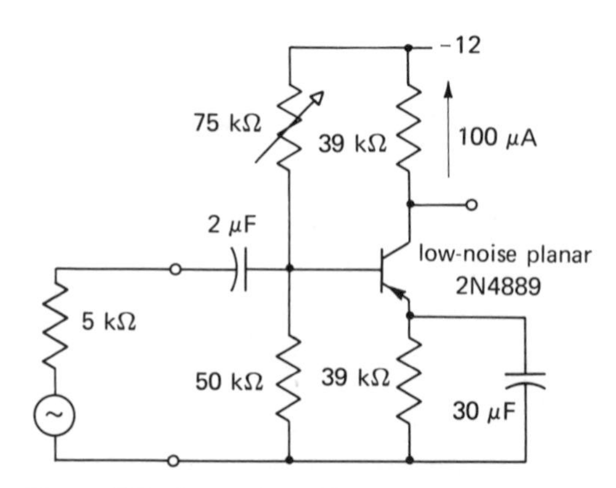

Figure 13.5. Low-noise, wide-band audio amplifier.

deposited carbon resistors are generally noisy, and alloy resistors should be checked to eliminate defective units.

The input capacitor should be relatively large, but the emitter capacitor should be sized to eliminate as much as possible of the $1/f$ noise spectrum. The amplifier in Fig. 13.5 should have an equivalent input noise voltage that is no more than 3 dB above the thermal noise voltage of the 5-kΩ source.

13.8 DIODE NOISE

The noise produced by point-contact diodes is high and unpredictable. However, most diodes produced today have an alloy or diffused junction, which makes their noise approximately the same as the noise of a transistor operating at the same current level. Zener diodes operate in a breakdown region, which produces a complex and variable Zener or avalanche noise. This noise may be controlled somewhat by using an adequate but not excessive diode current that depends on the normal current rating. A partial reduction of diode noise may be obtained by shunting the diode with a capacitor.

Some Zener diodes produce sudden random changes in the breakdown voltage, which are particularly disturbing at subaudio frequencies. Diode noise sources are available that produce white noise from the $1/f$ region up to 100 MHz.

13.9 TRANSISTOR BREAKDOWN

The reliability of a transistor depends partly on the designer's ability to build a circuit in which the transistors are operated within their capabilities. The damage or failure of a transistor is usually caused by part of the junction becoming overheated long enough to produce an irreversible change of the junction materials. To assure that the critical temperature limit is not exceeded, the manufacturers normally specify a maximum current, maximum voltage, maximum power dissipation, and temperature limits that a transistor can safely withstand. Transistor failure when the semiconductors change or melt because of inadequate heat dissipation is the topic of later sections of this chapter. A transistor may also become overheated by operation above the current or the voltage breakdown ratings.

13.10 CURRENT RATINGS

The maximum current rating of a transistor may have several meanings. The current rating may refer to the fusing current of an internal connection. The current rating may indicate the highest current at which the device is considered useful. At other times the rating may be fixed by an arbitrary minimum current gain, which is too low to be useful in design. Generally, the peak collector current used in design should be determined by referring to a β versus collector-current curve. In power applications the peak collector current may be determined by second breakdown considerations that make the safe current a small fraction of the manufacturer's maximum.

13.11 VOLTAGE BREAKDOWN RATINGS

A confusing variety of voltage ratings for transistors has developed from the need to indicate the limiting capability of different types of transistors under a variety of service conditions. Moreover, voltage ratings are given for the emitter junction, the collector junction, and for the collector-to-emitter voltage. The voltage breakdown ratings always refer to a junction with reverse bias. When the maximum voltage rating is exceeded, the reverse current, normally quite small, abruptly increases and the transistor is said to "break down." With breakdown, the current may be localized in a small area that becomes overheated and thus destroys the transistor.

A few of the commonly used breakdown ratings are represented by the collector current-voltage curves shown in Fig. 13.6. The low-voltage region of

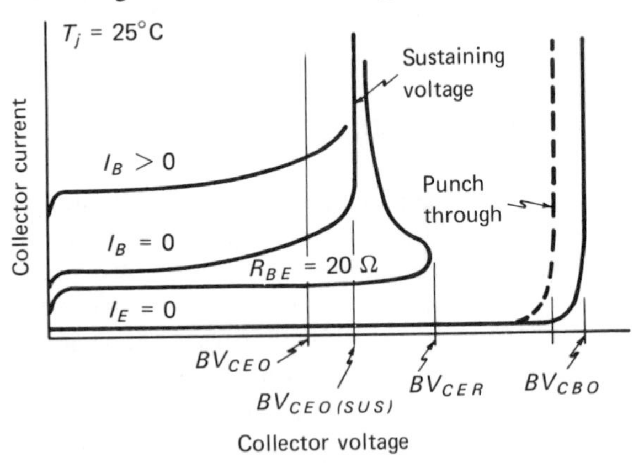

Figure 13.6. Collector current-voltage curves showing breakdown.

the collector characteristics have the normal current-voltage curves up to collector voltages that are about 80 per cent of the collector sustaining voltage $BV_{CEO(SUS)}$. The subscripts on the letter symbols show that the sustaining voltage is measured with the emitter as the common terminal and with the base open. As the collector voltage approaches the sustaining voltage, the base-current curves turn up, showing that the transistor current gain is higher than at low voltages. The increase of current gain with collector voltage is known as *avalanche multiplication* and is generally observable a little below the common-emitter breakdown. With low S-factors a small increase of current gain is not important by itself and may not be recognized.

Because the emitter is the most-used common terminal, the common-emitter breakdown voltage, BV_{CEO}, which is always specified, is the best all-purpose breakdown rating. The BV_{CEO} voltage is specified for a nominal increase of current gain and is a conservative rating, because with the base open the breakdown voltage depends partly on the transistor leakage currents. The effect of a higher temperature in lowering the BV_{CEO} voltage is offset by the design practice of using low-valued base resistors in high-temperature applications. The sustaining voltage, which is sometimes used in place of BV_{CEO}, tends to give a transistor a somewhat more favorable rating.

With the base terminated in a resistor R, the voltage at which the collector current begins to avalanche is a higher breakdown voltage VB_{CER}. The highest breakdown rating, BV_{CBO}, is obtained with the emitter open and with the voltage applied between the collector and base. Some types of transistors fail at a slightly lower BV_{CBO} because of punch through, or reach through. The BV_{CBO} ratings are not often useful and tend to give a transistor a too favorable rating, since the common-base rating is about twice the common-emitter rating.

The collector-emitter breakdown characteristics are complicated by the effects of the base circuit impedance and the signal. The manufacturers supplying transistors for different applications use different circuit conditions and different terminology for defining the boundary between an operating and a breakdown region. There is some justification for these differences, and the designer should obtain breakdown data from a manufacturer of devices that are intended for his application.

The collector breakdown ratings BV_{CEO} and $BV_{CEO(SUS)}$ may be used for class-A and class-B designs, provided the collector current is not too high when the collector voltage approaches the breakdown voltage. The BV_{CER} and BV_{CBO} ratings may be approached only when the collector current is small, as in some class-B and switching applications. Otherwise, there is danger of second breakdown.

The effect of breakdown with low collector currents is sometimes observed only as a faulty circuit operation, particularly with high junction temperatures. If the increased current is allowed to continue, overheating or second breakdown may permanently damage the transistor. A transistor generally recovers from breakdown if the current, or pulse energy, is sufficiently limited to prevent overheating. However, second breakdown is usually so difficult to control that a transistor is destroyed even in laboratory studies of the phenomenon.

13.12 SECOND BREAKDOWN

Second breakdown occurs when a transistor is stressed simultaneously by a high voltage and high current. On failure the collector voltage drops almost instantaneously to a small value. Second breakdown is caused by a concentration of current into such a small part of the collector diode that the junctions are fused into a permanent collector-to-emitter short circuit. Second breakdown is a regenerative process that is not initiated unless certain high voltages and currents are coincident for a finite length of time. If the transistor is switched through the high-current high-voltage region fast enough, the heat is not sufficiently localized to result in permanent damage. Second breakdown is most likely to occur in circuits with inductive loads when the base-emitter junction is reverse biased and in high-speed transistors that have a thin, active base region.

Power-transistor second-breakdown characteristics are generally specified by safe operating area curves similar to those shown in Fig. 13.7. The solid curves in the figure indicate the collector current and voltage limits below which a germanium 2N1073A transistor will not go into second breakdown. As second breakdown is independent of temperature and the duty cycle, these curves may be used, provided the average power-dissipation rating is not exceeded. The average power rating depends on the heat sink in use, and this rating is indicated by the dashed lines in the figure.

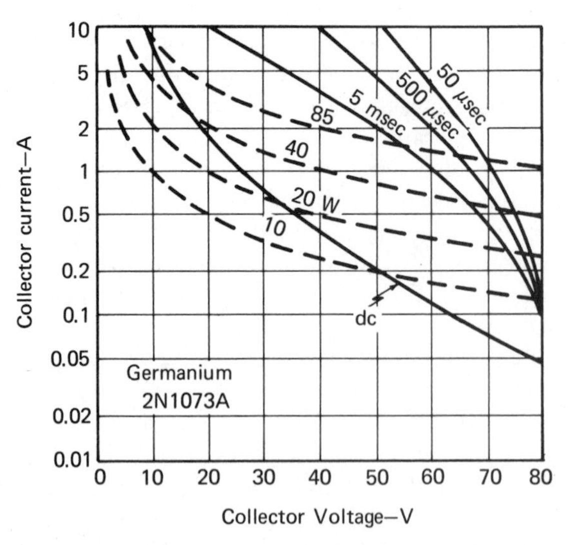

Figure 13.7. Safe operating area curves for power transistor.

The safe-area curves show the area for dc (Q-points) and for signal times of 5 msec, 0.5 msec, and 50 μsec. The curve for dc shows that with a 40-V

Q-point voltage the transistor is safe from breakdown only if the collector current is less than 0.4 A. This current is small compared with the 10-A collector rating, and the 16-W power input is small compared with the 85-W power rating of the device. The curves show that in switching applications higher peak currents and voltages may be used, provided the switching time is less than 1 msec. Thus, the danger of breakdown is considerably reduced by reducing the time the load line occupies the high volt-ampere region of the collector characteristics.

The region below the dc line is safe for all circuits in which the base-emitter diode remains forward biased under both transient and dc conditions. Reverse biasing of the base-emitter diode is generally to be avoided because the breakdown voltage of the emitter diode is very low, and safe-operating collector values are not generally defined for the reverse-biased emitter diode. The region near the collector-emitter breakdown voltage BV_{CEO} is safe only for relatively low currents, as in the OFF transistor in class-B operation or in low-power class-A operation.

Some reduction of the second breakdown hazard has been made by changing the internal design of power transistors, but in general any transistor can be destroyed by second breakdown. For a given type of transistor the safe area is larger for a high-voltage device and for a low f_T value. A transistor with a high thermal resistance may be limited by the transient temperature rise, which is produced by a high-current pulse signal before the signal is large enough to produce second breakdown. Thus, the various ratings must be carefully applied to ensure reliable operation.

13.13 POWER LOAD DIAGRAMS

Because power amplifiers are generally assumed to have a pure resistance load, the familiar load line extends from a current maximum with zero voltage to a voltage maximum with zero current. However, the reliability of a transistor depends more on the instantaneous volt-ampere product, which is generally a maximum when neither the power, the current, nor the voltage is a maximum. With reactive loads the instantaneous volt-ampere demand on the transistor may exceed by many times the collector input power.

Load diagrams that show the instantaneous collector current and voltage are commonly used to study the load. The diagrams are obtained as an oscilloscope display that presents the collector current on one axis and the collector voltage on the other. A voltage proportional to the current may be obtained by using a current probe or by inserting a small series resistor and picking off the voltage drop by a differential amplifier. The collector-to-emitter voltage is obtained similarly by a differential amplifier, to exclude the voltage drop in the emitter resistor.

With a resistance load the oscilloscope should display a straight diagonal

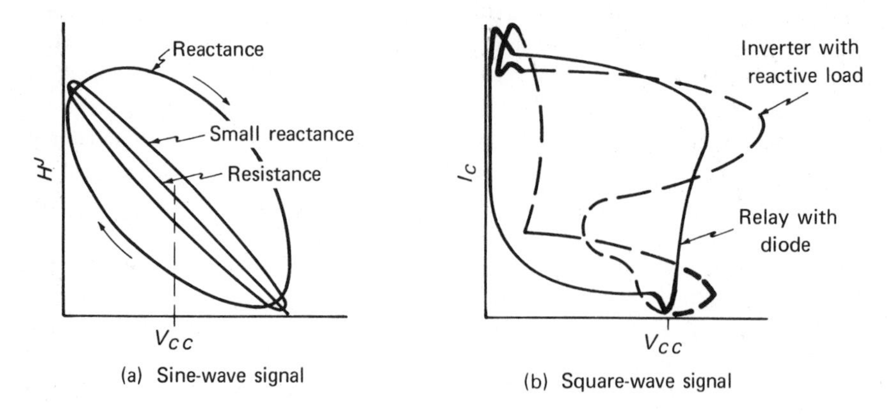

Figure 13.8. Collector load lines (dynamic).

line, as shown in Fig. 13.8a, similar to load lines drawn on a set of collector characteristics. However, even a well-built output transformer has enough inductance and capacitance to change the load diagram with a resistance load at either high or low frequencies. These changes always present the possibility of transistor breakdown.

The load diagram of a resistance load having a small reactance component is an ellipse, as shown in Fig. 13.8a, when the signal is a single frequency sine wave. If the output signal is distorted, either by saturation in the transformer or by an overdriven transistor, the load diagram tends to take on a very irregular shape and may even cross over itself. The load diagram of an inverter with a resistance load appears approximately as shown in Fig. 13.8b. With a square-wave drive and a reactance load, the load diagram may be so irregular and complicated as to defy interpretation, as in Fig. 13.8b.

The study of load diagrams shows that, even under relatively favorable conditions, the instantaneous volt-ampere load may constitute a severe test of a transistor's ability to resist second breakdown. A difficulty in the use of the load diagram is that the usual oscilloscope presentation does not show how long a given pulse lasts. Hence, it is difficult to correlate the load diagram with a second-breakdown chart of the transistor. However, load diagrams are helpful in appraising the effects of changes made to reduce the possibility of second breakdown.

Load-line excursions into the second-breakdown area may be reduced by any device that limits the peak collector voltage. With an output transformer and a resistance load, as in an audio amplifier, the danger of second breakdown is greater when the amplifier is driven at low frequencies with the load removed. Protection is obtained by shunting the primary with a Thyrite varistor, by limiting the drive voltage, or by reducing the collector supply voltage.

Transistors that are used to switch reactive loads, such as a power solenoid, may be protected by shunting the load with a diode or the transistor with a Zener diode. These diodes limit the peak voltage excursion and make the load diagram approximately rectangular (Fig. 13.8b).

Perhaps the most critical operation for power transistors is in a switching inverter in which the signal is a square wave and the load is a reactance or a motor. Voltage spikes caused by the primary leakage inductance may be reduced by reducing the leakage inductance, reducing the amount of drive, or by connecting Zener diodes across the transistors. The "ballooning" of the load diagram caused by the reactive load may be reduced by shunting the primary with a capacitor or by reducing the collector supply voltage. A capacitor load has the disadvantage that the inverter switches more slowly and turns on a higher current with increased dissipation. Generally, a self-excited inverter may be operated satisfactorily if the transistor BV_{CEO} rating is 2.5 times the collector supply voltage. For rapid switching, the transistor f_T rating should be 1000 times the inverter frequency.

Inverters that are required to supply large loads usually have two transformers. The input transformer is allowed to saturate while the output transformer operates in the linear mode. With linear operation the leakage inductance is reduced. Thus, the transient voltages generated by switching are reduced. The reliability of a transistor inverter depends mainly on the quality of the transformers and the care used to ensure that an unexpected load condition does not overstress the output transistors.

13.14 DESIGN CONSIDERATIONS

The breakdown characteristics of transistors are so varied and depend so much on the signal and the source that most practical designers use considerable overdesign to ensure a reliable circuit. Unless one makes a specialty of breakdown problems, a conservative design is perhaps the best practical procedure. A circuit intended for quantity production may be designed with the aid of an application-engineer representative for the transistor manufacturer.

Since the breakdown regions of transistors are fairly closely defined for common applications, the less skilled designer can produce credible designs by adding a small safety factor and making very careful performance tests. A resistance-coupled amplifier can be safely designed with the collector supply voltage as high as the BV_{CEO} rating, because the class-A Q-point is normally well below the supply voltage, and the collector current is low whenever the collector voltage is high, as at cutoff. When the Q-point dissipation is less than the transistor power rating, there is rarely any problem with the second breakdown in a class-A amplifier.

Power amplifiers with output transformers, either single sided or push-pull, are conservatively designed if the transistor BV_{CEO} rating is two to two and one half times the supply voltage, provided the collector load is a resistance and the load is always in place. With the resistance load the peak collector voltage approaches twice the supply voltage when the collector current is a minimum. However, if the load is accidentally removed or short circuited, the transistors may be damaged either by transient voltage spikes with the open circuit or by second breakdown with the short circuit. The spikes can be limited by Zener diodes, and short-circuit protection is generally shown for power circuits if necessary.

The greatest danger of breakdown generally occurs in circuits having an iron-core relay or solenoid as the load. Often, the peak voltage that occurs when the transistor is switched OFF can be limited to a reasonable value by shunting the reactance with a resistor or a diode and resistor.

The small physical dimensions of high-frequency transistors make these devices significantly different from low-frequency transistors. RF transistors have lower breakdown ratings because the junctions are relatively thin. Therefore, the operating Q-points are usually nearer the maximum voltage and power ratings of the transistors. The low voltage ratings make RF transistors sensitive to damage by transients or by high currents produced in testing a circuit. RF circuits should be protected by R-C filters in the power supply to prevent switching transients when the dc power is turned ON or OFF.

13.15 TRANSISTOR POWER DISSIPATION

The power converted to heat within a transistor must be removed fast enough to keep the junction temperature below an upper limit fixed by the desired transistor life and reliability. The power supplied to the transistor is the collector power $V_C I_C$ plus the base input power, which is between 1 and 10 per cent of the collector power. However, the collector power is not all converted to heat, as in Class-B service. Therefore, for the thermal calculations below we assume for simplification that the $V_C I_C$ product is always corrected to represent only the power that is actually converted to heat.

The heat produced within a transistor accumulates until there are temperature differences along which heat flows to the outside. When equilibrium is reached, the total temperature fall from the collector junction to the outside ambient temperature is high enough to remove the heat. The designer is primarily concerned with the junction temperature existing under thermal equilibrium. For commercially accepted reliability, germanium junctions are operated between 90 and 110°C, and silicon junctions are used between 150 and 200°C.

The junction temperature used as a design objective depends on many factors, which include costs, the kind of application, previous experience, and reliability requirements. As always, there is no real substitute for experience and tests in deciding what temperature to use, but for a starting point one is safer in using the lower of the two values. As a design is completed, competition may force an upward revision of the temperature, and reliability may require a lower temperature. The initial choice is never very important.

When reliability is desired, a good practical rule is that the life of most electrical components and transistors is doubled by each 8°C decrease of the operating temperature. The rule neglects effects like a boiling or melting point but is considered a useful guide in the design of electronic and power equipment.

13.16 THE HEAT-TRANSFER EQUATION

The power dissipated within a transistor eventually reaches the surrounding environment, which has an essentially fixed temperature. Inside the transistor case the heat is transferred mainly by conduction in metals. From the case to the ambient the heat is transferred by combined radiation and convection with, perhaps, some conduction through solids.

Because the temperature differences in the thermal circuit are less than 100°C, the heat transferred by a given temperature drop may be calculated by Newton's law of cooling; that is,

$$T - T_A = \Theta P \tag{13.2}$$

Newton's law, like Ohm's law, $E = RI$, states that the temperature difference equals the product of the thermal resistance Θ and the heat power P that is moved. The temperatures are in °C, the thermal resistance is in °C/watt, and the thermal power is in watts. Usually, the thermal power is known, the thermal resistances are known or calculated, and the resulting temperature drop is the product ΘP. For a power transistor with an input power $V_C I_C$, the temperature difference between the junction and ambient is

$$T_j - T_A = \Theta V_C I_C \tag{13.3}$$

The total thermal resistance between a junction and the ambient is usually a series of three resistances, as represented in Fig. 13.9. The resistance Θ_{TR} between the junction and the mounting flange is given by the manufacturer. The insulating washer between a power transistor and the heat sink usually has a thermal resistance Θ_W of 0.5 to 1°C/W. The lower value is obtained by using a lightly greased washer, the thermal resistance of which is often negligible. Typical resistance values for insulating washers are given in Appendix A.3.

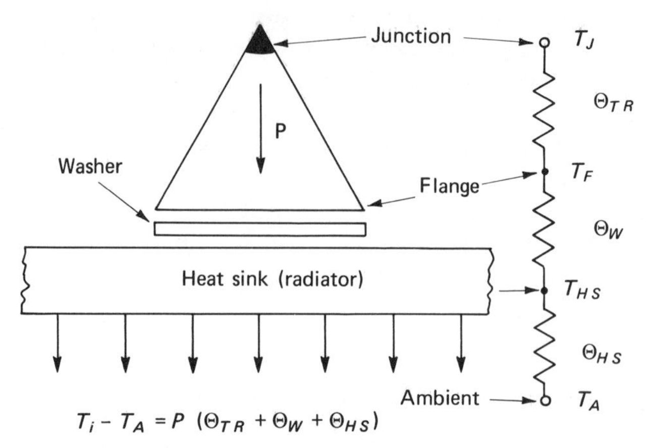

Figure 13.9. Transistor thermal resistances and temperatures.

13.17. THERMAL RESISTANCE OF HEAT SINKS

The heat sink generally used with power transistors is a flat plate or a finned heat sink. The thermal resistance Θ_{HS} of a heat sink in still air may be easily calculated, or is given by the manufacturer of a finned type. The thermal resistance of a flat plate is usually between 500 and 1200°C/watt for each square centimeter of exposed surface. While the 2.4:1 range of thermal resistance may seem large for practical use, the experienced designer can select a value that is within an accuracy of 20 per cent.

The thermal resistance of a flat-sheet radiator is essentially the same for each square centimeter of the surface. For a small sheet, 5 by 5 cm, the thermal resistance is a low 500°C/W/cm^2. With one side of the sheet radiating, there are 25 units of area radiating in parallel, so the net thermal resistance is 20°C/W. If both sides are exposed, the thermal resistance is only 10°C/W. If the sheet is large, say 25 by 25 cm, the thermal resistance per unit area is higher, perhaps 1200°C/W/cm^2, but the thermal resistance is only 2°C/W.

Radiating fins on a heat sink approximately double the effective area of the surface on which the fins are mounted. In still air the fins make only a small improvement, which may be equalled by bending a larger flat sheet into three sides of a box. Where there is insufficient space for an adequate heat sink, forced air can be used to reduce the thermal resistance to one half and, perhaps, to one tenth of the free convection thermal resistance.

A practical rule for calculating thermal resistance is satisfactory for many purposes. However, the thermal resistance varies with the size, shape, and color of a radiator, and a heat sink should always be evaluated by measuring the actual temperature rise in situ. The manufacturer always gives the total thermal resistance of small transistors from junction to ambient. The junction-to-case resistance, which is sometimes given, may be helpful when a small power transistor is used with a relatively small clip-on heat sink.

13.18 HEAT-SINK DESIGN– TWO EXAMPLES

Generally, the thermal resistance of the heat sink is the main loss in the thermal series and is, therefore, important in controlling the junction temperature and the transistor reliability. A heat sink must be as large as practical to obtain high power from a given transistor and must be small when designing compact equipment. A compromise is usually required in selecting the size and temperature of a heat sink. A good beginning compromise is obtained by assuming that the heat-sink temperature is about halfway between the junction temperature and the highest expected ambient. With a 40°C ambient this compromise makes the heat-sink temperature about 70°C for germanium transistors and 100°C for silicon transistors. These temperatures usually produce junction temperatures that are well below the rated maximums and definitely limit the maximum power obtainable from the transistor. However, the suggested compromise is useful for estimating the dimensions of a heat sink.

The first example of heat-sink design shows that the power obtainable from a given transistor may be greatly limited by the space available for a large heat sink. The second example shows that a small heat sink may be very helpful and that considerable error may be tolerated in estimating the thermal resistances.

Consider a 2N1073 germanium transistor that is rated capable of dissipating 85 W with a 25°C flange temperature. The 25°C flange temperature is usually impractical without refrigeration because equipment temperatures occasionally reach 40°C. Suppose that we assume a 40°C ambient and use the compromise heat-sink temperature of 70°C with a junction temperature of 100°C, as shown in Fig. 13.10. The thermal resistance of the transistor and washer may total 2°C/W. The heat-sink temperature is assumed to be halfway between the junction and the ambient; hence, the heat-sink thermal resistance must also be 2°C/W. As a first approximation, a thermal resistance of 800°C/W for each square centimeter of radiating surface implies that the heat sink must have an area of 400 cm^2. If we use both sides, a square heat sink must be at least 14 cm on a side. With a total thermal resistance of 4°C/W, the transistor, by equation (13.2), may be allowed to dissipate only 15 W.

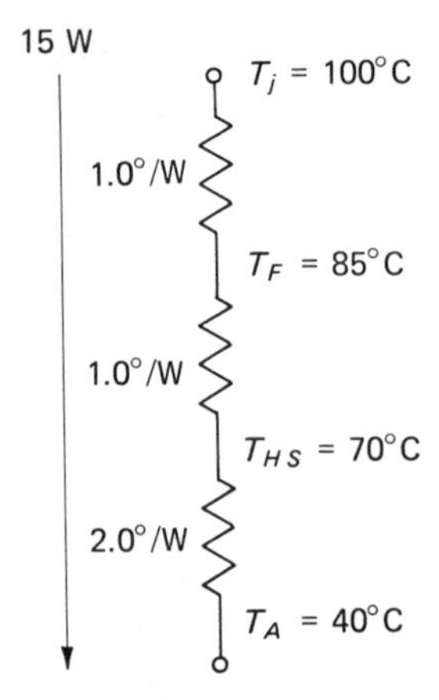

Figure 13.10. Thermal example with 70°C heat sink.

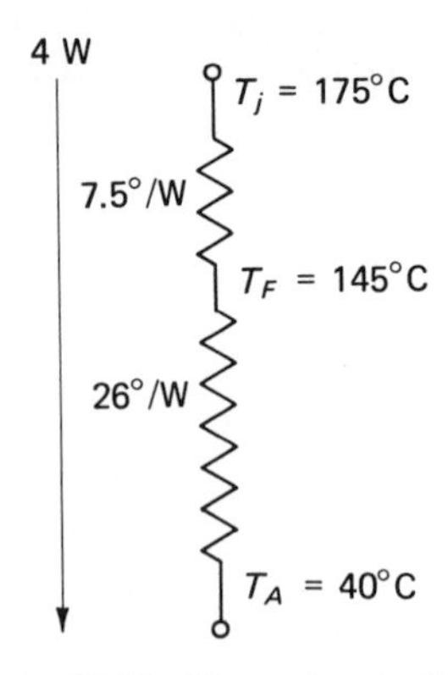

Figure 13.11. Thermal example with 175°C junction temperature.

The heat-sink estimate shows the designer that, even with a large flat heat sink, the transistor may dissipate only 15 W, a value far below the manufacturer's rating. Moreover, we show in Section 13.21 that the 2N1073 probably cannot dissipate more than 10 W with free-air cooling because of thermal runaway. If the design requires 15-W dissipation, there is probably little choice but to use forced air or liquid cooling.

With these heat-sink estimates at hand, the designer can proceed with the design only by making a difficult choice. The higher power can be dissipated only by using forced-air cooling. A small heat sink in still air can only be used by considerably reducing the power that must be dissipated. These decisions are often the most difficult step in a design. Once the decisions are made, the procedure is simple. The approximations can be replaced by careful calculations, and the design may be experimentally tested.

As the second example, consider the problem of cooling the 2N3739 transistor used in the 1-W power amplifier described in Section 7.11. The manufacturer's data suggest 175°C as a maximum junction temperature, and gives 7.5°C/W as the junction-to-flange thermal resistance. Because the transistor is required to dissipate 4 W, the junction-to-flange temperature difference is 30°C, as shown in Fig. 13.11.

If the highest expected ambient temperature is 40°C, the temperature difference available to move heat from the flange to the ambient is $175 - 30 - 40 = 105°\text{C}$, with a flange temperature of 145°C. A heat sink that dissipates 4 W with a 105°C temperature difference has a thermal resistance of $\Theta = 105°/4\ \text{W} = 26°\text{C/W}$. This estimate shows that the transistor may be operated for a short time without a heat sink, since the thermal resistance from the case to free air is 60°C/W. The heat sink for continuous operation need only to double the radiating area of the transistor itself.

Suppose, however, that the designer wishes to reduce the Q-point temperature drift by reducing the heat-sink temperature to 80°C, which is 40°C above the 40°C ambient temperature, as in Fig. 13.12. The heat-sink should have a thermal resistance of Θ = 40°C/4 W = 10°C/W, and the junction temperature is reduced to 40°C + 30°C = 70°C above the ambient temperature. This calculation shows that the heat sink may be 5 by 5 cm, using both sides or radiating fins on one side, or 7 by 7 cm using one side of a flat sheet.

With the small heat sink the thermal resistance of the insulating washer can be neglected. With the larger heat sink the use of an insulating washer may make the use of a slightly larger heat sink necessary. In the first example the total heat-sink resistance is only 2°C/W, and the washer, if used, accounts for 25 to 50 per cent of the total.

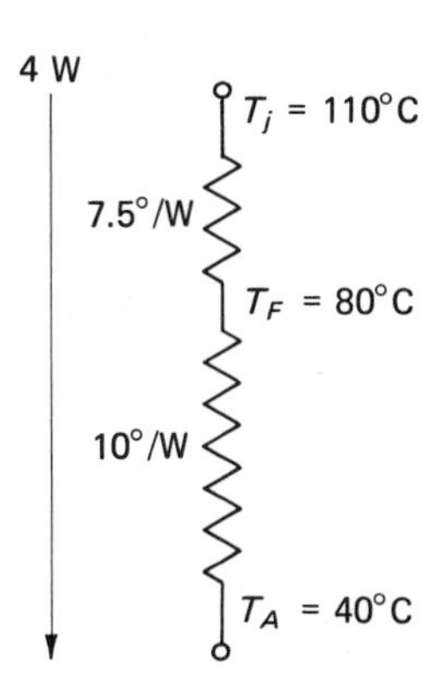

Figure 13.12. Thermal example with 108°C junction temperature.

13.19 THERMAL RUNAWAY

When a transistor is operated with a high junction temperature, the collector current and the temperature may suddenly increase, and the transistor is said to be in a state of *thermal runaway*. A transistor may operate for days with a perfectly stable junction temperature until an unusual overload or deterioration increases the junction leakage current and initiates thermal runaway. Once started, thermal runaway usually produces such a high junction temperature that the transistor is temporarily inoperative or is permanently damaged.

Thermal runaway is caused by a regenerative thermal feedback. All transistors have a collector-to-base leakage current I_{CO} that adds to the base bias current and increases the bias at high junction temperatures. The low-voltage I_{CO} value given for germanium transistors is a temperature-sensitive component of the leakage current, which doubles for each 10°C increase of the junction temperature. With high junction temperatures the exponential increase of I_{CO} causes enough thermal feedback to produce runaway.

Ordinarily, silicon transistors are free of thermal runaway, and the high-temperature, high-voltage I_{CO} given for silicon transistors is mainly a voltage-sensitive component of leakage current. Because the voltage-sensitive component of leakage current does not cause runaway, the I_{CO} given for silicon power devices is a measure of the transistor quality and is used only to estimate the Q-point shift at high temperatures and with moderate collector voltages.

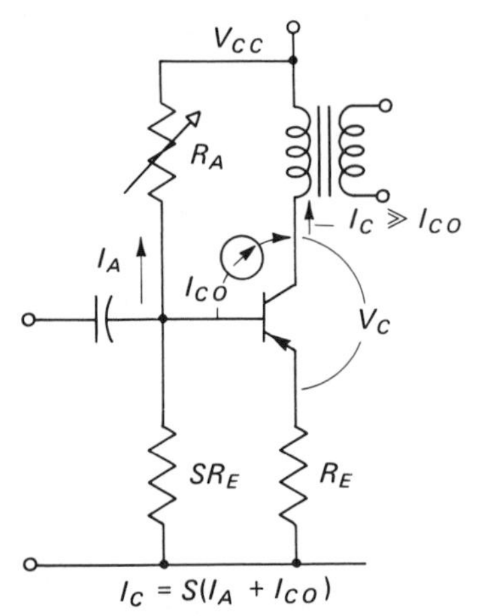

Figure 13.13. Power transistor bias with I_{CO}.

A circuit typical of power-transistor applications is represented in Fig. 13.13. The collector-emitter Q-point voltage is V_C, the low-temperature bias current is I_A, and the dc S-factor is S. The collector-to-base leakage current is shown as a constant current source I_{CO}, which may normally be neglected with low junction temperatures.

The effect of the leakage current in shifting the Q-point at high temperatures is calculated simply by adding I_{CO} to the bias current, and the high-temperature collector current is

$$I_C = S\,(I_A + I_{CO}) \tag{13.4}$$

This equation shows that the Q-point change caused by I_{CO} is reduced by reducing the S-factor, by making the bias current large, or by selecting a transistor with a small I_{CO}. Whether the transistor is susceptible to runaway is not indicated by the relative magnitude of I_A and I_{CO}, because runaway depends on the thermal-feedback rate and is not a Q-point drift problem.

13.20 CONTROL OF THERMAL RUNAWAY

Normally, any change in the collector current that increases the heat input to a transistor is offset by an increase of the junction temperature that is enough to move the increased heat away from the junction. If the heat input doubles, the junction-to-ambient temperature difference doubles, and the new temperature is a stable temperature. If the new value of I_{CO} causes an additional power input, the junction temperature must increase even more. Because the collector cutoff current increases exponentially with temperature, while the heat transfer away from the junction increases linearly with temperature, there is a value of I_{CO} above which there is thermal runaway. The thermal feedback loop is unstable when the power input, $SI_{CO}V_C$, exceeds the heat that is moved away from the junction by a 5°C junction temperature rise. By equation (13.2), the additional heat power moved by a 5°C temperature rise is 5°C/Θ. Therefore, the junction temperature is stable if

$$\Theta V_C S I_{CO} \leqslant 5^\circ\mathrm{C} \tag{13.5}$$

With this equation and the given values of Θ, V_C, and S, the designer may calculate the value of I_{CO} that produces thermal instability. The temperature that produces the calculated I_{CO} is found and compared with the junction temperature calculated by equation (13.3). If the junction temperature is below the temperature of instability, the design is stable. Although the temperature rise produced by I_{CO} is neglected, the stability test is conservative and may be used without an additional safety factor for several reasons. At high temperatures the rate of increase of I_{CO} tends to be less than the rate assumed in deriving equation (13.5). The I_{CO} value in the data sheets is a maximum, and some selection of the power transistors may be permitted.

The value of I_{CO} for germanium transistors is usually given as a maximum value of I_{CO} at 25°C, and values at higher temperatures may be obtained from Fig. 13.14. Since I_{CO} doubles for each 10°C increase of junction temperature, we may assume that I_{CO} is 8 times larger at 55°C, 64 times larger at 85°C, and about 100 times at 90°C.

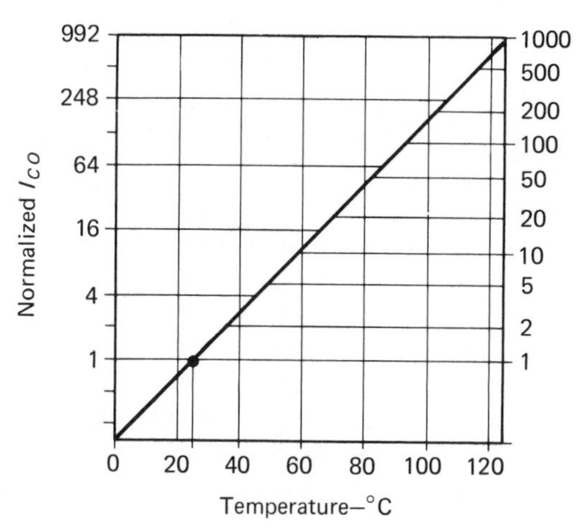

Figure 13.14. Normalized I_{CO}-temperature curve.

13.21 A THERMAL-STABILITY EXAMPLE

As an example of a thermal-stability calculation, consider the 2N1073 germanium transistor discussed in Section 13.18. The thermal I_{CO} is given as 0.3 mA at 25°C with $V_C = 2$ V. Assume that the transistor is used in a power amplifier with $V_C = 25$ V, $S = 5$, and the total thermal resistance from the junction to ambient is 4°C/W. The maximum junction temperature for thermal stability is found as follows:

Substituting the given values in equation (13.5) and solving for I_{CO}, we find

$$4(25)5\, I_{CO} = 5 \qquad \textbf{(13.6)}$$

and

$$I_{CO} = 0.010\text{A} \tag{13.7}$$

The limiting value of I_{CO} is only 32 times the value at 25°C. Therefore, runaway may be expected when the temperature rise doubles I_{CO} five times. Five 10°C increases make the amplifier unstable with a junction temperature that is only $25 + 50 = 75$°C. If the maximum ambient temperature is 40°C, and the total thermal resistance is 4°/CW, the transistor can dissipate only $(75 - 40)/4 = 8.7$ W and remain thermally stable. This power rating is only one tenth the manufacturer's rating.

Suppose that the designer wishes to use the transistor with a higher junction temperature so that higher output power may be obtained. The junction temperature can be increased 10°C by a 2:1 reduction of any factor in equation (13.5). The thermal resistance Θ can be lowered to about 3°C/W, the collector voltage is usually fixed by other considerations, and an S-factor below 5 severely reduces the power gain. Except with forced air or conduction cooling, the 2N1073 is limited to a continuous power dissipation of about 10 W.

The previous example illustrated the influence of the collector cutoff current on a design and shows also a disadvantage of overdesign. The selection of a transistor with an 85-W power rating has actually reduced the power that can be dissipated by contributing a high-value I_{CO}. A germanium 2N2869 rated for only 11 W with an 80°C flange temperature has an I_{CO} of 0.1 mA at 25°C. This 3:1 reduction of I_{CO} allows an additional 16°C advantage that may be used to increase the junction temperature from 70 to 86°C for a heat transfer of $46°\text{C}/4°\text{C/W} = 11.5$ W. Since this power transfer gives a 63°C flange temperature, the transistor is operated well within the power rating.

13.22 SUMMARY

The smallest signal that can be easily observed with an amplifier is approximately equal to the random noise generated by the resistors and transistors of the amplifier. The resistance noise power is proportional to the absolute temperature, the resistance, and the bandwidth. The fundamental nature of resistance noise makes it a useful and precise reference for comparing noise. Transistor noise—a combination of thermal, $1/f$, shot, and cutoff noise—is generally compared to the theoretical thermal noise by a spot noise or a wide-band noise figure. Noise figures are noise-power ratios expressed in decibels.

A wide-band audio amplifier may be designed with a broad-band noise figure as low as 2 dB. A video amplifier may have a 3-dB, or higher, noise figure. A UHF broad-band amplifier may have a 3- to 10-dB noise figure. Low noise is obtained in transistor amplifiers by using a low-value emitter current, a narrow band, an optimum source impedance, and a transistor chosen for low noise in the frequency range.

The breakdown ratings of a transistor designate voltage and circuit conditions at which the collector current may increase abnormally, especially with high junction temperatures. The increased current may cause faulty circuit operation or permanently damage the semiconductors. The voltage breakdown ratings are generally given for operation with low collector currents, which may be unrealistic in power-amplifier and switching applications. Second breakdown is caused by the simultaneous existence of a high collector current and voltage that are sustained long enough to create a permanent junction-to-emitter short circuit. Second breakdown is avoided by limiting the collector current and voltage to the safe area of a second-breakdown chart. Failure by second breakdown is most likely to occur in switching amplifiers with motor or reactive loads.

The power converted to heat in a transistor must be transmitted to the room ambient. Otherwise, the junction materials are damaged by the high temperature. The temperature difference produced by heat flow equals the product of the thermal resistance and the power transmitted as heat. The thermal resistance of transistors is generally given by the manufacturer. The thermal resistance of a heat sink in free air is approximately 500 to 2100°C/W for each square centimeter of the radiating area. Small areas of about 25 cm^2 have the lower thermal resistance, and large areas of about 600 cm^2 have the higher thermal resistance per unit area. Thermal fins approximately halve the thermal resistance in still air and are two to four times more effective in moving air.

Reliable operation of transistors may be obtained by operating transistors with adequate protection from breakdown and by limiting the junction temperature rise. Runaway, a problem with germanium power transistors, is caused by regenerative heating of the junction. Runaway may be prevented by ensuring that the temperature rise produced by the temperature-sensitive component of I_{CO} does not increase the junction temperature rise by more than 5°C. The temperature rise may be reduced by reducing the power input, $SI_{CO}V_C$, and by reducing the thermal resistance between the junction and the ambient.

PROBLEMS

13-1. Refer to the low-noise amplifier shown in Fig. 13.5 and assume that the upper cutoff frequency is at 12 kHz. (a) What is the rms-equivalent resistance-noise voltage of the 5-kΩ signal source? (b) If the amplifier has a 3-dB noise figure, what is the equivalent input-noise voltage? (c) If the amplifier has a voltage gain of 150, what is the approximate peak-to-peak noise voltage observed at the collector terminal?

13-2. Assume that the transistor represented by Fig. 13.3 is to be operated with 0.1-mA emitter current. (a) What is the smallest signal that can be observed with a 100-Ω source resistance? (b) If the 100-Ω source is coupled by an input transformer that presents 5 kΩ to the transistor, what is the smallest signal that can be observed from the same 100-Ω source?

13-3. A transistor with a thermal resistance of 1.0°C/W is mounted on a flat plate 8 by 14 cm. The plate is 4 mm thick, which makes the temperature drop in the plate negligible. (a) What is the thermal resistance from the plate to ambient, assuming radiation and convection from both sides of the plate? (b) What is the junction temperature rise above ambient when the transistor is dissipating 11 W? (c) What is the highest permissible ambient temperature if the transistor is a germanium type?

13-4. A germanium transistor is to be operated with $V_C = 12$ V, $I_C = 0.92$ A, $S = 20$, and a 90°C junction temperature in a 65°C ambient. (a) What is the required junction-to-ambient thermal resistance? (b) If I_{CBO} is given as 0.1 mA at 25°C, is the design thermally stable? (c) How large a square flat-plate heat sink would you recommend?

13-5. Assume that a push-pull class-A hi-fi amplifier is to be operated with 10-W output and use the safe-area curves shown in Fig. 13.7. (a) What is the maximum collector voltage that can be safely used? (b) Is it safe to use a higher collector voltage if the signal is a 200-Hz sine wave? (c) What is the maximum safe collector voltage if a transformerless amplifier is operated class B with a reactive low-frequency load?

DESIGNS

13-1. Select a germanium transistor that will dissipate at least 15 W. Mount the transistor on an aluminum plate that is 8 by 12 cm and bias the transistor to make it dissipate 10 W with a 12-V collector supply. Use a dc S-factor of about 5 so that there is no problem of runaway. Measure the sink temperature rise and calculate the thermal resistance per unit area. The temperature is best measured with a thermocouple-type thermometer, but a small thermometer will give reasonable accuracy if mounted in close thermal contact.

13-2. Using the heat sink and transistor described in Design 13-1, increase the S-factor and the collector supply voltage until runaway is observed. Confirm the validity of the runaway equation. The room temperature value of I_{CBO} is easily measured with the base open, using a 2- or 3-V collector supply. The transistor may be protected from runaway damage by connecting a power resistor in series with the collector supply. Use a resistor value that reduces the collector voltage to about 80 per cent of the supply voltage just before runaway begins.

NOISE REFERENCES

13-1. Miller, J. R., ed., *Communications Handbook*, Part II. Dallas, Tex.: Texas Instruments, Inc., 1965.

13-2. Cowles, L. G., *Analysis and Design of Transistor Circuits.* New York: Van Nostrand Reinhold Company, 1966.

THERMAL AND BREAKDOWN REFERENCES

13-3. RCA Silicon Power Circuits Manual, SP-51. Harrison, N. J.: Radio Corporation of America, 1969.

13-4. *Motorola Power Transistor Handbook,* 4th ed. Phoenix, Ariz.: Motorola Semiconductor Products Division, Inc., 1969.

13-5. *GE Transistor Manual,* 7th ed., pp. 32-42, 107-110, 144-145, 458, 516-532. Electronics Park, Syracuse, N. Y.: The General Electric Co., Inc., 1964.

14

DIODES AND MICROWAVES

Although best known for its use at low frequencies, the diode has always been in the vanguard of progress at high frequencies and is perhaps the most versatile of all semiconductors. Even before the vacuum tube, the galena detector made radio communication a practical reality. In 1948 a pair of coupled diodes became the point-contact transistor. Today, at microwave and millimeter-wave frequencies, the diode is the oscillator, the amplifier, and the detector. At UHF and microwave frequencies, the junction diode is a frequency multiplier, a voltage-controlled capacitor, a mixer, or a voltage-controlled resistor. The two-terminal diode has a surprising versatility and many interesting applications.

The last decade has seen great progress in the development of new diodes and in the application of varactors, but the data needed for design are widely scattered and often concealed in papers written for technical specialists. The purpose of this chapter is to outline the high-frequency characteristics and applications of diodes so that the reader can undertake simple designs for VHF and UHF applications. Some of the material is applicable above 1 GHz, which is a rapidly developing field for semiconductor circuit design and applications.

14.1 DIODES

Most diodes are *pn* junctions of germanium, silicon, or gallium arsenide that are made by different processes, depending on the characteristics desired. For microwave frequencies a metal semiconductor contact is used in the Schottky-barrier diode, and a "cat whisker" of metal is used in the point-contact diodes, which have similar characteristics. Alloy junctions are used to make high-current rectifiers, tunnel diodes, and varactor diodes. Planar diffused junctions are used to make fast computer diodes, high-voltage low-leakage rectifiers, and step-recovery frequency multipliers. Three microwave devices, which are not used as diodes in the ordinary sense, should be considered separately: The *PIN* diode is formed as a *p*-intrinsic-*n*, diffused-junction, silicon "diode," but is operated forward biased as a voltage-variable resistor. The avalanche (IMPATT) diodes are biased in the reverse breakdown region to produce high-frequency signal power or noise. The Gunn-effect diodes are cubes of semiconductor material that produce microwave oscillations when biased at relatively low dc voltages. The Gunn devices are called transferred electron devices or diodes because they have a preferred polarity and are damaged when subjected to a reverse polarity.

For the designer the characteristics of the various devices as circuit elements are of first importance. Hence, they are discussed here as detectors, mixers, oscillators, varactors, etc. Moreover, because diodes are particularly important at microwave and millimeter frequencies, emphasis is on UHF and microwave applications. The reader may easily find similar applications at VHF and lower frequencies.

The most common uses of signal diodes are for RF current and power measurements, for detection of modulated signals, as in a radio or a TV receiver, and for converting high-frequency signals to lower frequencies, as in a mixer. These applications of diodes use the diode as a high-frequency rectifier, and the low-frequency or dc components of the diode current are the desired low-frequency signal.

14.2 STORED CHARGE AND JUNCTION DIODES

The nonlinearity that makes a diode so useful for detection and frequency conversion is seriously impaired at high frequencies by stored charge. When the voltage applied to a junction diode is changed from forward to reverse polarity, a current flows in the reverse direction for a significant time, which is called the reverse-recovery time t_{rr}. Thus, when the half period of the input signal approaches the reverse-recovery time, the rectification efficiency falls off until the diode is useless as a rectifier. Point-contact and Schottky diodes are better high-frequency detectors than junction diodes because the former are free of charge storage in the semiconductor. However, when the input frequency is so high that the diode capacitance acts as a bypass, the rectification efficiency decreases with frequency, thus setting an upper frequency limit for efficient detection by even a point-contact diode.

The problems of using diodes at high frequencies may be understood by constructing the circuit shown in Fig. 14.1 and examining the reverse-current characteristics of typical diodes. The circuit has a signal source with a general-purpose, junction diode D, a load resistance R_L, and a low resistance R_I across which an oscilloscope is connected for observing the current wave forms. The oscilloscope should have a low input capacitance and a flat frequency response to at least 1 MHz. Also, to keep the oscilloscope input capacitance from affecting the current wave form, the resistors R_L and R_I should be changed together so that R_L is always 10 times R_I. The signal source is adjusted to supply an open-circuit voltage of about 10 V rms.

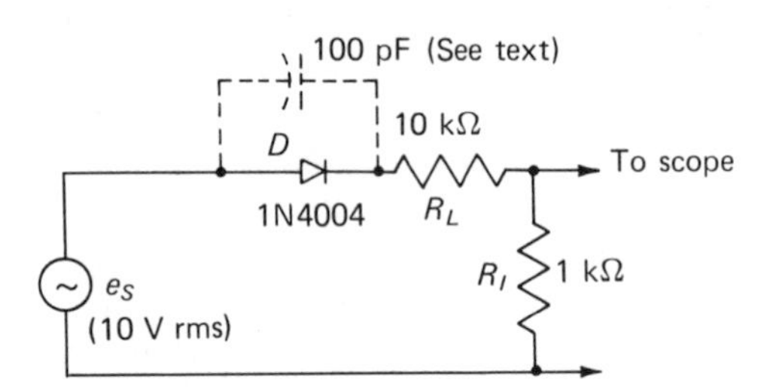

Figure 14.1. Circuit for diode current test.

With low frequencies and a 10-kΩ load the diode current has the familiar wave form of a half-wave rectifier, as shown in Fig. 14.2. As the frequency is increased, a current pulse appears in the reverse-current direction, and the oscilloscope trace may show a slight upward slope, which is caused by the diode capacitance. Shunting the diode by a 100-pF capacitor will increase the slope and permit an estimate of the diode capacitance. The reverse-current

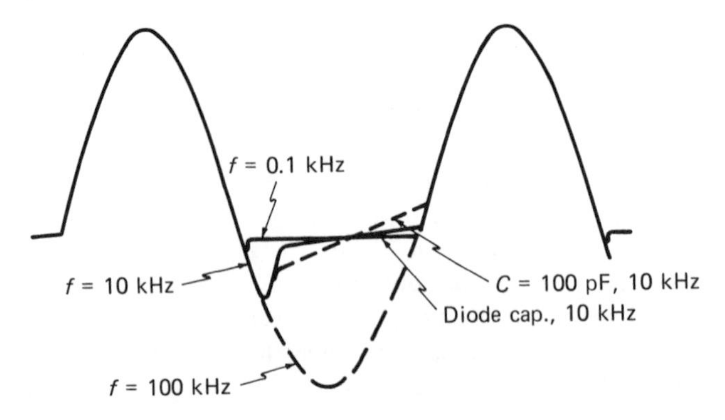

Figure 14.2. Junction-diode current wave forms.

pulse is produced by stored charge, which in junction diodes is usually a more severe high-frequency limitation than diode capacitance.

With a 10-kHz input frequency the diode conducts during the OFF time for approximately 10 μsec and has a peak reverse current of about one tenth the peak forward current. The wave form shows that when the input voltage reverses to turn the diode OFF, the diode at first remains ON, since the current rises rapidly in the direction of reversed current. This phenomenon of conduction, when the diode should be OFF, is caused by minority carriers, holes, which are stored in the junction during forward conduction. The stored charge makes the diode conduct until the reverse voltage returns the charge as a current pulse. The time that the diode conducts this relatively high reverse current is the reverse-recovery time t_{rr}.

By decreasing the resistors R_L and R_I downward by factors of 10, the reader may observe that the reverse-recovery time is relatively independent of the circuit impedance. However, as the input frequency is increased to 100 kHz, the recovery time occupies an increasingly longer part of the half-cycle during which the diode should be OFF; and the peak of the reverse current increases until the current wave form is approximately sinusoidal. Moreover, the reader may confirm that the rectification efficiency, as indicated by the dc voltage across R_L is reduced to 50 per cent when the reverse-recovery time just fills the lower half-cycle.

Since a diode conducts until the stored charge is removed, the rectification is impaired when the reverse-recovery time equals the half-period of the input signal. Thus, the frequency at which the rectification efficiency is 50 per cent may be estimated by the equation

$$f_c = \frac{1}{4t_{rr}} \tag{14.1}$$

The charge storage times and equivalent cutoff frequencies of typical diodes are given in Table VI. These frequencies may be used as guideposts in applications, but the calculated values at times may deviate from the cutoff values observed in some circuits. The reverse-recovery time depends on the magnitude and wave form of the applied signal. Thus, the rectification efficiency of a diode is determined in a complicated way by the circuit and the signal.

TABLE VI

Charge Storage Times and Equivalent Cutoff Frequencies

Diode Type	Recovery Time (μsec)	Frequency Estimate (MHz)
Rectifiers, 1 to 10 A	3. - 10.	0.05 - 0.2
General-purpose planar (1N459)	0.25 - 1.	0.5 - 2.
Ultrafast planar (1N4610)	0.001 - 0.02	25. - 500.

14.3 MICROWAVE DIODES

The diodes used for detection and measurements at microwave and millimeter frequencies have always been the point-contact crystal diodes. Some microwave diodes are rated for use at 30 GHz, which is a factor of 100 higher than the useful frequency range of junction diodes. Point-contact and tunnel diodes utilize electrons for conduction and thus are not limited by charge storage. Schottky diodes, which employ a similar metal-semiconductor junction and use electrons as the carrier, are called *hot-carrier* diodes. These devices, like point-contact and tunnel diodes, are limited in frequency mainly by the connecting leads, the protecting package, and the capacitance of the contact. Schottky diodes have the high breakdown voltage and temperature characteristics of silicon, nearly the low turn-on voltage of germanium, and the speed of a majority carrier device. Since they are replacing point-contact diodes in many high-frequency applications, a comparison of the static characteristics is shown in Fig. 14.3.

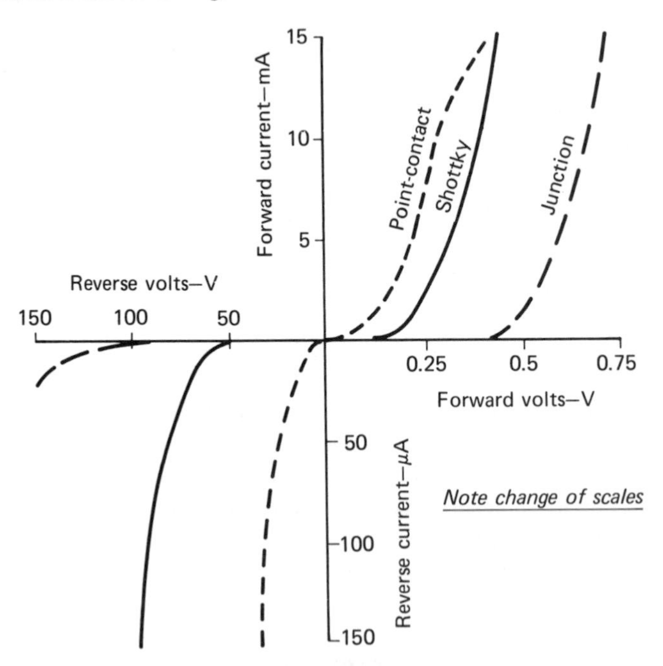

Figure 14.3. Silicon-diode current-voltage characteristics.

Germanium diodes are inexpensive and useful at UHF frequencies, but have a high reverse leakage and are temperature sensitive. An important advantage of germanium diodes is their low forward voltage drop and their ability to rectify low-voltage signal levels. Under comparable conditions germanium diodes begin to rectify signals of about 0.05 V. Point-contact silicon diodes and hot-carrier diodes rectify at 0.15 V, and low-power silicon diodes rectify at 0.5 V. Gold-bonded silicon diodes rectify at about 0.3 V but are now surpassed by inexpensive hot-carrier diodes. Germanium diodes are said to operate more efficiently than silicon diodes at low temperatures (important in space probes), whereas silicon diodes operate easily at high temperatures (50 to 150° C).

Tunnel diodes are hot-carrier devices with a negative resistance that may be used in high-frequency amplifiers and oscillators. When the negative-resistance characteristic is suppressed so that the diode may be used more easily for mixing and detection, the diodes are called *back diodes.* Back diodes are manufactured in both germanium and silicon and are used when turn-on is required at very low voltage levels. However, back diodes operate over such a narrow voltage range that they are satisfactory only when the signal level is relatively fixed.

14.4 SHUNT CAPACITANCE AND HOT-CARRIER DIODES

Shunt capacitance is perhaps the most limiting parameter of the hot-carrier devices at microwave frequencies. Schottky and point-contact diodes have an essentially negligible reverse-recovery time and, therefore, should make better rectifiers, mixers, and detectors at high frequencies than the junction diodes, which exhibit charge storage. In practice, however, hot-carrier devices are useful at frequencies only 10 to 100 times the frequency limits of a junction diode, because the diode capacitance shunts the nonlinear junction and, thereby, limits the open-circuit impedance when the diode is OFF. A high-frequency, low-capacitance diode may have a shunt capacitance of only 1 pF, but at 1 GHz this reactance is only 160 Ω (a reactance value worth remembering). Thus, a diode with a 1-pF capacitance cannot be an efficient rectifier when the circuit impedance exceeds 160 Ω. For this reason, high-frequency rectifier and detector circuits always have low impedances, and appreciable power is required to drive the diode to a low impedance.

Point-contact silicon and germanium diodes selected for microwave applications have shunt capacities as low as 0.3 pF, and in low-impedance circuits (50 Ω) may be used as mixers and detectors at 30 GHz and above. The silicon Schottky diodes have a capacitance of about 1 pF and require two to five times the input power for the same efficiency. Thus, Schottky diodes are used above 1 GHz only when needed for their low noise, uniformity, and reliability. Gallium arsenide hot-carrier diodes have noise figures as low as 6 dB and may be used to and above 10 GHz.

The effect of shunt capacitance as a limitation in the application of diodes is illustrated by the ac voltmeters commonly used for UHF signal measurements.

14.5 DIODE VOLTMETERS AT UHF FREQUENCIES

The application of diodes for voltage measurements at high frequencies is illustrated by the circuit in Fig. 14.4. This circuit uses a Schottky diode to charge the 0.01 disc capacitor. The meter is generally calibrated to read rms voltage, but the deflection is actually proportional to the peak-to-peak signal voltage. The variable resistor at the left of the meter is necessary to provide a circuit for dc current and provides a convenient means for calibrating the meter reading. The input capacitor protects the diode from dc voltages, provided the capacitor is not too large.

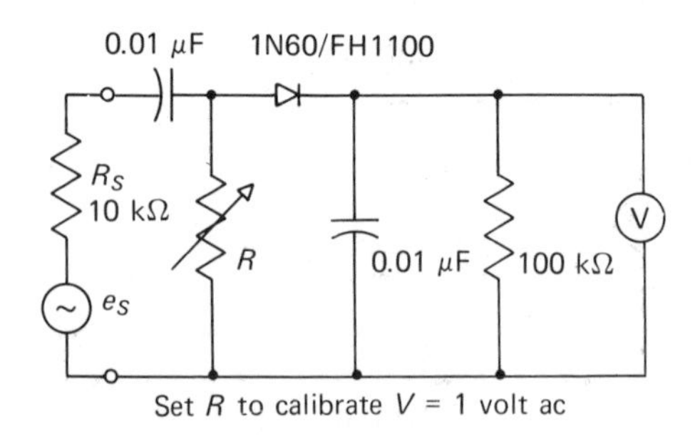

Figure 14.4. High-frequency ac voltmeter.

The voltmeter calibration is independent of frequency until the diode capacitance shunts the nonlinear junction and reduces the ac voltage across the diode. Thus, the cutoff frequency varies with the impedance R_S and with the input voltage. When the input signal is sinusoidal and a few volts rms, the rectification efficiency is reduced to 50 per cent when the reactance of the diode capacitance equals the source impedance R_S. For example, if the Schottky diode shown in Fig. 14.4 is shunted by 100 pF, the upper cutoff frequency is reduced to 250 kHz with a 2-V rms input signal. The reactance of 100 pF is 10 kΩ at 160 kHz, the calculated cutoff frequency.

The capacitance of the Schottky diode is given as 1.2 pF, so the calculated cutoff frequency for the UHF voltmeter is 14 MHz. The observed cutoff frequency is about 25 MHz. With a germanium point-contact diode the cutoff frequency is 40 MHz.

The cutoff frequency of the voltmeter may be increased approximately 1 decade for each factor-of-10 reduction of the source impedance. When R_S is 1 kΩ, the cutoff frequency with a germanium diode is about 400 MHz. With lower impedances the rectifier provides useful indications at microwave frequencies. Decreasing the source impedance forces the diode to rectify at a lower impedance level and reduces the effect of the capacitance, but increases the amount of RF power required to produce a given signal. Probes for high-frequency TV servicing use similar circuits and with germanium diodes give accuracies of ± 2 dB (26 per cent) up to 250 MHz. The 1N21B point-contact, silicon diode provides rectification efficiencies exceeding 90 per cent at 500 MHz and 70 per cent at 1 GHz.

The range of input voltages that is handled best by a crystal voltmeter is between 0.1 and 2 V. When the applied voltage exceeds 20 V rms, an attenuator must be inserted to reduce the voltage actually applied to the crystal. Capacitance attenuators are preferred at UHF and GHz frequencies.

Because appreciable power is required to operate high-frequency voltmeters, the diode that converts the ac signal to dc tends to clip the signal on one or both peaks and loads the circuit being tested. The load on the circuit is determined in part by the dc load; hence the dc meter should have a high sensitivity.

A voltage-doubling detector suitable for detecting low-level, high-frequency signals is shown in Fig. 14.5. The detector circuit is similar to that of the voltmeter in Fig. 14.4, except that the shunt diode improves the rectification efficiency and loads the signal source on both peaks of the signal.

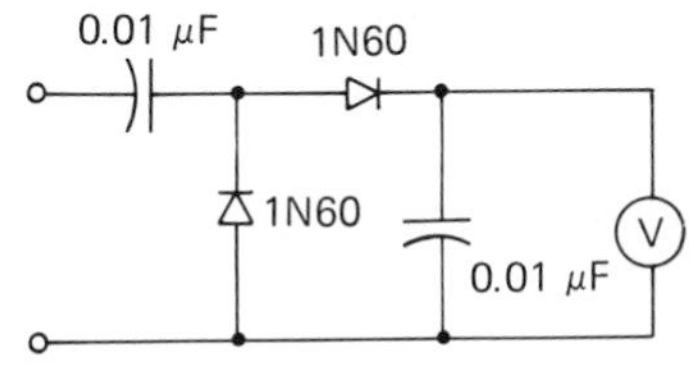

Figure 14.5. Voltage-doubling UHF voltmeter.

14.6 SQUARE-LAW VOLTMETERS

If the input voltage to a crystal voltmeter is small, say a few hundredths of a volt, the rectified current is proportional to the square of the input voltage, and the current may be used for voltage or power measurements at low levels and high frequencies. In some instances the square-law relation holds up to several tenths of a volt. Operated as a square-law meter, a crystal rectifier has a low input impedance because the diode current is measured directly, as shown in Fig. 14.6. The impedance of a germanium diode at low signal levels is approximately 20 kΩ in parallel with 2 pF. The resistance component of a silicon diode is several orders of magnitude higher, but the shunt capacitance is the same, 2 pF, and the diode does not follow the square law as well. For measurement applications the germanium diodes are more nearly ideal diodes.

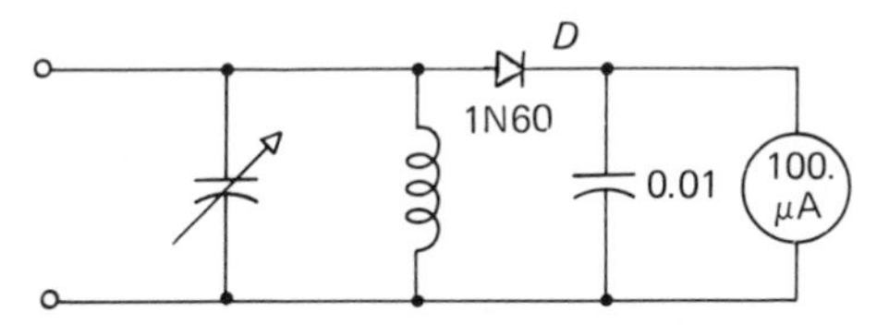

Figure 14.6. Square-law detector.

14.7 DIODE DETECTORS

Detectors use the nonlinear characteristic of diodes to convert the amplitude changes of a high-frequency signal to a corresponding low-frequency or dc signal, which is the information originally modulated on the high-frequency signal. The detectors used at microwave frequencies are small-signal detectors in which the dc output signal is proportional to the square of the high-frequency amplitude. The detected low-frequency signal is easily amplified for presentation as the transmitted information. A linear detector uses the diode to rectify a large signal. The low-frequency output is proportional to the high-frequency amplitude and is more linear than a square-law detector.

Because tunable amplifiers are expensive and complicated for frequencies above 100 MHz, signals at these frequencies are often detected without prior amplification. Point-contact diodes are used extensively for low-level detection in the UHF and microwave frequency range, but Schottky diodes are more uniform and less subject to burnout by overloads. Both types operate as square-law detectors for input power levels from 0.001 μW, −60 dBm, up to about 1 μW, −30 dBm. Both types require a few microamperes of forward bias current to perform as square-law detectors.

The lower useful limit of power sensitivity depends on the noise in the device and on the capacitance that shunts the junction at microwave frequencies. Junction diodes are generally unsatisfactory as mixers or detectors above 100 MHz, depending on the circuit impedance. Schottky diodes are essentially free of $1/f$ noise and make satisfactory detectors even above 5 GHz. Schottky diodes are replacing the back diodes formerly used in this frequency range. Selected point-contact germanium and silicon diodes are required above 10 GHz and are still the best millimeter-wave detectors because of their low-noise and low-capacitance advantages.

For large-signal detection, Schottky diodes operate as linear, peak-voltage detectors when the peak-to-peak signal exceeds 0.25 V. The hot-carrier diode has the advantage of a decided nonlinearity between forward and reverse voltages of 0.25 V, which are a factor of 2 or 3 lower than for a silicon junction diode. Moreover, hot-carrier diodes have the advantage that their high-reverse-breakdown voltage, at least 70 V, allows the diode to be used at high signal levels and over a wide amplitude variation of the modulated carrier. Thus, these newest members of the diode family offer a distinct improvement for linear detection in many applications.

14.8 DIODE MIXERS

A mixer is a circuit used to convert an incoming high-frequency signal to a lower modulated frequency. Conversion is accomplished by driving a diode simultaneously with the high-frequency input signal and a local oscillator (LO). The diode is driven by the sum of the two signals, and the nonlinearity of the diode produces a current that includes sum and difference frequencies. Thus, mixing a 470-MHz UHF TV signal and a 515-MHz local-oxcillator signal produces a 45-MHz signal that can be amplified in the IF amplifier of a TV receiver. By this means a UHF tuner is able to convert UHF signals at power levels down to 0.01 μW, depending on the diode conversion efficiency and noise level.

The Schottky diode is a highly efficient mixer. It has less conversion loss and thereby produces lower mixer noise than for any other diode. Its lower impedance level provides better impedance matching, and the high efficiency at high signal levels reduces adjacent channel interference and distortion problems.

The circuit of a UHF TV tuner that employs a diode mixer is shown in Fig. 14.7. The local oscillator is a transistor CB stage tuned by the capacitor C_3. The antenna is connected across the parallel-tuned circuit C_1 and L_1. Because the diode presents a low impedance, the diode loop is a series-resonant circuit C_2 and L_2 inductively coupled to the input circuit. The LO is loosely coupled to the diode, and the power input to the diode is only a few milliwatts. The capacitor C_4 and inductance L_4 are resonant at the 45-MHz IF frequency, so the capacitor has a low impedance at UHF frequencies. The diode may be either a UHF germanium diode or a Schottky diode.

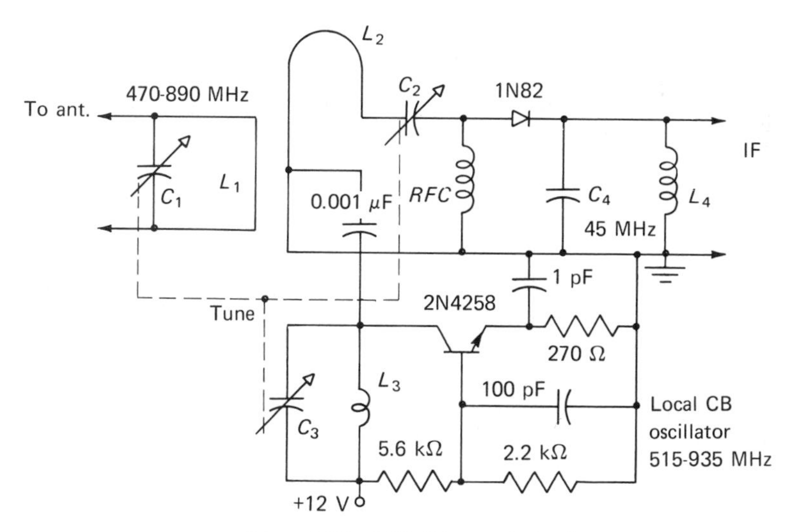

Figure 14.7. UHF TV tuner (mixer).

By fabricating the inductors as thin copper strips and enclosing the tuned circuits in copper- or silver-plated steel enclosures, the tuned circuits have high no-load Qs, which provide a high degree of selectivity and a signal bandwidth of about 10 MHz.

UHF TV tuners that have a transistor-tuned input stage and a transistor mixer have the advantage of about 10-dB gain, instead of a 10-dB loss, and a little lower noise figure. A similar TV tuner has been made as a thin-film microcircuit that uses diode varactors as tuning capacitors. In this way the tuning can be remotely controlled, and the physical size of the tuner is considerably reduced.

A diode mixer constructed in a coaxial line for single-frequency operation at 2 GHz is shown in Fig. 14.8. The diode and the center conductor make up a half-wave line that is short circuited at one end and loaded by the diode at the other. The input is inductively coupled by inserting a short length of the input line into the detector cavity. The capacitance corresponding to C_4 in Fig. 14.7 is formed by the length of copper tubing slightly smaller than the inside diameter of the coaxial line to which the diode is connected for the IF output.

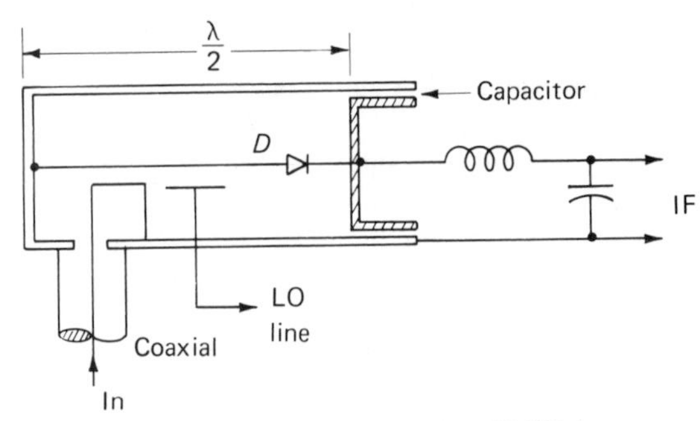

Figure 14.8. Microwave mixer (2 GHz).

14.9 HIGH-FREQUENCY OSCILLATORS

The power available from a transistor oscillator decreases rapidly with frequency in the UHF range, and only a few tens of milliwatts are available at 5 GHz. Examples of CB-transistor oscillators are readily found in the transistor manuals with descriptions of coaxial lines or cavities for tuning.

Tunnel-diode oscillators have an element of simplicity that has made them useful as the signal source in hand-held short-range radios, in remote-control devices, and as the mixer oscillator in receivers at microwave frequencies. A number of tunnel-diode circuits may be found in reference 14-1.

Above 10 GHz, Gunn-effect and avalanche diodes convert dc power directly to microwave energy, thereby eliminating much complex circuitry. The avalanche diodes oscillate, or amplify, when biased in the reverse-breakdown region using a phenomenon that was predicted by W. T. Read in 1958. The diodes are known as Read diodes, avalanche transit-time diodes, or as impact avalanche and transit-time (IMPATT) diodes. The continuous-wave

(CW) avalanche diode has a negative resistance produced by a combination of the avalanche process and carrier drift. The diode accomplishes a single-step conversion of dc power to microwave energy, and may be operated either as a CW oscillator or as a pulsed power source of higher power output.

A typical silicon IMPATT diode can generate a watt at 10 GHz, and power has been produced at frequencies to 300 GHz. Gallium arsenide IMPATTS are expected to yield higher power with higher efficiencies and lower noise than silicon IMPATTS. An IMPATT diode may be inserted in a coaxial line, as shown in Fig. 14.9. The diode has inductance that is resonated by the capacitor C_1 so that the 50-Ω line impedance is transformed to a 1- or 2-Ω load for the diode. The bias power is connected to the center conductor of the coaxial line, as shown, with a capacitor C_2 to isolate the power source from the outgoing line. The conversion efficiency is low, so with 1-W output the heat sink must dissipate about 15 W. The thermal resistance of the diode is high because of the low areas required in a high-frequency device. Therefore, the oscillator must be provided with a low-temperature liquid cooling system. For systems operating with pulsed output signals, as in radars, IMPATT devices produce pulsed power outputs of 30 W at 3 to 4 GHz.

Gunn diodes produce microwave power when a small cube of gallium arsenide is connected to a 6- to 10-V dc source at currents of 0.2 to 1. A. Gunn devices are called diodes because they are two-terminal devices and are polarity sensitive. A Gunn oscillator produces up to 0.2 W at frequencies of 12 GHz. Because the diode is small and requires a high dc power density, thermal design for heat removal is a prime consideration and an engineering problem. A Gunn diode may be inserted in a microwave cavity in essentially the same way the IMPATT diode is made a part of the coaxial cable in Fig. 14.9.

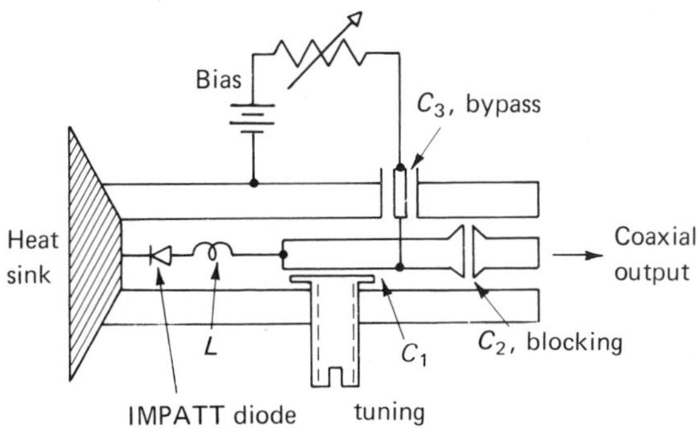

Figure 14.9. IMTATT Microwave-oscillator.

Avalanche diodes are used in microwave radars, parametric amplifiers, and oscillators. Because these devices are characterized by a relatively high noise, their applications may be more limited than Gunn diodes. IMPATT and Gunn diodes produce power outputs that are 1 to 2 orders of magnitude higher than a transistor or a tunnel diode.

Throughout the high-frequency spectrum, semiconductor devices are still surpassed by vacuum-tube devices. Above 10 GHz, traveling wave tubes, magnetrons, and backward-wave oscillators produce kilowatts of power. However, semiconductor devices have important advantages of simplicity, small size, and high reliability.

14.10 VARACTOR DIODES

The capacitance of a reversed biased diode may be reduced a factor of 3 or more by increasing the applied voltage. For silicon diodes the capacitance is given by the equation

$$C_t = \frac{C_O}{(V_R + 0.7)^n} \tag{14.2}$$

where C_O is the capacitance when the applied dc voltage V_R is 0.3V. The exponent n in the denominator may be 0.5 or 0.33, depending on whether the junction is abrupt or graded. For approximate calculations equation (14.2) may be simplified to

$$C_t = \frac{C_O}{\sqrt{V_R}} \tag{14.3}$$

which shows that the capacitance varies inversely with the square root of the applied voltage. The control of capacitance by a voltage is particularly attractive at UHF and microwave frequencies, where a variable air capacitor is many times the size of a microcircuit.

Silicon tuning diodes provide a capacitance variation of 3 to 1, and in a few low-frequency low-voltage types, a capacitance variation of 10 to 1. The voltage-variable capacitors offer advantages in automatic frequency control and for amplifier tuning from 1 MHz through microwave frequencies.

A high-Q capacitor is desirable in most applications. The Q of tuning diodes varies inversely with frequency, and the Q value is generally given at 100 MHz. The Q of a typical diode capacitor varies over the intended frequency range from 20 to 200, and at 100 MHz is between 50 and 500.

The advantages of a diode for tuning include simplified circuits and construction, a higher reactance range, faster response times (vital in frequency sweeping circuits), and the ability to control a multiplicity of devices simultaneously.

14.11 VARACTOR DIODE TUNING

Voltage-variable capacitors, VVC, are commonly used as the voltage-variable tuning element in tuned amplifiers, for automatic frequency control of oscillators, and for frequency modulating an oscillator. In these applications, the amplitude of the ac signal across the varactor is relatively small, so the capacitance is only a function of the dc, or low-frequency, control voltage. Tuning varactors are designed to have high Q values and large capacitance variations with voltage.

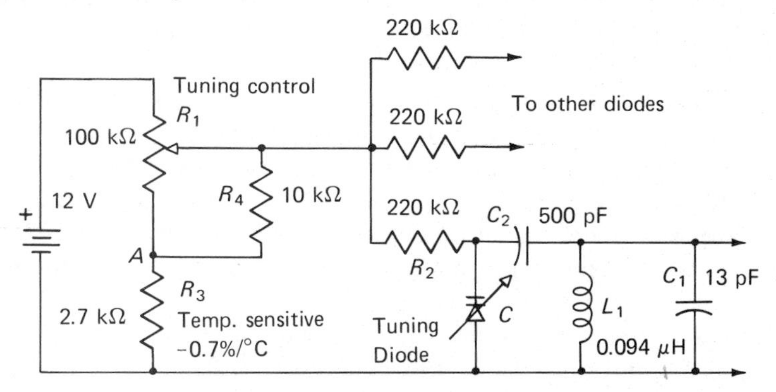

Figure 14.10. FM varactor-diode tuning circuit, 88-108 MHz.

The circuit used for tuning an L-C resonant circuit is illustrated in Fig. 14.10. The tuned circuit is formed of L_1, which is tuned by C_1 and the diode capacitance C. The diode voltage is varied by the potentiometer R_1. The resistor R_2 isolates the diode from other diodes that may be connected to the potentiometer, and the resistor has a large value to prevent loading of the tuned circuit. There is a negligible current in R_2. The capacitor C_2 is used to block dc from the tuned circuit, and may be a small capacitor if the capacitance change in the diode is more than needed for frequency control. For the following example, C_2 is large enough to have a negligible effect on the tuning.

Suppose that the L-C circuit is to tune over the FM frequency range from 88 to 108 MHz, using 12-V dc for the maximum reverse voltage. Suppose, also, that the designer wishes to tune the circuit at the upper frequency f_h, using a total capacitance ($C_1 + C$) of about 25 pF. A reactance chart shows that the lower frequency f_l may be obtained by increasing the capacitor about 10 pF. We call this capacitance change ΔC.

The designer finds a diode with a 12-V capacitance, $C = 10$ pF, which may be increased to 24 pF with 0.3-V bias. Assume that the useful change $\Delta C = 12$ pF.

For small frequency changes the capacitance required at the high frequency is given by

$$(C_1 + C) = \frac{\Delta C}{(f_h/f_l)^2 - 1} \tag{14.4}$$

Substituting $\Delta C = 12$ pF and the frequency limits gives $(C_1 + C) = 23$ pF. Since the minimum capacitance of the diode is $C = 10$ pF, $C_1 = 13$ pF.

For resonance, with a 23-pF total capacitance, the reactance chart suggests trying $L_1 = 0.01\ \mu$H. A check shows that with $L_1 = 0.094\mu$H, the circuit tunes from 108 MHz, with $C_1 + C = 23$ pF, to 88 MHz, when $C_1 + C + \Delta C = 35$ pF. Because the VVC diodes have a temperature coefficient of -2mV/C°, a negative temperature coefficient silicon resistor is used for R_3. The resistance changes -0.7 per cent/C°, and the potentiometer is selected to make the voltage 0.3 V at the point A so that the control voltage changes -2.1 mV/C°. The temperature compensation is adjusted by changing the value of R_3, and the shape of the voltage-frequency curve is changed by adjusting the resistor R_4.

If the tuned circuit determines the frequency of an oscillator, the voltage applied to the diode may be used to control the oscillator frequency, as in an automatic frequency control, or may be used to frequency modulate the oscillator. With VVCs, tuned circuits inside a microcircuit or capacitors in an integrated circuit may be readily varied from outside the package. VVCs are useful for remote control of receivers, for adjusting circuits by telemetry, and for signal seeking and sweep-frequency applications.

14.12 VARACTOR FREQUENCY MULITPLIERS

Varactors used for frequency multiplication are operated in the region between forward conduction and reverse breakdown, and thus are not used as rectifiers. A low-frequency diode operated as a rectifier produces harmonics, but the harmonic amplitudes and power are relatively low, particularly at high frequencies. Varactor diodes produce harmonics by a capacitance change or by the sudden return of stored charge, with the advantage that a high percentage of the input power is recovered as harmonic power.

Although a nonlinear device, a varactor diode is a capacitance with a Q value between 20 and 200 at the recommended operating frequency and is, therefore, a low-loss circuit component. A varactor designed for high-power VHF and UHF applications is driven with a peak-to-peak input signal that is nearly equal to the reverse-breakdown rating, and is biased so that the peak signal in the forward direction does not produce significant rectification.

The circuit of a reactance frequency doubler is shown in Fig. 14.11. The

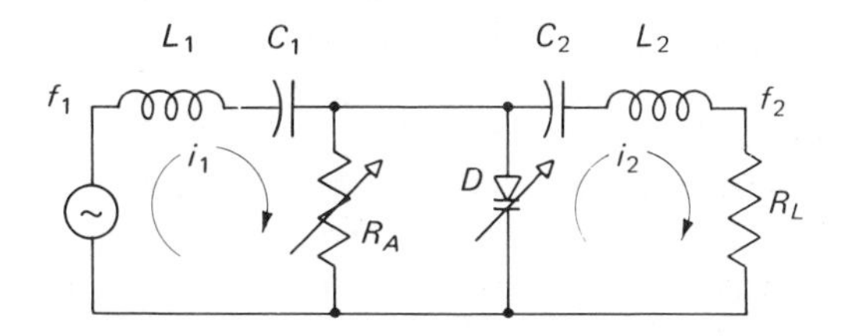

Figure 14.11. Varactor frequency doubler.

signal input is connected to the diode through a series-resonant circuit L_1 and C_1 tuned to the input frequency f_1. The load is connected through C_2 and L_2, which are tuned to the second harmonic frequency $2f_1$. The diode bias is adjusted, as desired, by the resistor R_A (100 kΩ), which is relatively high valued compared with the circuit impedance. The bias current is produced by rectification when the varactor is driven slightly into forward conduction on the signal peaks.

If the desired load current is the third or a higher harmonic, the circuit in Fig. 14.12 is used. The input circuit is tuned to the input frequency, as before, and the output is tuned to the desired harmonic. The series-resonant circuit L_2 and C_2 across the diode is tuned to the second harmonic and is called the *idler*. The idler permits a second-harmonic current to flow in the diode along with the fundamental. Thus, by mixing, the varactor produces the frequencies $f_1 + 2f_1 = 3f_1$, $2(2f_1) = 4f_1$, etc.

Although the variable-capacitance varactor offers second-harmonic output efficiencies of 80 per cent at VHF and UHF frequencies, diodes with step recovery of stored charge are more efficient at high frequencies and for high-order harmonic generation.

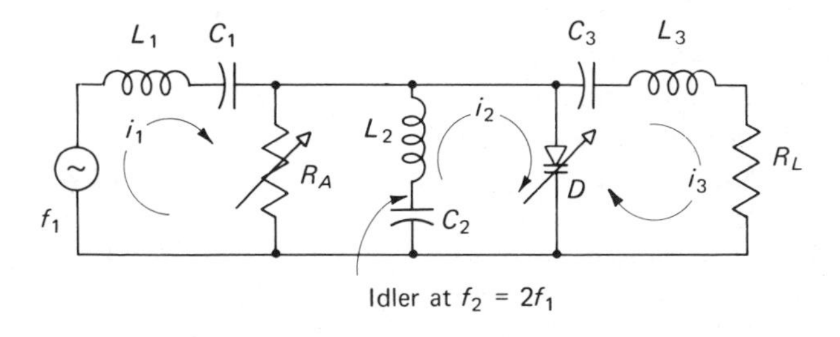

Figure 14.12. Varactor frequency multiplier.

14.13 FREQUENCY MULTIPLIERS

Frequency multipliers are commonly used to produce high-frequency power because more power can be obtained this way than from transistors. When a low-frequency signal is used to overdrive an amplifier, harmonics are produced that may be selected by tuned circuits. In this way a single-step frequency multiplication of 2, 3, 4, and up to 8 or more times is often used. By a frequency multiplication of $8 \times 4 \times 3$, a crystal operating at 104 MHz may control a signal source at 10 GHz.

Frequency multipliers for high-frequency applications generally use step-recovery diodes that produce an exceedingly steep wave front.

14.14 STEP-RECOVERY DIODES

Charge storage in a diode, which is a detriment in detection, is used to advantage for harmonic generation in diodes known as step-recovery varactors, snap diodes, or power varactors. These diodes turn OFF in times of 0.1 nsec, and produce harmonics that can be used at frequencies beyond the reach of transistor and vacuum-tube oscillators. The snap diode, when driven into forward conduction, stores charge and presents a low impedance. When the drive current reverses, the stored charge is returned in a compact bunch and conduction suddenly ceases, thereby generating a voltage spike, or pulse. By feeding these pulses into a resonant load, harmonics are selected that generate the desired output frequency.

A diode that has charge storage in the ordinary sense conducts a high current in the forward direction and stores charge in the junction capacitance. Immediately following a reversal of the applied voltage, as shown in Fig. 14.2, the stored charge returns as from a capacitor, and the reverse current gradually returns to zero.

A step recovery diode has the reverse-current characteristic shown in Fig. 14.13. The steep wave front at the sudden cessation of reverse current is used to enhance multiplier action and is a rich source of high-order harmonics. For high conversion efficiency the harmonic output frequency should be one-tenth the reciprocal of the recovery (snap) time T_S. Thus, an efficient multiplier requires the output frequency to meet the condition

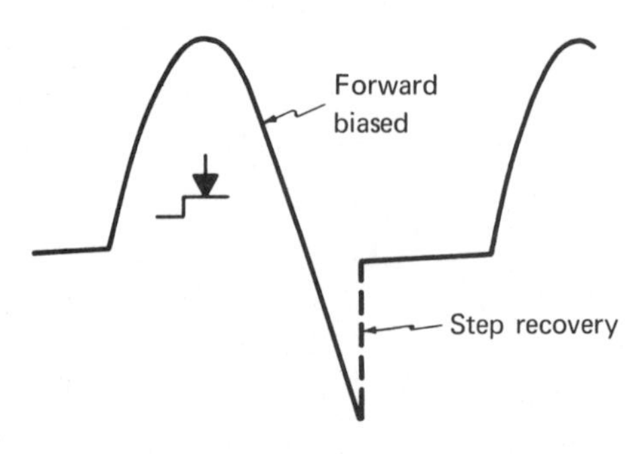

Figure 14.13. Step-recovery diode current wave form.

$$f_{\text{out}} << \frac{1}{T_S} \tag{14.5}$$

Diodes for microwave frequency multipliers may have step-recovery times of only 0.1 nsec. Equation (14.5) shows that efficient conversion may be obtained at 1 GHz, and we may expect useful outputs up to about 10 GHz.

On the other hand, if a diode is built with a fast recovery, the junction volume is small and the charge can be stored for only a short time, known as the *minority carrier lifetime* T_L. Thus, the charge lifetime places a lower limit on the useful input frequency (for a given diode), which is given approximately by the equation

$$f_{\text{in}} > \frac{2}{T_L} \tag{14.6}$$

A diode with a snap time of 0.1 nsec, as above, has a minimum lifetime of about 10 nsec, and equation (14.6) shows that the minimum input frequency is about 0.2 GHz. For this diode we may expect a maximum efficient frequency multiplication of about 5, and with reduced efficiency, a factor of perhaps 20.

Step recovery produces a pulse wave form that is approximately independent of the signal level. Therefore, the percentage of harmonic generation is independent of the power input. The circuit used with a step-recovery diode is essentially the same as shown in Fig. 14.12, except that an idler may not be necessary. If the final frequency is above 1 GHz, the diode output may be coupled into a microwave cavity, as shown in Fig. 14.14.

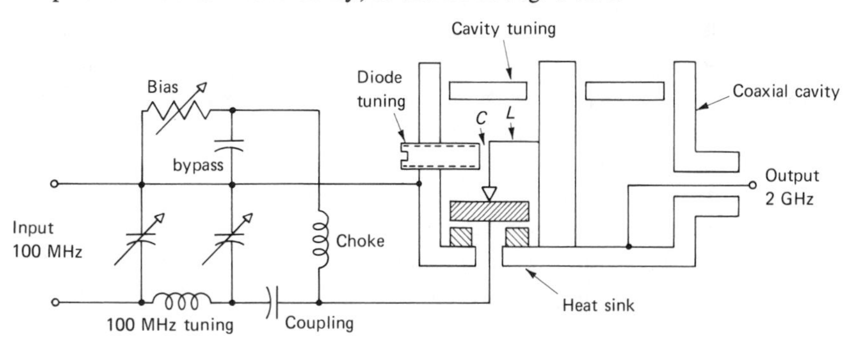

Figure 14.14. Step-recovery X20 frequency multiplier.

For low-frequency applications the diodes are designed with a high capacitance and low thermal resistance. With an input power of 25 W at low frequencies a multiplier produces a single harmonic with efficiencies approaching $160/N$ per cent, where N is the harmonic number. With a 50-MHz input frequency a multiplier may use $N = 8$ for an efficiency of 20 per cent. However, with the same overall frequency increase, a chain of three multipliers with $N = 2$ has an efficiency of nearly 50 per cent. Because of the higher efficiency, multipliers used above 1 GHz are usually frequency triplers or doublers. Thus, parallel-connected transistor drivers followed by a chain of varactors have achieved power outputs of 20 W at 2 GHz, 1 W at 13 GHz, and a few milliwatts at 50 GHz. This is more power than can be obtained from a single transistor.*

14.15 *PIN* DIODES

The *PIN* diode is a *pn* diode with an intrinsic layer of silicon between, which acts as a dielectric barrier. The barrier reduces the diode capacitance and increases the stored charge. At microwave frequencies the amount of stored charge cannot be significantly changed, so the diode resistance is independent of the microwave signal but may be varied by changing the forward bias. By varying the bias from reverse bias (no stored charge) to about 10-mA forward bias, the resistance at microwave frequencies may be varied from 10 kΩ to 10Ω. The diodes may be turned ON or OFF in times measured in tens of nanoseconds.

When the control current is varied continuously, the *PIN* diode is useful for attenuating, amplitude control, or for modulating an RF or microwave signal. When the control current is switched ON, OFF, or in steps, the diode is useful for circuit switching, pulse modulation, or phase shifting. The small size and high switching speed of the *PIN* diodes make them an ideal component for RF and microwave applications.

An attenuator may be constructed by inserting a single *PIN* diode as a variable resistance in series with or across a line. The attenuation is achieved partly by reflection and partly by dissipation in the diode, and the attenuators are called reflective switches or attenuators. In some systems the reflected power cannot be tolerated, and the attenuator must be designed to present a constant input or output impedance, or both.

The circuit of a bridged-T attenuator is shown in Fig. 14.15. The bias on a pair of diodes, shunt and series, is increased on one and decreased on the other so that the circuit has a constant impedance with the attenuation changed by 20 dB. By spacing diodes along a transmission line at quarter-wavelength intervals and by controlling the diodes differently, the attentuation may be varied, with the input impedance essentially constant. These structures are used in RF and microwave circuits as attentuators and modulators, which are particularly desirable because of their wide bandwidth.

*For higher power than can be produced at present with semiconductors, vacuum-tube and klystron oscillators produce several hundreds of watts (CW) power at 10 GHz and more than 1 W at 100 GHz. Pulsed magnetrons produce peak powers 1000 times higher. Traveling-wave oscillators, which are designed for use above 10 GHz, produce CW power exceeding 100 W up to 100 GHz.

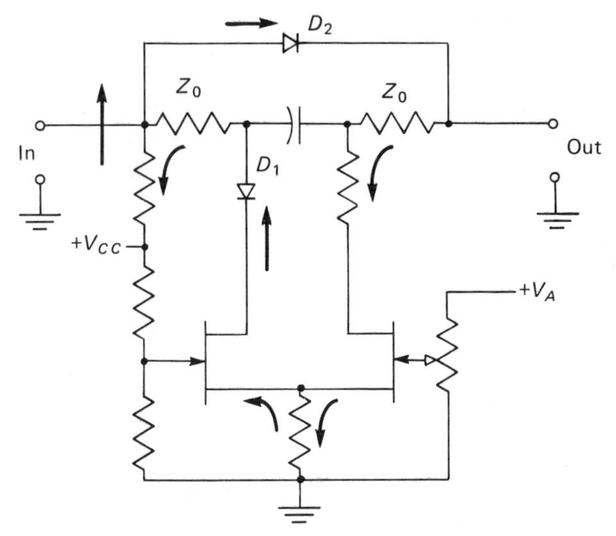

Figure 14.15. Bridged-T *PIN* diode attenuator (arrows show current increments with increase of V_A).

The diodes are capable of switching peak powers measured in kilowatts with switching times between 10 and 500 nsec. The carrier lifetime limits the application of *PIN* diodes to frequencies above a minimum frequency that is between 1 and 100 MHz, depending on the diode.

14.16 MICROWAVE CIRCUITS

Microwave circuits are constructed very differently from the lumped-constant circuits used at RF and low frequencies. The dimensions of the components and connecting leads must be as small as practical, the active elements often dissipate considerable heat and must be carefully heat sinked, and the circuit impedances must be confined to a relatively narrow range of values. A microwave package is small and carefully shielded. At low microwave frequencies a tuned circuit may use lumped capacitors and inductances, and at high frequencies the tuned circuits are sections of transmission lines and cavities.

At a frequency of 5 GHz, one wavelength in free space is only 6 cm and in a transmission line may be less than 3 cm. A wire 0.25 cm long has an inductance of 1 nH, which resonates with a capacitance of 1 pF at 5 GHz. Thus, a microwave circuit must have such small dimensions that the designer works with a magnifying glass and microscope, or the circuit is constructed by photolithographic techniques, as for an integrated circuit.

The circuits for low-microwave frequencies are constructed by combining small coils and capacitors with coaxial lines and microwave cavities. This, the oldest method of extending the low-frequency techniques into the microwaves, is useful up to 2 GHz.

One way of making smaller circuits uses the photographic and etching processes that have been developed in fabricating integrated circuits. By these techniques *thin-film microcircuits* are constructed, using photoetching and the vapor deposition of metals to make resistors, capacitors, coils, and lines. This process may extend the lumped coil and capacitor circuits to 10 or 30 GHz, but the process is economical only for mass production, and the circuits require considerable hand adjustment of the components. *Thick-film* techniques employ metal-bearing inks or pastes that are printed on a substrate and fired to about 800°C to produce conductors, resistors, and capacitors. The thick-film techniques have the advantage over thin films of lower production costs, which permit the manufacture of smaller lots and give greater design flexibility. These microcircuits, using uncased transistors to reduce lead lengths and shunt capacitance, have pushed the conventional form of discrete circuits to frequencies of 6 GHz or higher.

Transistors that are manufactured for microwave circuits are small discs or squares with strap-like leads for the emitter, base, and collector. Some devices have a shield between the base and collector and two emitter connections to reduce the effect of the emitter-lead inductance.

A third method of constructing microcircuits employs strip transmission lines laid down as a single conductor on an insulating substrate with a ground plane conductor on the opposite side. *Strip-line circuits* use sections of lines for tuning or impedance transformation along with lumped coils and capacitors. An integrated circuit with directly attached lumped components is called a *hybrid circuit*. Hybrid strip-line circuits have the advantage of being useful up to 10 GHz, but they are difficult to design and adjust. The high dielectric coefficient of the substrate may make a section of line very short and temperature sensitive.

Because the reactance of 1 pF is only 160 Ω at 1 GHz and is 16 Ω at 10 GHz, the impedance of microwave circuits is generally between 100 and 1 Ω. This rather limited range of impedance values and the need to use transmission lines with a 50-Ω impedance tend to restrict circuit designs and the gain that can be obtained in a single stage. Gain is expensive at microwave frequencies, and to conserve gain the circuits and lines must be carefully matched.

14.17 MICROWAVE APPLICATIONS

The applications of microwaves have been growing very rapidly since 1964. Microwave systems are particularly attractive for aircraft, satellites, and space probes, because microwave antennas are small and are easily constructed to confine the radiated energy in a small conical beam. Microwave systems are attractive in communication systems because very wide bands of modulated signals may be transmitted and because signals can be transmitted in a coaxial cable or by radio relay networks.

Microwave systems are used in commercial and military radars, communication networks, in satellite communications (Comsat), and in telemetry, as from the moon. Most frequencies in use or now planned extend to about 10 GHz, and research is moving the frequency limit higher year by year.

14.18 SUMMARY

The high-frequency limitations of semiconductors generally stem from stored charge, capacitance, and the low efficiency of dc-to-ac power conversion. The relative simplicity of diodes has advantages at high frequency that are exploited with varactors, step-recovery diodes, mixer diodes, and the tunnel diode. IMPATT and Gunn diodes have negative-resistance characteristics that permit direct conversion of dc power to microwave and millimeter-wave powers.

The rectification and conversion efficiencies of junction diodes are impaired at UHF frequencies by stored charge and, sometimes, by shunt capacitance. Hot-carrier diodes, which are free of charge storage, are useful as rectifiers at low UHF frequencies, but they are generally surpassed by germanium point-contact diodes, which have lower capacitance and higher detection efficiencies.

Point-contact crystal diodes are used in square-law voltmeters, detectors, and power meters from the low microwave frequencies to millimeter-wave frequencies. As mixer diodes, the Schottky and point-contact devices are often used to convert high-frequency signals to a low-frequency IF for easier amplification and large-signal rectification.

Transistor oscillators are used up to about 5 GHz, and IMPATT and Gunn diodes are being used increasingly as signal sources at higher frequencies. Vacuum tube and klystrons still produce several orders of magnitude more power at high frequencies than is possible with semiconductor sources.

Varactor diodes are being used increasingly for frequency multiplication, for tuning, and in diode amplifiers. High conversion efficiency may be obtained because charge stored in the diode capacitance is returned to the circuit without being dissipated, as in a rectifier. The step-recovery diode similarly returns stored charge in exceedingly short times, thus converting low-frequency power with high efficiency into power at frequencies 5 or more times the input frequency. Devices that use stored charge possess a low-frequency operating limit because the stored charges have a finite lifetime.

The chapter concludes with a brief outline of microwave techniques and applications.

PROBLEMS

14-1. What is meant by (a) a diode detector, (b) a diode mixer, (c) a rectifier, and (d) a varactor? Sketch a typical circuit for each.

14-2. If the source and load impedances shown in Fig. 14.12 are both 50 Ω, what values for L_1, L_2, and L_3 would you use for a first trial of the frequency multiplier? Assume that the input frequency is 100 MHz.

14-3. Using the values given in the text for the varactor-tuned FM circuit in Fig. 14.10, calculate the high- and low-frequency limits for the given design. Find how to change C_1 and L_1 if the frequency range is to cover only 88 to 98 MHz.

14-4. What are the characteristics and the frequency range which makes each of the following diodes useful? (a) Germanium; (b) point contact; (c) planar diffused; (d) tunnel diode; (e) Schottky; (f) PIN, IMPATT, and Gunn diodes.

DESIGNS

14-1. Design and test a frequency doubler that operates with a 100-MHz input signal. See reference 14-2.

14-2. Construct and test the varactor tuning circuit shown in Fig. 14.10.

14-3. Design and construct an oscillator that operates at 1 GHz. See reference 14-2.

REFERENCES

14-1. *GE Transistor Manual*, 7th ed., "Tunnel Diode Circuits," pp. 349-374. Electronics Park, Syracuse, N. Y.: The General Electric Co., Inc., 1964.

14-2. *RCA Silicon Power Circuits Manual*, SP-51, "Microwave Amplifiers, Oscillators, and Frequency Multipliers," pp. 352-378. Harrison, N. J.: Radio Corporation of America 1969.

14-3. *Solid State Devices and Application Notes.* Palo Alto, Cal.: Hewlett Packard Co., 1967.

14-4. *Microwave Journal.* Dedham, Mass.: Horizon House-Microwave, Inc. A monthly journal of technical articles describing circuits and devices for microwave applications.

15

INTEGRATED-CIRCUIT AMPLIFIERS

Linear integrated circuits may be used in the design of circuits and instruments that would be impractical with discrete components. The designer has in a small package a high-gain, high-quality dc amplifier that may be adapted to many purposes simply by adding a feedback circuit. Because of its high gain and inherent stability, the IC amplifier with feedback has a very stable and predictable transfer characteristic. With low-cost ICs, designers are producing many new circuits that have a reliability and quality of performance unobtainable with discrete-component circuits.

The input stages of high-gain IC amplifiers are usually differential amplifiers. The advantage of a differential IC amplifier is that with positive and negative collector supplies, the amplifier has zero dc input and output voltages, so the output is easily direct-coupled to the input or to other circuits. The ability to reject common-mode input signals makes the differential amplifier useful for rejecting the dc and ac signals so often encountered in system design.

This chapter has a brief description of the design and construction of a high-gain IC, followed by selected applications to illustrate typical uses of ICs in circuit and instrument design.

15.1 INTEGRATED-CIRCUIT CONSTRUCTION

The active and passive components of monolithic ICs are formed on a single small silicon substrate by the same techniques used in making planar transistors. All components and interconnections of the circuit are formed in a batch process in which several hundred nearly identical ICs are fabricated on a single substrate.

The construction process begins with a thin slice of silicon that is protected by oxidizing the entire surface to form a silicon oxide coating. The circuit components and the connections are then photographed on the oxide surface so that the oxide may be selectively removed by chemical etching. Where the oxide is removed, semiconductors or resistors are formed by diffusing impurities into the silicon substrate. On the exposed silicon surface, conductors are formed by evaporating metals that are deposited, forming thin-film metallic resistors and contacts for the external connections. When finished, the entire surface of the chip is protected either by silicon oxide or by the metallic films.

Following the diffusion process, the ICs are cut apart and each is mounted in a package and connected by fine wires to external terminals. Monolithic ICs have the economy of a batch process except for the high initial cost for the circuit design, the photographic masks, and the tooling up for production.

Hybrid ICs are microcircuits in which the passive components are formed on an insulating substrate either by thick- or thin-film techniques, while the active devices, normally in chip form, are added separately. The active chips, which may include monolithic integrated circuits, are connected to the metalized circuit by wires, and the entire circuit may be sealed in a plastic package. The assembly and interconnection of the many components make the hybrid circuit more expensive and less reliable than a monolithic circuit.

Hybrid circuits can incorporate better devices than are available in integrated circuits, and they allow for greater control over the precision of the component values. Hybrid circuits are more versatile, as they can be tailored to a specific application and made in smaller quantities than monolithic ICs. They are capable of operation at frequencies and power levels not yet achieved with monolithic ICs. High-performance operational amplifiers (op amps) are commonly manufactured as hybrid circuits and supplied in an epoxy-filled package.

15.2 INTEGRATED CIRCUITS FOR DESIGN AND APPLICATIONS

Monolithic structures can economically include many active devices, so it is practical to use differential amplifiers and novel circuits to provide superior performance. Thus, the monolithic ICs make excellent performance characteristics possible by feedback and limiting, by internal power supply regulation and decoupling, and by internal temperature compensation.

Integrated circuits greatly reduce the number of components and the skills required to design a high-performance circuit. The ICs have the obvious advantages of small size and weight, but the most important advantages are, perhaps, the reliability and quality of performance that can be obtained as easily as if using a single transistor. Moreover, low-cost ICs have a guaranteed uniformity for replacement and a wide-range temperature characteristic that is difficult to surpass with discrete components.

Many specialized circuits are available as integrated circuits. These include voltage regulators, IF and video amplifiers, FM sound systems, color TV chroma demodulators, audio-frequency amplifiers, multipliers, and many circuits for digital computers. All these circuits are readily available with characteristic data and application information for the designer.

Linear amplifiers specifically intended for analog computers are generally known as operational amplifiers, or op amps. An op amp is used with feedback in low-frequency analog studies. The amplifiers are often supplied with internal capacitors to shape the high-frequency cutoff so that the feedback may be easily applied. High-gain general-purpose differential amplifiers for instrument and audio-frequency applications usually have connections brought out so that the designer has the option in controlling the high-frequency cutoff. These amplifiers, which have a differential input stage and considerable common-mode rejection (CMR), generally offer the designer considerable flexibility in their use.

Integrated-circuit arrays are available that have unconnected active devices for circuit design. These arrays have groups of diodes, Darlington pairs, transistors, and complete stages. These devices match each other and have similar temperature characteristics because they are manufactured in close proximity on a single chip. The arrays may be used to replace discrete components in special-purpose and low-volume circuit designs.

Linear ICs may be adapted to perform a variety of circuit functions in data processing, communications, and instrumentation. Sometimes the terminals on an IC, which are intended for external compensation, can be connected to external circuits that adapt the amplifier to a special purpose. Thus, for any application needing a complex, high-gain circuit, the designer has the opportunity to save much time and hard work by adapting an integrated circuit to serve his purposes.

A typical IC presents interesting opportunities for comparing a microcircuit design with that of the discrete circuits described in preceding chapters.

15.3 MICROPOWER INTEGRATED CIRCUITS

An IC amplifier should be designed to use as little dc operating power as is practical. An IC substrate may be only 1 by 1 mm, so dc power is converted to heat in a very small area. Thus, even low-power input stages may be subjected to a relatively high temperature rise, which complicates the control of Q-point drift and reduces the allowable ambient temperature range. Moreover, many applications require a low standby-power consumption. Others require low amplifier noise and high input impedance characteristics, which are generally aided by using a low Q-point power. For these reasons, ICs have been developed that require very little dc power.

The description of a low-power op-amp design lends considerable insight into the design of a multistage integrated circuit. For reference, the circuit of a μA735 micropower op amp is shown in Fig. 15.1. This amplifier operates over a supply voltage range of ±3 to ±18 V, and on a ±3-V supply requires only 100-μW standby power. With a ±15-V supply the standby power is 6 mW. In comparison, a μA709 op amp is designed to operate on ±15 V and requires about 100 mW. The input resistances of these amplifiers are approximately 10 and 0.1 MΩ, respectively.

The power required by an op amp is determined mainly by the output power requirements and the output stage. If the load requires signal power, there is, of course, no way to obtain this power except from the power supply, and with a class-A output stage the standby power is over three times the required load power. The micropower amplifier uses a class-B complementary-symmetry power amplifier, which is comprised of transistors Q_{18}, Q_{19}, Q_{20}, and Q_{21}. By means of the resistor R_{14}, shunt feedback is applied over the last three stages to minimize the crossover distortion and to reduce the internal output impedance. The location and the design of the diodes Q_{14}, Q_{15}, Q_{16}, and Q_{17} determine how well low standby power and fixed Q-points can be maintained with temperature and load changes.

High input impedance may be obtained by using high-β transistors operated at low collector currents (for high h) and by using a Darlington input connection (for high β). Low noise is obtained by improved processing techniques and by using input transistors that have high β with collector currents as low as 1 μA. Thus, the input resistance of one side of the amplifier is $\beta h = 2.5$ MΩ, which may be increased by emitter feedback.

The first two stages of the μA735 present an imposing array of transistors and diode-connected transistors. Since the amplifier uses two differential stages ahead of Q_{13}, one side of the amplifier may be neglected in tracing the circuit. Thus, the input stage has a direct-coupled *CE-CC* transistor pair Q_1 and Q_6, and the second stage is the *CE-CC* pair Q_8 and Q_{10}. The differential stages couple to the output amplifier through a unit-gain phase inverter Q_{12}. The output amplifier, which has feedback, is comprised of Q_{13} and two complementary *CC-CC* pairs, Q_{18}, Q_{20} and Q_{19}, Q_{21}, which are biased class B.

The amplifier has a total of three *CE* stages and four *CC* stages. A separate *CE* stage Q_3 is used to increase the common-mode feedback of the input

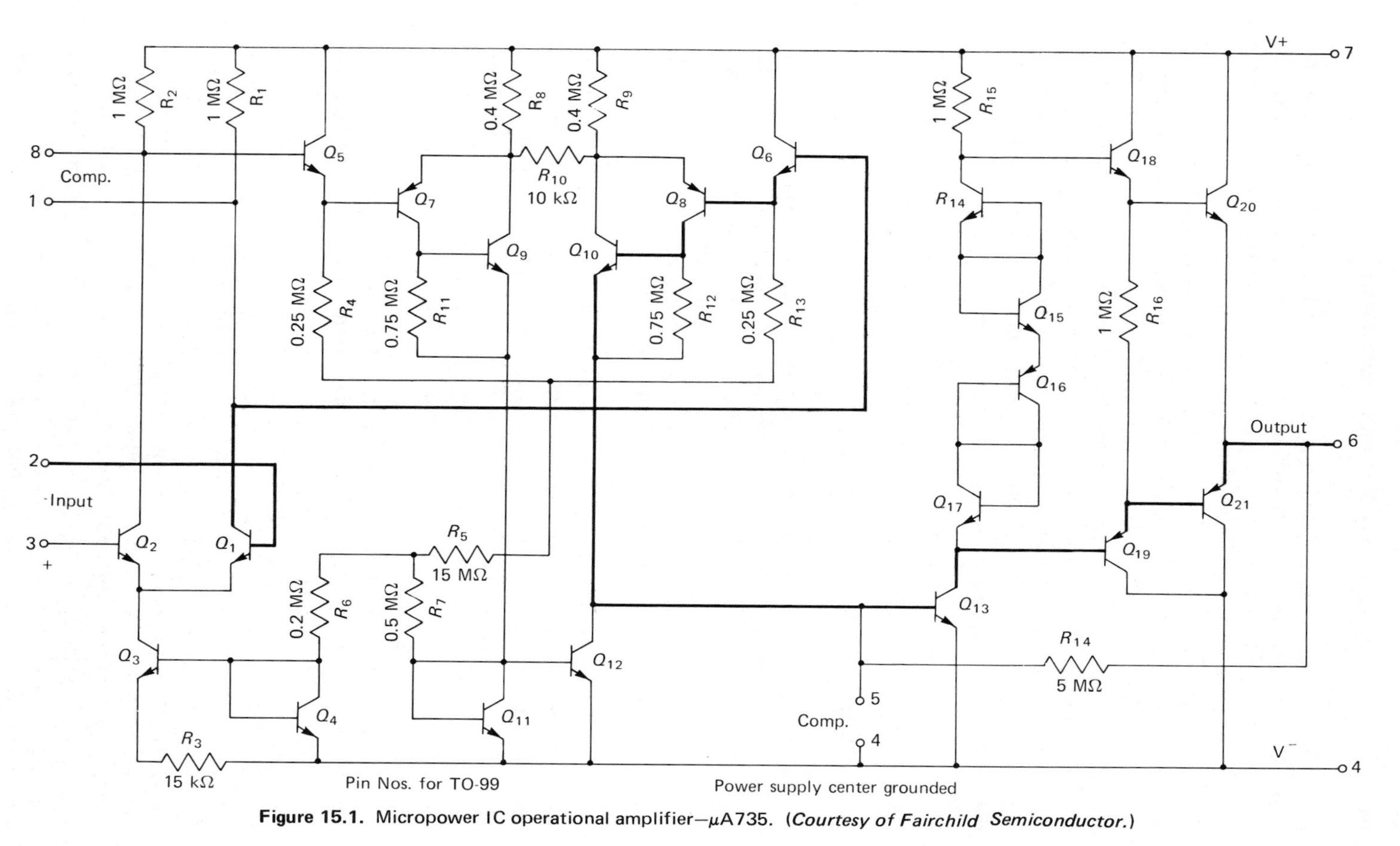

Figure 15.1. Micropower IC operational amplifier—μA735. (*Courtesy of Fairchild Semiconductor.*)

stages, and six transistors are used as diodes. Q_{12} which appears to be a *CE* stage, does not add to the overall voltage gain because the collector voltage is only the base-emitter voltage drop of Q_{13} and Q_{12} operates as a phse inverter to aid conversion of the differential gain to a single-sided stage.

It is interesting to observe that three stages of the amplifier use *pnp* transistors that only a few years ago could not be manufactured economically along with *npn* transistors. Also, the amplifier uses thin-film resistors with one value R_{14} as high as 5 MΩ deposited on the silicon wafer. The amplifier has a voltage gain of 20,000, using a ±3-V supply and a 5-kΩ load. The amplifier is intended to be used with two external compensating capacitors and resistors. Because of the low values of emitter currents, the transistor f_T values limit the amplifier gain-bandwidth product to about 1 MHz.

15.4 INTEGRATED-CIRCUIT AMPLIFIER DESIGN

The balanced differential amplifier is generally used in the high-gain stages of an IC. Differential stages may be direct-coupled easily because the common-mode (emitter) feedback minimizes the *Q*-point drift and unbalance. A differential stage has a high differential gain without requiring an emitter-bypass capacitor, and the process by which ICs are fabricated tends to match similar components and to match their temperature coefficients over a broad temperature range.

One of the designer's first problems is to decide how many *CE* and *CC* stages are required in a high-gain amplifier. Here, as is often the case, simplicity is the key to good design, and overdesign only adds to the design problems, the manufacturing difficulties, and the cost. The designer should use *CC* stages wherever possible because they are easily biased, easily direct-coupled, and the *Q*-point drift of a series of *CC* stages is small compared with the drift in a single *CE* stage.

The voltage-gain requirements of the amplifier must be met by using *CE* stages with sufficient feedback and bias control to limit the *Q*-point temperature drift. Because of the difficulty of applying feedback over more than three *CE* stages, the voltage-gain requirements should be met by using two, or at most three, *CE* stages. The minimum number of *CE* stages that give the required voltage gain may be found by equation (3.15).

According to equation (3.15), two *CE* stages with a 3-V supply voltage may have a voltage gain of 6000, and the data sheet shows a typical gain of 20,000. Sometimes the gain may be slightly increased by using a transistor collector as an active load, but this technique is not employed in the μA735. However, one additional stage is required for feedback in the power stages, and one *CE* stage Q_3 is required in the input amplifier to provide common-mode feedback. These considerations explain the use of three *CE* stages between the input and output of the amplifier. We may assume that the output amplifier provides some additional gain even with the internal feedback.

The *CC* stages of an IC amplifier may be considered the means by which the input impedance is coupled to the load impedance without a loss of voltage gain. The μA735 has a minimum input impedance of approximately 10 MΩ and is specified to operate with a 5-kΩ load impedance. This impedance ratio, 10,000/5, implies a current gain of 2000, which may be obtained with two *CC* stages if the current gains are 45. Two additional *CC* stages may be expected because of the *S*-factor control by R_1 and R_9 in the *CE* stages. Thus, we may expect a total of four *CC* stages and three *CE* stages in the main amplifier. A desirable simplicity of design is indicated because the μA735 has only four *CC* stages and three *CE* stages. The published descriptions of the amplifier indicate that the *pnp* stage Q_8 has a current gain of only about 2, but that it is used for additional isolation between the collector of Q_1 and the *CE* transistor Q_{10}.

By equation (3.15), a two-stage *CE* amplifier with a ±15-V supply may be expected to have an open-loop voltage gain of 150,000 to 300,000. The second gain figure assumes that the entire 30-V supply is available to the second stage, but that the input-stage emitters are essentially at the midpoint voltage. These gain values are typical for ordinary OP amps. Substantially higher gains require a third *CE* stage or higher supply voltages.

The relatively low voltage gain of the μA735 operated on a ±15-V supply shows that the design maintains a fixed overall gain independent of the supply voltage. Further discussion of the μA735 amplifier may be found in reference 15-1.

15.5 INTEGRATED-CIRCUIT DESIGN PROBLEMS

Because the components of an integrated circuit are manufactured in a batch process by diffusion and metallic-deposition techniques, the circuit components have characteristics that are different from the discrete components used in ordinary circuits. The transistor characteristics designed for an integrated circuit tend to determine the diffusion schedule, so the range of values of the diffused resistors tends to be somewhat limited. The practical range of resistance is generally considered to be 20 Ω to 20 kΩ for a diffused resistor. Similarly, the limited area available for deposited resistors limits their range of values. Thus, a high resistance may be manufactured as the collector of a transistor, and a low resistance may be the emitter of another transistor. The maximum practical value for a MOS capacitor in a monolithic construction is about 100 pF, so larger capacitors are connected externally.

On the other hand, because the components of an integrated circuit can be located very close together physically, the resistors and transistors have closely matched electrical parameters and temperature coefficients. Since the microelectronic parts have relatively high absolute tolerances, the matching characteristics must be emphasized. Thus, high-gain amplifiers use balanced circuits, such as differential amplifiers, and use considerable common-mode feedback to limit the Q-point drift with temperature. Moreover, the power dissipation must be reduced to a minimum, because the silicon chip is so small that the internal heating may cause too much drift or reduce the allowable ambient temperature range.

Perhaps the most serious limitation of monolithic ICs is the difficulty of integrating a variety of semiconductor devices on a single silicon chip. Complementary transistors, Zener diodes, resistors, capacitors, and FETs are produced with difficulty and by a large number of processing steps. Hybrid circuits offer greater design flexibility in choosing the circuit components and in using high-performance devices.

An integrated circuit may be designed by using a breadboard and components manufactured specifically for an integrated-circuit breadboard. The breadboard circuit may be used to evaluate the dynamic performance of a circuit, but the temperature and Q-point characteristics cannot be accurately observed because of the high thermal resistances between separate parts.

Integrated circuits must be very carefully designed and tested before manufacturing is commenced. The design cost may be divided between thousands of units, but an error in the design or even a minor circuit change involves expensive retooling and production losses. The major costs of monolithic ICs are the large initial cost of the masks and design, and the production costs of mounting, bonding, and testing. Generally, the circuits use a large number of active devices and relatively few resistors, a practice that tends to optimize the gain and the performance. Novel circuits can be used to produce superior performance characteristics because the addition of semiconductors does not add greatly to the cost.

Diodes constructed as the collector-base or base-emitter junctions of a transistor are freely employed in monolithic ICs. A diode-connected transistor that has the collector and base in common generally provides the highest conductivity and is best anywhere the low voltage breakdown of the base-emitter junction can be tolerated.

The early types of ICs were susceptible to latch-up, or to damage by short circuits, overloads, and momentary abuse. Most of the recently developed ICs are reasonably immune to these problems, have built-in temperature and phase compensation, and have performance characteristics that are very attractive as amplifiers or parts of complex electronic systems or instruments.

15.6 STABILITY PRECAUTIONS

The high available gain of an IC tends to make an IC amplifier seem impractically unstable, especially when operated open loop. Generally, an OP amp is not intended to be operated open loop, and with feedback the amplifier should be easily stabilized. Most oscillation problems are caused by poor circuit layout and inadequate power-supply bypassing.

Linear ICs generally have a very wide band frequency response and are potentially unstable up to the unity-gain frequency given by the gain-bandwidth in Hertz. Instability or oscillations are produced either by improper compensation to control the loop gain-frequency response or by parasitic feedback caused by stray capacitance between the output and input.

The best way to compensate an amplifier is generally given by the manufacturer. The parasitics are reduced by making a neat circuit layout by placing leads and components near a ground plane, and by using a grounded metal shield between the output and input components.

Many ICs can be operated on an unregulated supply or a battery, especially when the amplifier is used with considerable feedback. However, even with a well-regulated power supply, high-frequency bypassing may be important. Positive feedback through the power-supply leads may cause very high-frequency oscillatons of a single amplifier or may cause instability when the supply is shared by several amplifiers. The remedy for these problems is generally to bypass both sides of the power supply to ground, using an 0.01- or 0.1-μF capacitor connected as close to the amplifier as possible. High-frequency oscillations sometimes cause an unreasonable low-frequency or dc performance of the amplifier, or displace an oscilloscope trace without causing a perceptible ac signal. Low-frequency blocking is caused more often by RF feedback than by inadequate low-frequency bypassing or poor regulation of the power supply.

15.7 IC APPLICATIONS

The IC or OP amp manufacturers furnish circuits suggesting many applications of these devices, both linear and nonlinear. Generally, an IC should be used only when there is an advantage of economy or reliability in using such high-gain devices. Many everyday applications need only one or two discrete transistors and fewer circuit components than are required by an IC. On the other hand, the low-cost and simplicity of the IC cannot be equaled when the designer needs a direct-coupled high-gain differential amplifier. The following examples are selected to illustrate applications for which an IC amplifier seems particularly well adapted.

15.8 HIGH-INPUT-IMPEDANCE ADJUSTABLE-GAIN AMPLIFIER

The high input impedance of an IC cannot always be used to advantage with a source having a high dc resistance. The difficulty is that the bias current of a direct-coupled transistor input stage must flow through the signal source or a resistor provided for that purpose. A typical transistor IC needs a bias current of about 1 μA. If the external resistor is 1 MΩ, the bias current produces a 1-V voltage drop. If the amplifier has a voltage gain of 10, the output Q-point is shifted 10 V, unless the opposite input is connected to a similar 1-MΩ external resistor. Thus, the external circuit must have a resistance less than 1 MΩ, although the input impedance of the IC may be greater than 10 MΩ. This example shows that an IC amplifier generally cannot be operated from a high source impedance and simultaneously used for high gain.

The inverting side of a differential amplifier may be biased either by current supplied through the feedback resistor R_f, as in Fig. 15.2, or by current through the input resistor R_1. A good practical rule is to assume that the voltage drop of the bias current through R_f should not exceed 1 V. This rule makes the maximum feedback resistor $R_f = (1 \text{ V})/I_B$. If the bias current is given as 0.5 μA, the rule requires that R_f be 2 MΩ, or less. The designer may now use any desired value for R_1 less than R_f when gain is required.

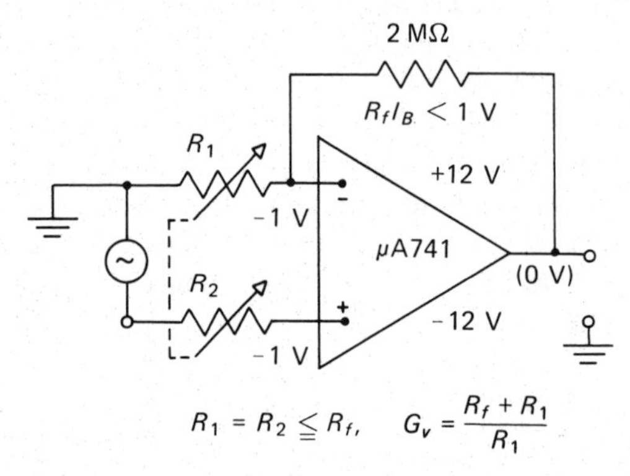

Figure 15.2. High-input-impedance adjustable-gain IC amplifier.

For the same reason, the input resistor R_2 on the noninverting input should not be larger than the feedback resistor R_f. Moreover, when the amplifier has gain because R_1 is less than R_f, the voltage drop in R_2 must be correspondingly reduced by making R_2 equal to or less than R_1. Thus, the rules for selecting the external bias resistors may be written as

$$R_2 \leqslant R_f \leqslant \frac{1}{I_B} \tag{15.1}$$

and

$$R_2 \leqslant R_1 \tag{15.2}$$

Observe that the bias rules are for a circuit in the form shown in Fig. 15.2 and that R_2 must meet both conditions. This circuit is limited in application because a low-resistance signal source is required, the gain control cannot be calibrated easily, and the resistors R_1 and R_2 should be varied simultaneously. However, the signal source does see the high input impedance of the amplifier.

For an application in which the bias current cannot be required to flow through the source, an IC amplifier may be used as a unity-gain stage, as shown in Fig. 15.3. The bias current required for the noninverting side of the amplifier is supplied from a battery via the bootstrapped 0.1-MΩ resistor. The bias may be adjusted by the potentiometer so that the input terminal is close to ground potential, which makes the output Q-point independent of the source resistance. Bootstrapping the bias resistor makes the input impedance approximately the rated "input impedance" of the amplifier.

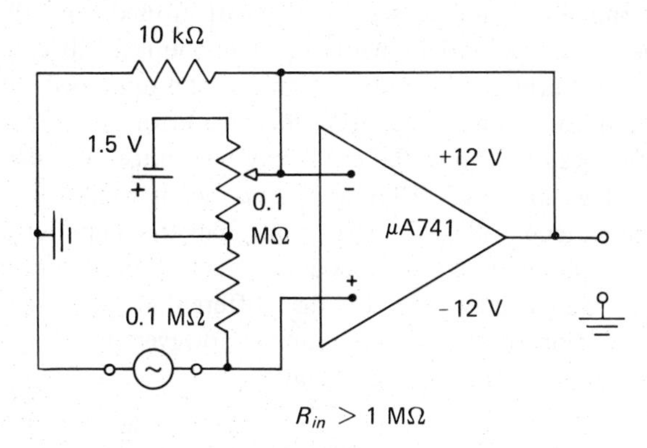

Figure 15.3. High-input-impedance unity-gain amplifier.

The circuit shown in Fig. 15.4 may be used when a high input impedance, variable-gain amplifier is required. In this example the bias resistor cannot be bootstrapped, and bias is supplied instead through an external high-value resistor. The signal source may be direct-coupled or capacitor-coupled to the

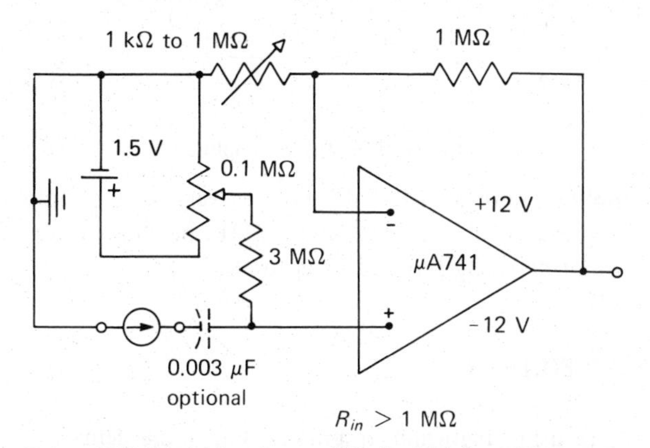

Figure 15.4. Variable-gain amplifier for high impedance source.

noninverting input. The gain may be varied either by changing the feedback resistor or the variable resistor shown in Fig. 15.4. If high gain is required and the feedback resistor is impractically large, a T network may be used in the feedback path.

15.9 HIGH-GAIN BRIDGE AMPLIFIER

A differential amplifier with a high degree of common-mode feedback provides a simple and convenient way for amplifying and measuring an off-ground difference signal. The bridge shown in Fig. 15.5 may, at balance, have both terminals *A* and *B* several volts off ground even though they are both at the same potential. With a direct-coupled differential amplifier connected to the bridge, the output signal is the amplified difference signal and is balanced to ground. The bridge-balance error is less than 1 per cent, even for low values of R_f, and the error may be reduced by increasing R_1.

The input stage of an IC differential amplifier is generally able to operate with a common-mode voltage of nearly one half the amplifier supply voltage, provided the amplifier is used at a moderate gain. With high gain a part of the common-mode signal may add to the differential signal and produce an inaccurate indication of the bridge balance. However, in many applications a fixed error signal is not a serious problem.

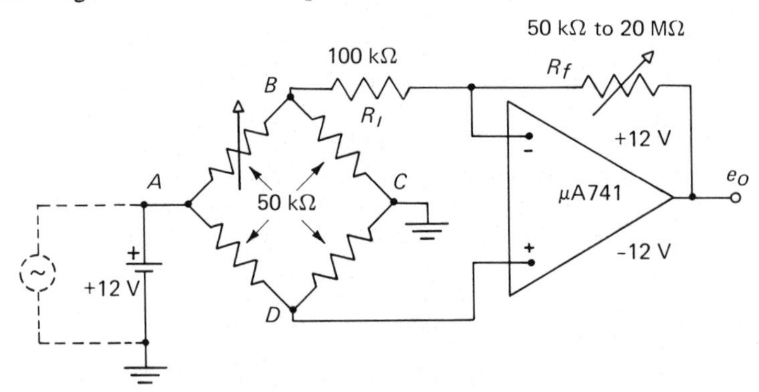

Figure 15.5. Bridge amplifier.

An IC amplifier with a high CMR ratio is very useful for observing small signals in the presence of common-mode 60-Hz pickup, or for observing any signal that cannot be easily referred to ground.

15.10 ACTIVE FILTERS

Active filters using high-gain amplifiers and a mathematically determined feedback network produce very precise filter characteristics. The active filter has the advantage that characteristics that otherwise would require high-Q inductors can be obtained by using only resistors and capacitors. Active filters that require a high-gain amplifier are limited to low-frequency applications, but in many applications these filters have significant weight, size, and cost advantages. Active filters can be constructed for higher frequencies if a moderate gain is acceptable. Otherwise, the internal high-frequency roll-off network may adversely control the frequency response.

The circuit of a high Q notch filter is illustrated in Fig. 15.6. The filter uses the standard twin-T filter, and the notch frequency is that at which the reactance of the capacitor C equals the resistance R. By connecting the amplifier as a unity-gain inverter, the voltage gain from the noninverting input is 1, and the terminal of the twin-T, which is normally grounded, is driven in phase with the filter output signal. When connected as shown, the Q of the filter may be adjusted from a maximum of about 100 down to 0.25. The null balance and the Q adjustment are essentially independent.

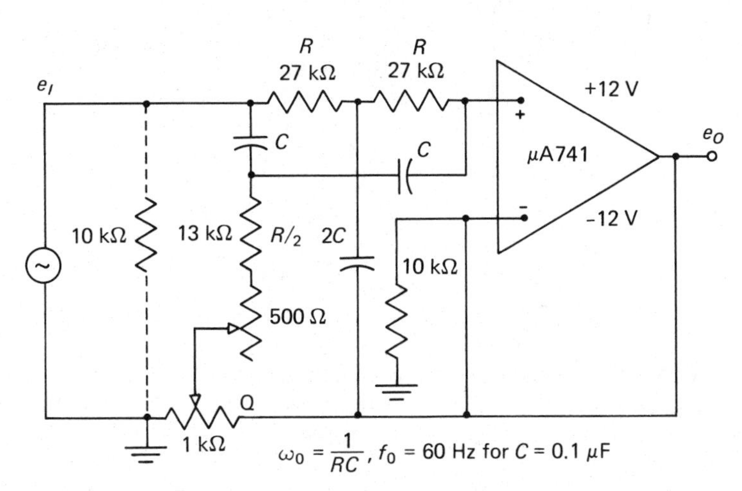

Figure 15.6. Active 60-Hz notch filter.

The characteristics of some active filters are relatively independent of the amplifier gain, and a high-gain IC may not be required for satisfactory performance. In the present example the filter requires a high loop gain for stable operation with a high Q. However, a stable Q of 4 may be obtained with a single-stage emitter follower. Therefore, the use of an IC is not generally justified for a low-Q application.

15.11 AUDIO OR TAPE PREAMPLIFIER

The high gain of an IC amplifier may be used to advantage in an audio preamplifier when the designer wants a precisely determined frequency response. The RIAA phono equalization requires a frequency response that decreases a total of 49 dB when the frequency increases from 50 Hz to 20 kHz. To shape the frequency response accurately with feedback, an amplifier should have at least 60-dB open-loop gain. The amplifier and feedback network shown in Fig. 15.7 are capable of reproducing the RIAA phono response with a high degree of accuracy, because the nominal 80 dB of open-loop gain is more than adequate to ensure that the overall response is independent of the amplifier.

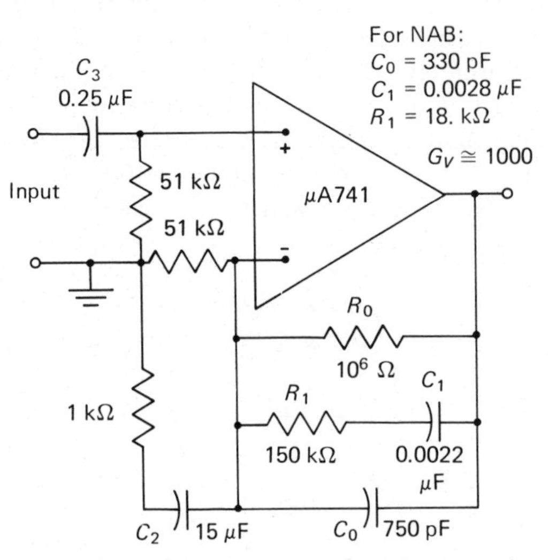

Figure 15.7. Phono (or tape) preamplifier (RIAA).

The RIAA frequency response is represented by the straight-line approximations shown in Fig. 15.8. When the characteristic is produced by a single feedback network of the form shown in Fig. 15.7, the R and C values can be found easily by using a reactance chart and the adjusted corner frequencies given in Table VII. The table also includes equivalent corner frequencies for the NAB frequency response used in audio tape recorders. The resistor R_O is picked to give the amplifier the desired low-frequency gain. The other network elements are found according to the table. The capacitors C_2 and C_3 may be used at the option of the designer to reduce the amplifier response below 20 Hz as a means of reducing turntable rumble.

TABLE VII

Compensating Network Corner Frequencies

C and R	RIAA	NAB
C_1, R_0	73	57
C_1, R_1	500	3200
C_0, R_1	1460	26500

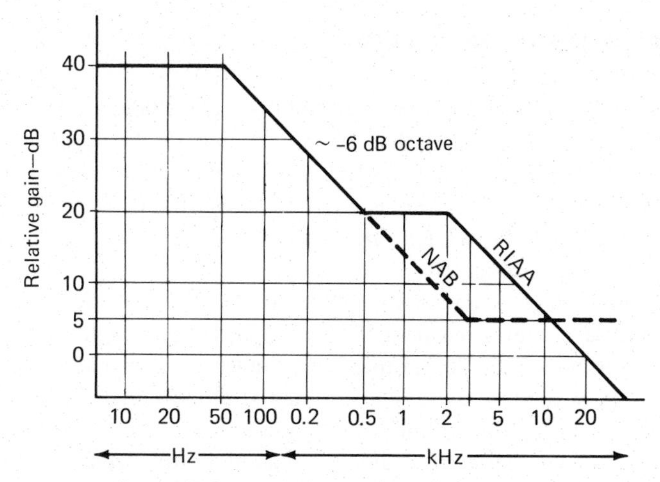

Figure 15.8. RIAA phono frequency response.

15.12 NONLINEAR-FEEDBACK APPLICATIONS

Several interesting applications of the IC amplifiers use nonlinear elements in the feedback path to produce exceedingly useful circuits. By connecting a dc meter and a full-wave rectifier in the feedback path, as shown in Fig. 15.9, a dc meter is caused to respond with a linear indication that is proportional to the ac voltage e_1. Although the rectifiers are nonlinear, the amplifier forces the voltage across R_1 to equal e_1 so that the current in the meter is e_1/R_1. As long as there is adequate feedback, the meter reading is proportional to the average ac voltage e_1. With hot-carrier diodes, ac meters of this type generally operate down to about 10 mV full scale. An advantage of this circuit is that the meter may be linearly calibrated for direct current, and with a suitable shunt may use the dc scales.

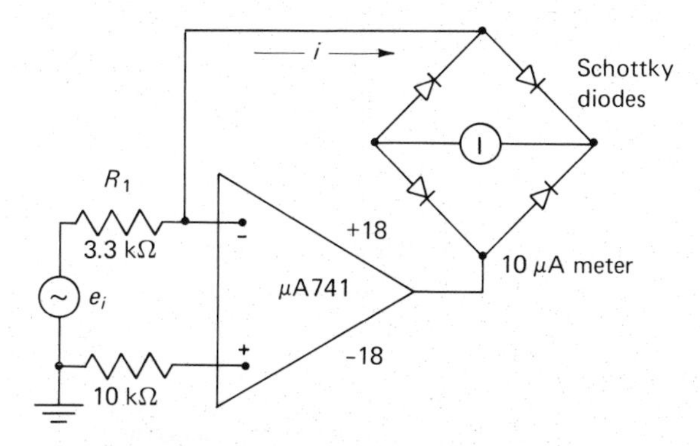

Figure 15.9. Linear rectifier-type ac meter, 100 mV rms.

15.13 LOGARITHMIC AMPLIFIERS

Logarithmic amplifiers compress an input signal, which varies over many decades, into an output signal that is proportional to the logarithm of the input signal. Logarithmic amplifiers are used in measuring audio signals and noise when a readout in decibels is required. Logarithmic amplifiers are also used for measuring low-level ionization-chamber currents or for logarithmic-frequency and pulse-rate meters.

The current-voltage characteristic of a semiconductor diode, which is the most predictable nonlinear element known to physics and engineering, may be used for converting a linear signal to a logarithmic output signal.

The forward voltage across a silicon diode is proportional to the logarithm of the current, and the logarithmic conversion is accurate over a current range of about 6 decades. Thus, if a diode is placed in the feedback path of a high-gain amplifier, the output voltage varies with the log of the input current and voltage.

For a wide range of logarithmic response, a low-β planar transistor, connected as shown in Fig. 15.10, is generally better than a diode. With the base grounded, the emitter is forward biased and the collector voltage is zero, relative to the base. For the circuit shown, the output voltage is proportional to the log of the input voltage from about 100 μV to 10 V, or 5 decades.

For an extremely wide range of the input voltage, the input offset current and voltage must be carefully compensated as in the circuit shown. Low-cost

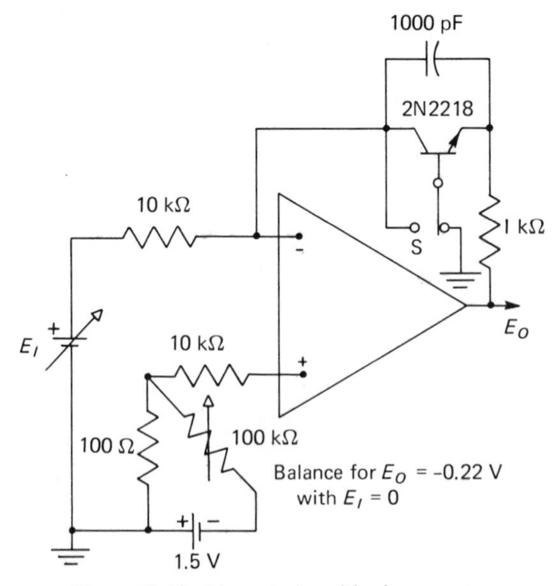

Figure 15.10. Linear to logarithmic converter.

packaged circuits for multiplication and division are available, and high-performance circuits need not be designed. However, logarithmic converters may be easily designed for less demanding applications.

With a pair of diodes parallel-connected back to back, as shown in Fig. 15.11, an amplifier responds logarithmically to both positive and negative inputs. In this form an amplifier may be used, when the signal to be observed varies over a very wide range, as in the output of a bridge. The gain control R_1 permits adjusting the gain of the amplifier, although the range of the logarithmic response is reduced as the gain is increased.

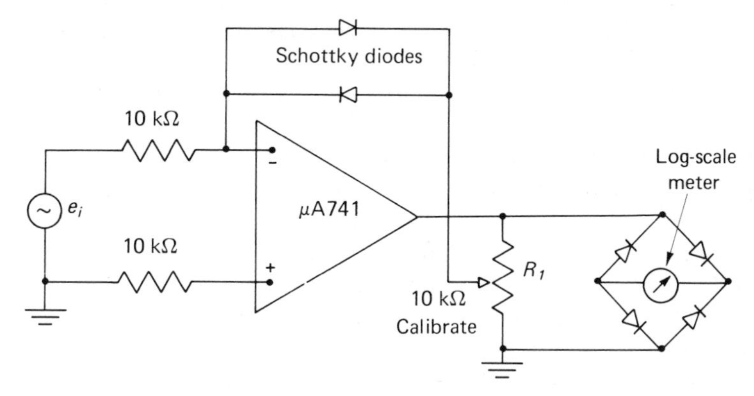

Figure 15.11. Log amplifier with 4- to 5-decade compression.

15.14 IC POWER AMPLIFIERS

The power output of a linear IC is generally between 10 and 100 mW. The peak-to-peak output voltage, the voltage swing, may be nearly equal to the total supply voltage when equal plus and minus supplies are used. When higher power outputs or lower load impedances are required, a power amplifier may be direct-coupled to the output of an IC. If the peak voltage output is adequate but the load has a low impedance, the complementary-symmetry power amplifiers described in Chapter 8 may be used as direct-coupled power boosters. Generally, with a minor adjustment of the compensation network, the booster amplifier can be included in the overall feedback loop.

ICs are now available that have power transistors in the package with metal ears for attaching heat sinks. These are hybrid amplifiers in which the power transistors are manufactured separately from the IC and enclosed in a common package. Because the entire package may operate at nearly 150°C, these amplifiers have a low overall voltage gain and include resistors to limit the S-factor of most of the stages. These hybrid amplifiers are available with power output ratings up to 100 W.

15.15 SUMMARY

Integrated circuits give the designer high performance, superior reliability, and known temperature characteristics at a small cost, as compared with a circuit that uses discrete components. A high-gain IC with feedback provides characteristics that are predictable and relatively independent of the amplifier. These characteristics are important advantages in systems requiring exact reproducibility or a mathematically accurate frequency response and gain.

Monolithic integrated circuits are produced economically only for applications that use thousands of identical circuits. Hybrid circuits have advantages either in small-quantity applications or in circuits that require low-noise, high-performance devices that are not available in ICs. Integrated-circuit arrays may be used for very small quantity designs, and some monolithic ICs can be adapted to special purposes by connecting external components.

A typical high-gain IC is a direct-coupled, two-stage differential amplifier followed by a quasi-complementary output amplifier. The differential amplifier simplifies the application of feedback, rejects common-mode signals, and has a low *Q*-point drift with temperature. The amplifier voltage gain is generally produced by two *CE* stages, and the input-output impedance step-down is produced by *CC* stages or Darlington pairs. Additional *CE* stages may be used to produce a high common-mode rejection, additional gain, or internal feedback. However, an amplifier with more than two *CE* stages in the main amplifier may be expected to require a second frequency-compensating network.

The applications described in this chapter illustrate the use of linear ICs as high-gain and as high-input-impedance amplifiers. The phono preamplifier illustrates the use of high gain and feedback to provide an accurate frequency response. An ac rectifier-type voltmeter illustrates the use of nonlinear feedback elements to produce a linear meter response. Similarly, a high-gain IC with a diode feedback element is shown to produce a logarithmic output signal with a 5- to 7-decade change of input signal. Hybrid-type OP amps that have FET input stages, chopper drift control, or similar advantages may be used to serve many applications besides the analog computer.

The designer has the opportunity to save much time and hard work by using ICs in systems requiring complex circuits. The small size and weight of ICs make them useful in watches, in small instrument packages, and for implantation in the body. The superior reliability of a complex IC is an advantage in any application. In consumer equipment the ICs are proving economical mainly to replace substantial parts of TV, radio, of hi-fi circuits. Recently developed ICs include the sound, IF, and AFC sections of color receivers, remote TV control circuits, and the IF and audio sections of FM receivers. These circuits will undoubtedly be used by designers for many applications besides their stated purposes.

16.1 JUNCTION-TYPE FETs

The junction-type FETs are formed as a semiconductor filament of silicon called the *channel* and an associated diode called the *gate.* As shown in Fig. 16.1a, the channel is reduced to a narrow gate-like region by forming the diode on opposite sides of the channel. Junction FETs are formed either as *n*-channel or *p*-channel devices and are represented by the symmetrical symbols shown in Fig. 16.1b. A voltage that reverse biases the gate diode creates a field capable of controlling or *pinching off* current flow through the channel. Because the gate diode is reverse biased, the driving source sees a high input impedance. The FET can be operated without bias (zero bias) unless the peak input signal forward biases the gate by more than 0.5 V, the diode *threshold voltage*. In order to obtain high-voltage output signals, we must often let the input exceed 0.5 V and, therefore, we are forced to use bias and reduced voltage gain.

The junction FETs are readily available with a wide range of characteristics and in either *p*-channel or in similar *n*-channel types. Because the channel is a simple semiconductor filament, both ends are alike and the circuit symbols are symmetrical end to end. Whether an FET is used as a follower or an amplifier is found by noting the channel polarities. The *n*-channel device is used with the battery connected as if for a vacuum tube. The channel terminal connected to the negative side of the battery, as shown in Fig. 16.1c, is called the *source* (*S*), and the terminal connected to the positive side is the *drain* (*D*). A *p*-channel device requires the opposite polarities, negative side to the drain. We can remember the drain polarity by observing that the arrow on the symbol points in to a positive drain or away from a negative drain. The source and the drain should always be indicated on circuit diagrams by the letters *S* and *D*.

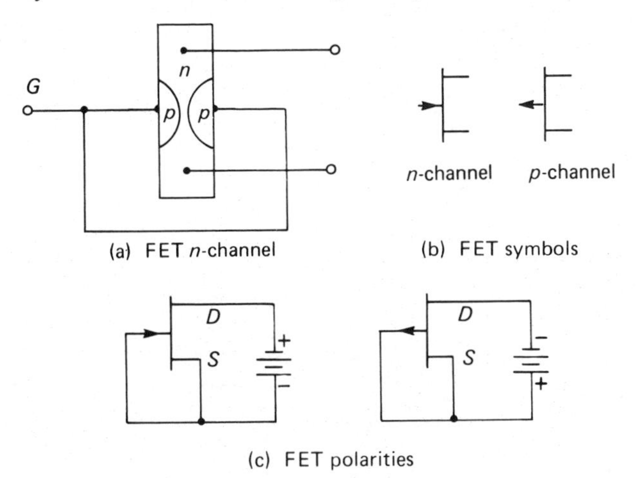

(a) FET *n*-channel

(b) FET symbols

(c) FET polarities

Figure 16.1. The junction FETs.

The static characteristics of a switching-type JFET are shown in Fig. 16.2. If this device is operated with zero gate bias [$V_{GS} = 0$], a 1-V change of the gate voltage for constant V_{DS} produces a drain current change of 3.0 mA. The zero bias mutual conductance g_0 is the ratio of the drain current change to the corresponding gate voltage change, which is 3000 μmho.

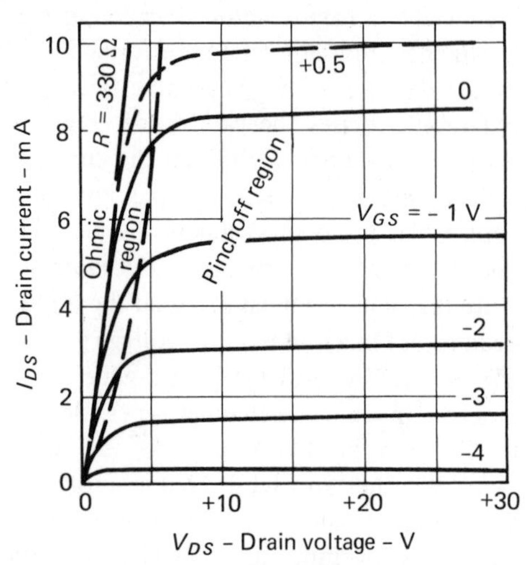

Figure 16.2. Junction field-effect transistor drain characteristics, *n*-channel general purpose.

The static characteristics of an amplifier-type JFET are shown in Fig. 16.3. This device has a low zero bias drain current I_{DON} and a mutual conductance of 1000 μmho. The figure of merit g_m/I_{DON} is 3.0 mho/amp for the amplifier type

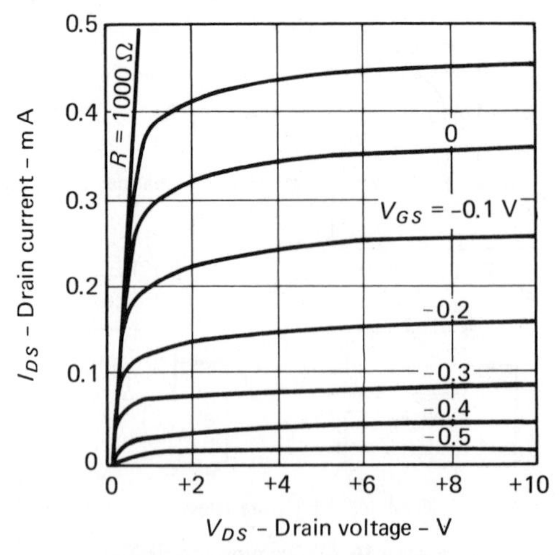

Figure 16.3. Junction field-effect transistor drain characteristics, *n*-channel amplifier.

and is only 0.35 mho/amp for the switching type. The advantage of a high figure of merit is explained in Section 16.6. Observe that the low g_m device has the higher figure of merit, and both have higher figures of merit at low drain currents.

The gate voltage required for channel current cutoff is called the pinchoff voltage V_p. The pinchoff voltage of FETs is usually between 2 and 8 V. The switching-type FET illustrated in Fig. 16.2 has a pinchoff voltage of 4 V and the amplifier-type pinchoff voltage is 0.6 V. For similar constructions, the figure of merit varies inversely with the pinchoff voltage.

The FET offers the advantages of simple biasing techniques, relative insensitivity to temperature, very low noise level, and useful power gain, even above 500 MHz. Because FETs have impedance and gain characteristics much like vacuum-tube pentodes, most tube circuits are easily converted to use FETs. They are also useful in many applications as constant current resistors and as voltage-controlled relays (switches), choppers, and modulators.

16.2 INSULATED-GATE FETs

The insulated-gate semiconductors are similar to the junction FETs except that the gate is formed (see Fig. 16.4a) as a small high-quality capacitor, thus permitting the control of current flow in the channel without the disadvantages of the gate diode. For the control of the channel current by an electric field, the capacitor must have exceedingly small dimensions and the silicon dioxide dielectric must be such a thin layer that it has a low breakdown rating—about 30 V. The gate capacitance is low, and the leakage resistance so high, that the gate is easily charged to a voltage above breakdown.

Gate damage is prevented only by using a grounded soldering iron and by protecting the gate from static discharges, as from one's body in dry weather. Insulated-gate FETs should be picked up by the case rather than by the leads, and the leads should be shorted together during handling by foil or a wire loop. Some devices have built-in Zener diodes to protect the gate from low energy sources, but reasonable precautions are still required in the application of

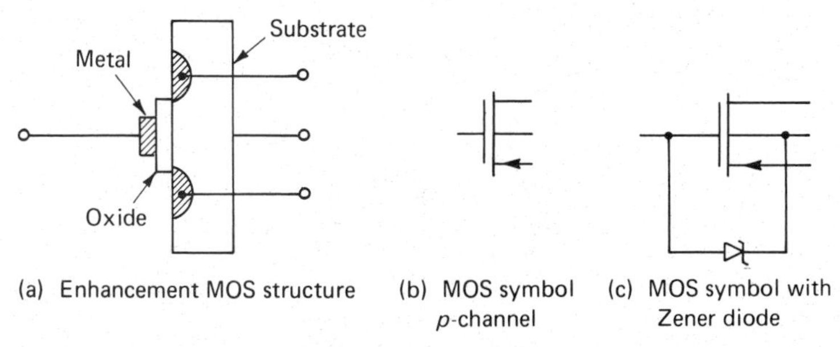

(a) Enhancement MOS structure (b) MOS symbol p-channel (c) MOS symbol with Zener diode

Figure 16.4. Enhancement MOS FET.

insulated-gate devices. However, the Zener diodes have the disadvantage of a high leakage current that cannot be tolerated in electrometer applications.

The MOS devices are available in two types, characterized by two different modes of operation—the *depletion mode* and the *enhancement mode*. In the depletion type the channel is a conductor and current normally flows in the conducting channel. A voltage applied to the gate reduces the current and may cut off the current completely. In the enhancement type the channel is normally cut off and is turned on by applying a gate voltage or bias.

Enchancement MOS devices are presently constructed as shown in Fig. 16.4a with a *p*-channel substrate as one side of the gate capacitor. The symbol for these devices is shown in Fig. 16.4b, and some devices, as shown in Fig. 16.4c, have a protecting Zener diode formed as a part of the structure. The *p*-channel enhancement MOS requires a negative drain voltage and a negative gate bias that allows the use of very simple bias techniques. As shown in Section 16.15, the enhancement MOS amplifier stages may be operated with the gate and the drain at the same potential, making it possible to direct couple a series of stages, even though the stages operate from a common power supply.

The static characteristics of an enhancement-mode MOS are shown in Fig. 16.5. This insulated-gate device is operated with a negative voltage on both the gate and the drain. With a gate bias of −7 V, the MOS has a g_m of 3000 μmho with a 4-mA drain current. The channel is cut off for gate voltages below −5 V. Observe that the MOS can be operated with equal gate and drain voltages, as indicated by the dashed line in Fig. 16.5. This line is the locus of *Q*-points that may be used for a direct-coupled amplifier operating with equal gate and drain voltages.

Depletion-type MOS devices are constructed by depositing a thin-film

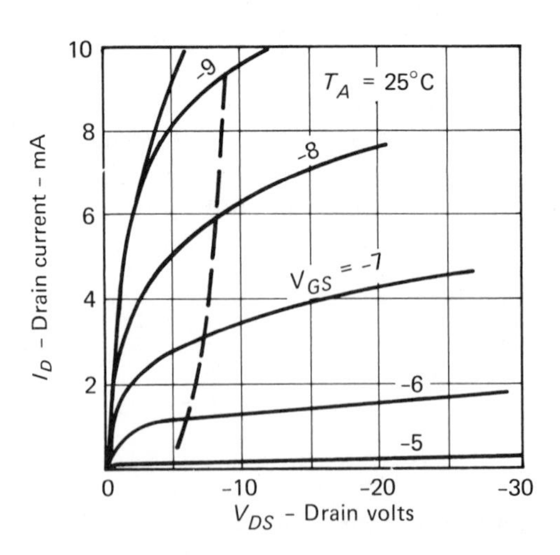

Figure 16.5. Transistor drain characteristics, *p*-channel enhancement type.

n-channel on a semiconductor substrate and by depositing the oxide dielectric and metal plate of the gate, as shown in Fig. 16.6a. The symbol for the n-channel MOS, given in Fig. 16.6b, shows a substrate terminal that is usually connected to the source. A depletion MOS operates with positive, negative, or zero bias and is, therefore, approximately interchangeable with the JFETs. The insulated-gate devices offer advantages at high frequencies (200 MHz) and in integrated circuits, but IGs presently have higher noise levels and are not as reliable as the JFETs. Exceedingly high gate resistance and low input capacitance make these devices attractive for high input impedance applications, as in the input stages of electrometers.

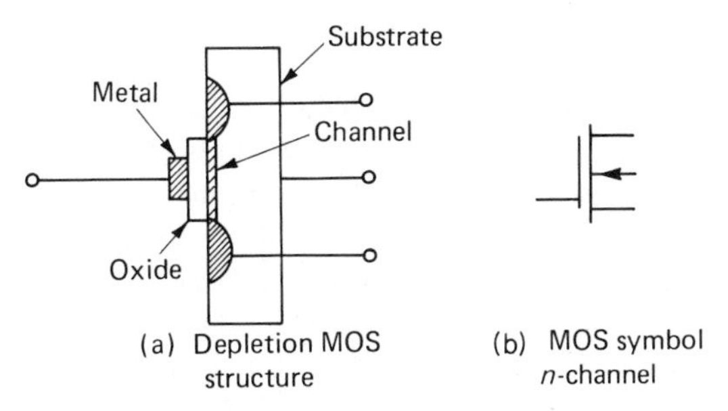

Figure 16.6. Depletion MOS FET.

The static characteristics of an n-channel depletion MOS are shown in Fig. 16.7. This device may be operated with zero bias and both positive and negative gate voltage excursions. The insulated gate prevents gate current. Observe that the static characteristics show that the MOS device represented in Fig. 16.7 is nearly interchangeable with the JFET represented in Fig. 16.2.

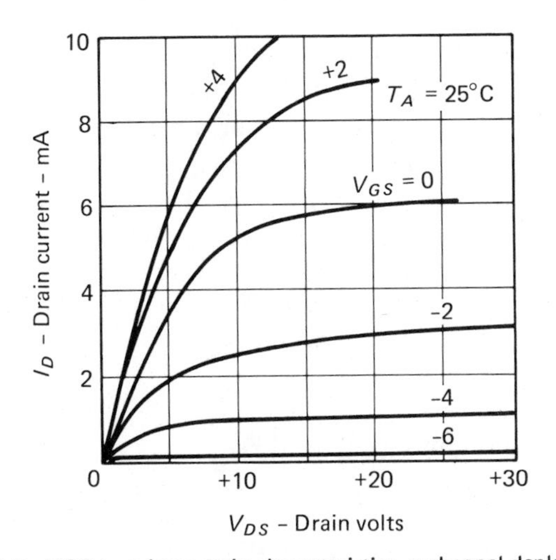

Figure 16.7. MOS transistor drain characteristics, n-channel depletion type.

16.3 RESISTANCE-COUPLED FET AMPLIFIERS

A simple resistance-coupled (R-C) *n*-channel FET amplifier is shown in Fig. 16.8. With a signal applied to the gate, as indicated by the sine wave, the source follows with the same signal polarity but at about six tenths the gate signal amplitude. Because the gate resistance is 300 times the source load resistance, the source follower has a power gain of $(0.6)^2$ (300), or about 100 times (20 dB). The signal at the drain terminal is of opposite phase and is four times the input signal; hence, the power output is about 700 times the power input (28 dB).

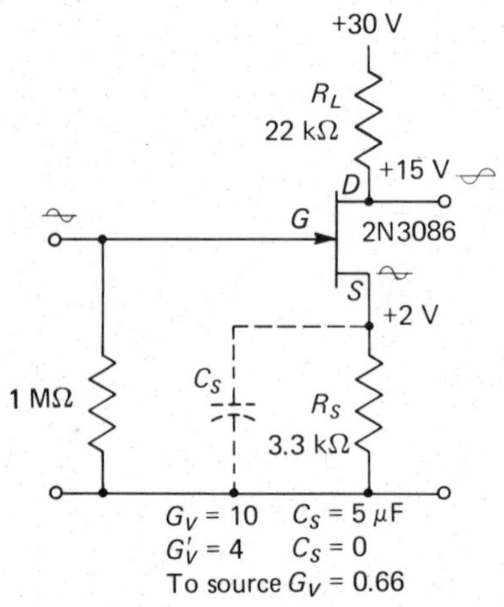

Figure 16.8. FET amplifier for *n*-channel FET or depletion MOS

The source resistor R_s furnishes a negative gate bias that lowers the drain current and raises the drain voltage V_D to about one half the drain supply voltage V_{DD}, thus permitting large ac output signals. The source resistor also causes negative feedback, which stablizes the *Q*-point and the ac gain, thus offsetting a part of the difference between one FET and another of the same type number. If used, a 5-μF capacitor connected across R_s reduces ac feedback and raises the voltage gain to 10.

Zero-bias operation is effected by shorting R_S and lowering the drain resistor value enough to offset the effect of increased drain current and restore the drain *Q*-point voltage. For many applications the simplicity of zero-bias operation justifies the selection of FETs or an adjustment of the load resistor. A 2N3086 with zero bias and a 10-kΩ resistor will give a voltage gain of about 7.

Amplifiers using depletion-type MOS devices may be constructed in the same way as FET amplifiers, e.g., as shown in Fig. 16.8. The characteristics of the two devices are similar enough to make them essentially interchangeable. The MOS, like the FET, gives greatest voltage gain when used with a bias and a high resistance load. With zero bias and a low-value load resistor the gain is low, but the MOS amplifier tolerates larger input signals.

16.4 ENHANCEMENT MOS AMPLIFIERS

A single-stage amplifier using an enhancement MOS 2N3608 is shown in Fig. 16.9. This amplifier uses a *p*-channel device so that the drain side of the battery is negative. Bias is supplied through a 20-MΩ drain-to-gate resistor, and a drain

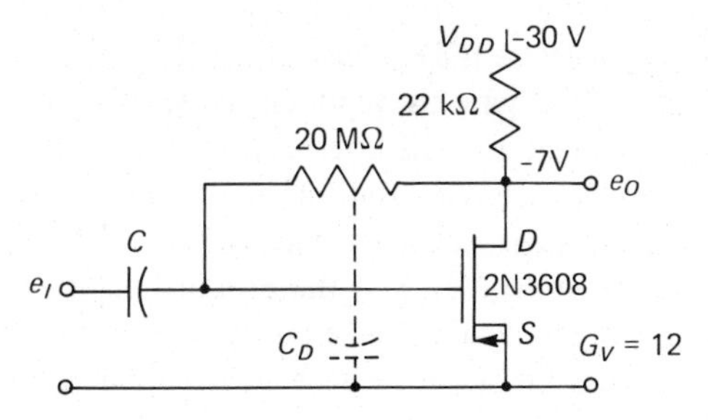

Figure 16.9. MOS amplifier; *p*-channel, enhancement MOS.

voltage of at least 20 V is required. Because the gate is practically an open circuit, the gate and drain voltages are equal (7 V) and about 1 V above the gate threshold voltage. Because the bias resistor R_f causes shunt feedback, the input impedance is:

$$R_I = R_f/(G_v + 1) \tag{16.1}$$

where G_v is the gate-to-drain voltage gain. For this amplifier G_v = 12; hence, the input impedance is 1.5 MΩ. As long as the driving generator impedance is at least a factor of 5 smaller than the input impedance, the generator is not loaded and full voltage gain is obtained. This requirement is easily met in most amplifiers; thus, the effect of feedback in reducing the input impedance can be neglected except when selecting the interstage coupling capacitor. If the midpoint of R_f is bypassed to ground, the input impedance becomes 10 MΩ.

16.5 FET AND MOS AMPLIFIER GAIN CALCULATIONS

Field-effect devices are high input impedance and high internal output impedance devices, similar to a vacuum-tube pentode. Amplifier gain calculations are greatly simplified by assuming that the input and output impedances are so high that they can be neglected. The gain calculated under the assumption is sufficiently accurate for practical circuit analysis. However, a method that takes the internal output impedance into consideration is described also so that we may understand both the exact and the approximate formulas.

Field-effect devices, like small-signal pentodes, may be represented by the equivalent circuit shown in Fig. 16.10. The dc gate and drain supply

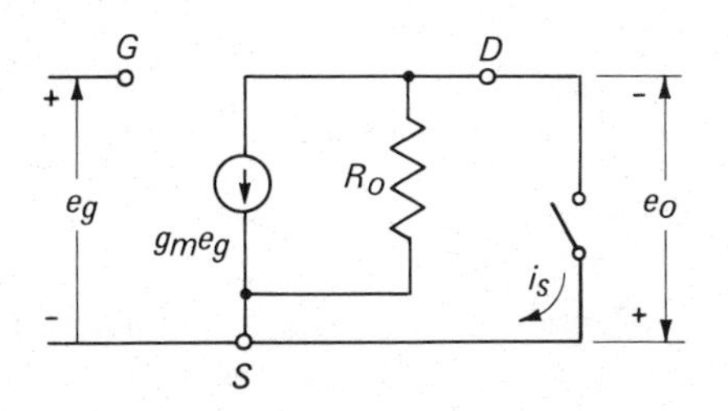

Figure 16.10. Norton equivalent circuit.

voltages are not shown because the equivalent circuit is used only to calculate ac currents and voltages. The gate is represented simply as an open-circuited terminal having an ac voltage e_g. The internal output impedance of the device is represented as an ac resistance R_O. The internal generator (the arrow enclosed by a circle) is represented as constant current source $g_m e_g$. This current is proportional to the gate voltage e_g, and the proportionality factor g_m is called the *transconductance* of the device. The equivalent circuit means simply that an ac input voltage e_g across the input resistor creates a constant ac current $g_m e_g$ through any connected load. When the drain is externally shorted (ac short) to the source, the entire generator current flows through the short and is

$$i_s = g_m e_g \tag{16.2}$$

so that

$$g_m = \frac{i_s}{e_g} \tag{16.3}$$

As equation (16.3) indicates, the mutual conductance is the short-circuit current per volt input.

When the drain is open, the generator current flows through R_O, and the drain ac voltage becomes

$$e_O = g_m R_o e_g \tag{16.4}$$

By definition, the *amplification factor* μ of a device is the open-circuit voltage gain, e_O/e_g. Therefore, equation (16.4) gives Van der Bijl's well-known (vacuum-tube) relation

$$\mu = g_m R_o \tag{16.5}$$

Field-effect devices are sometimes represented by the Thévenin equivalent circuit of Fig. 16.11 where the generator is a constant voltage (rather than constant current) source μe_g and the internal output impedance of the device is in series with the generator. Both equivalent circuits lead to the same result, but with a high internal impedance the Norton equivalent is simpler to use. One should be able to use either equivalent, whichever is more convenient.

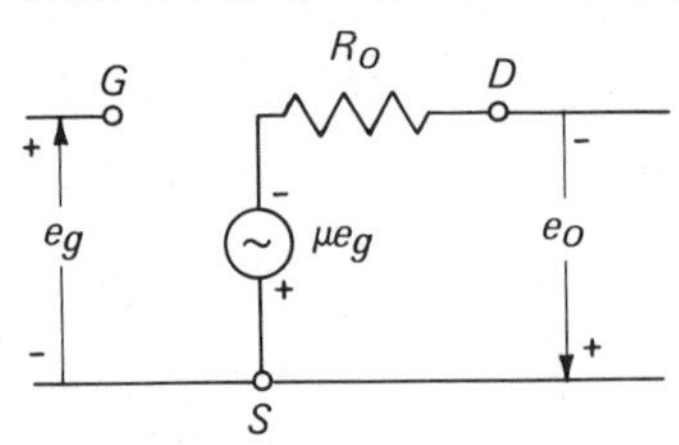

Figure 16.11. Thévenin equivalent circuit; $\mu = g_m R_O$.

Consider an amplifier represented by the Norton equivalent shown in Fig. 16.12 with a load resistor R_L connected externally from the drain to the source.

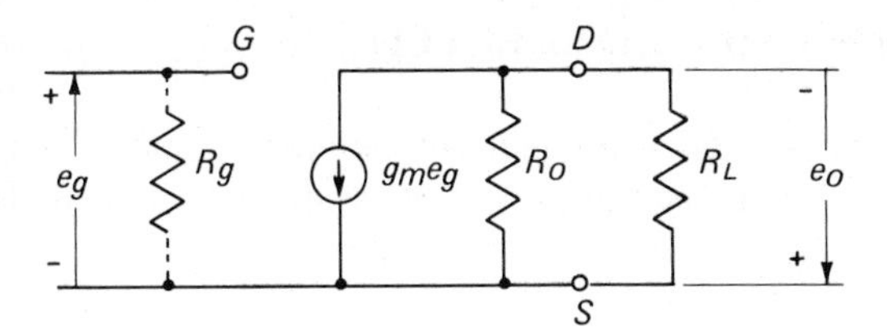

Figure 16.12. Norton equivalent of FET amplifier with load R_L.

The load R_L is in parallel with the internal impedance R_O, so instead of equation (16.4), we write

$$e_O = g_m \frac{R_o R_L}{R_o + R_L} e_g \qquad \textbf{(16.6)}$$

Field-effect devices usually have an internal impedance that is at least five times larger than the load; that is,

$$R_o = 5R_L \qquad \textbf{(16.7)}$$

Combining equations (16.7) and (16.6), and solving for the voltage gain, we find

$$G_v = 0.83 g_m R_L \qquad \textbf{(16.8)}$$

If it can be assumed that the internal output impedance is very large and negligible, then

$$G_v = g_m R_L \qquad \textbf{(16.9)}$$

A comparison of equations (16.8) and (16.9) shows that a voltage-gain calculation, neglecting the internal impedance of a field-effect device, will indicate a voltage gain that is 10 to 20 per cent too high. This small error is not the real difficulty in a gain calculation. Rather, the difficulty is that the transconductance of field-effect devices is rarely given within a 2:1 or 3:1 limit by the manufacturer, and the Q-point conditions existing in a practical amplifier are rarely as favorable as those given on the data sheet. Equation (16.9) is accurate enough for practical gain calculations providing we are careful in selecting a value of g_m appropriate for the Q-point and the nontypical device in use. Let us see how the operating Q-point affects the transconductance of a field-effect device.

16.6 THE TRANSCONDUCTANCE OF FETs

The transconductance g_m of FETs in the pentode or pinchoff region is found experimentally to vary as the square root of the drain current I_D. This fact may be written as

$$g_m = g_0\sqrt{\frac{I_D}{I_{DON}}} \tag{16.10}$$

where g_0 is the transconductance at the zero bias drain current I_{DON}. The manufacturer usually gives typical values for g_0 and I_{DON}. The theory of FETs indicates that g_0 is related inversely to the gate voltage V_P required to pinch off the drain current; that is,

$$g_0 = \frac{2I_{DON}}{V_p} \tag{16.11}$$

From equations (16.9) and (16.10) the voltage gain of a zero-biased amplifier is

$$G_v = g_0 R_L \tag{16.12}$$

and substituting g_0 from (16.11) we have

$$G_v = \frac{2I_{DON}R_L}{V_p} \tag{16.13}$$

In a resistance-coupled amplifier the dc voltage drop in R_L will be approximately one half the drain supply V_{DD}; hence, equation (16.13) may be written as

$$G_v \cong \frac{V_{DD}}{V_p} \tag{16.14}$$

Equation (16.14) shows why high supply voltages are desirable and the advantage in using devices having low V_P values. The advantage is expressed commercially by calling g_0/I_{DON} the *figure of merit* of an FET. Equation (16.11) implies that high figure-of-merit devices have low V_P values. Using equation (16.10), we can show that biasing an FET to reduce the drain current tends to improve the figure of merit. Hence, an amplifier will be satisfactory if biased to about one half the zero-bias drain current. The calculated voltage gain is usually found to be reasonably in agreement with the measured gain if the g_m of the device is assumed to be one half to two thirds the zero-bias value.

MOS devices may not conform exactly to the relations given above, but equation (16.10) approximates actual operating conditions sufficiently for most gain calculations (and also for vacuum-tube pentodes). We can assume, therefore, that the gain calculations outlined here for FET amplifiers are reasonably applicable also to insulated-gate and pentode applications.

16.7 FET GAIN CALCULATIONS: AN EXAMPLE

The amplifier shown in Fig. 16.8 will be used as an example for a gain calculation. We begin by drawing the equivalent circuit showing the externally connected resistors. With the bypass capacitor C_s in place, the source resistor is ac short circuited and the equivalent circuit is that of Fig. 16.12. The amplifier has a resistor R_g from gate to ground that is usually placed across the input in order to fix the input impedance at a known value, independent of temperature and humidity. Usually this resistor is so high it can be neglected as a load on the previous stage or generator. Hence, we assume that the gate is driven by a known signal voltage e_g.

The manufacturers give the zero drain g_0 as between 400 and 1200 μmho. The typical value is given as 800 μmho. However, the manufacturers give the typical zero-bias drain current I_{DON} as 1.5 mA, whereas the amplifier is biased and the drain current is only 0.7 mA. Using the typical g_0 in equation (16.10), we find that g_m is about 560. Equation (16.9) then predicts a voltage gain of 12. For practical purposes, the calculated gain agrees with the measured gain as closely as we may expect. However, because g_0 may vary by a factor of 3, so may the voltage gain. When a more uniform voltage gain is required, we are forced either to select the FETs or to use gain-stabilizing negative feedback.

16.8 SIGNIFICANT FEEDBACK

The variations in gain caused by the differences between FETs of the same type number are so great that local feedback must be used, as with transistors, to stabilize the gain of an amplifier. In a single-stage amplifier, as shown in Fig. 16.8, the voltage drop in the source resistor R_S causes a part of the drain voltage to appear in series opposition to the input voltage.

The effectiveness of feedback in improving an amplifier is approximately proportional to the gain reduction produced by the feedback. If the distortion or the gain variation of an amplifier is to be reduced by a factor of 4, we must reduce the gain of the amplifier to one fourth the gain without feedback. These statements imply that the gain without feedback is known or is measured. In actual practice, we need to know how to determine if a given stage is operating properly and that there is enough feedback to ensure a significant improvement in the gain stability of an amplifier—without knowing or measuring the gain without feedback. We shall show that the amount of local feedback in a stage may be easily estimated by measuring the ac input and feedback signals.

As an example of significant feedback consider the amplifier shown in Fig. 16.13, designed to have just enough feedback to reduce the voltage gain G_v by a factor of 4, 12 dB of feedback. Expressed algebraically, significant feedback exists because the gain with feedback G_v' is just one fourth the gain without feedback; that is,

$$G_v' = \frac{G_v}{4} \tag{16.15}$$

The ac voltages shown in Fig. 16.13 represent ac signals existing with a 4-mV input signal, this signal being chosen to make the gate-to-source voltage a normalized 1 mV. The 1-mV gate-to-source voltage produces a 12-mV signal across the load R_L; the resistor R_S was selected by the designer to establish 3-mV ac from the source to ground. To maintain these signals, we must supply a 4-mV gate-to-ground input signal; hence, with feedback the stage gain is 3. The 1-mV gate-to-source voltage represents the input signal required if the source feedback signal is bypassed; hence, the input signal is four times larger with feedback than without.

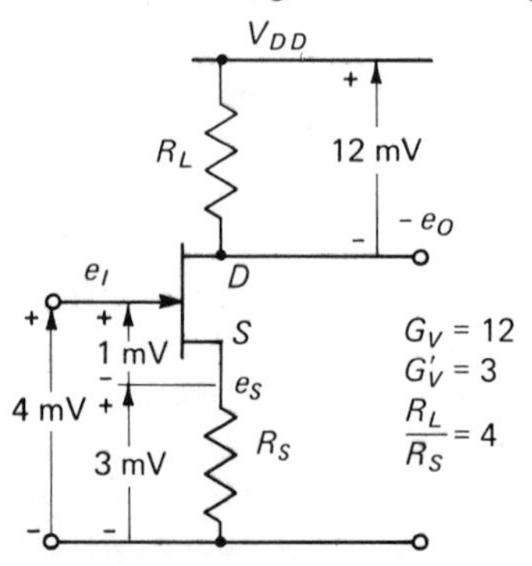

Figure 16.13. Feedback amplifier with minimum significant feedback; $e_s = 0.75e_I$ and $G_V' = 0.75\,(R_L/R_S)$

The important facts to be recognized are: (1) the gain is reduced because of the signal conditions existing in the input circuit; (2) the 4-to-1 relationship is shown by the input signals; and (3) the gain reduction is 4 because the feedback signal is three fourths the input signal. Statement (3) says that significant feedback exists if the ac source voltage is at least three fourths the amplifier input signal:

$$e_S = \frac{3}{4}\,e_I \tag{16.16}$$

Feedback in the amplifier may be removed by using a capacitor to bypass the signal existing across the source resistor. With the capacitor, a signal input of 1 mV maintains a load voltage of 12 mV; therefore, the no-feedback voltage gain is

12. Because the resistor ratio required to establish effective feedback varies with and is determined by the amount of voltage gain available without feedback, we often need to know how to set the resistance ratio when the voltage gain without feedback is known. The present example shows that significant feedback exists if the resistance ratio is one third the gain without feedback; that is,

$$\frac{R_L}{R_S} = \frac{G_v}{3} \tag{16.17}$$

Eliminating G_v from equations (16.15) and (16.17) produces a useful relation for the gain with feedback in terms of the resistance ratio:

$$G_v' = \frac{3}{4}\frac{R_L}{R_S} \tag{16.18}$$

The algebraic generalization of equations (16.17) and (16.18) is usually written as

$$G_v' = \frac{G_v}{1 + B_v G_v} \tag{16.19}$$

where

$$B_v = \frac{R_S}{R_L} \tag{16.20}$$

The voltage loss factor B_v is the reciprocal of the resistance ratio R_L/R_S. We should observe that equation (16.19) is simply a mathematical representation of the voltage relations existing in the amplifier input circuit. Our definition of significant feedback is equivalent to a requirement that the feedback loop gain $B_v G_v$ be 3, which is the meaning of equation (16.17).

Equation (16.17) tells the designer that significant feedback requires a resistance ratio that is one third of the gain without feedback. Equation (16.18) tells the experimenter that significant feedback actually exists if the measured gain with feedback is three fourths the resistance ratio. If the gain is lower than three fourths the resistance ratio, the feedback is not of practical significance. Equation (16.16) is a similar test for significant feedback in terms of the input voltage and the feedback voltage. All three tests apply equally well to transistor amplifiers.

16.9 SOURCE FEEDBACK: AN EXAMPLE

As a practical example of source feedback, consider the amplifier in Fig. 16.14 with the measured signal voltage normalized to a 1-mV gate-to-source voltage. The 1-mV gate-to-source signal may be easily set up by using a differential oscilloscope connected to show this voltage differentially. A comparison of the input circuit voltages shows that the source voltage is only 0.6 of the input voltage. Hence, the amplifier fails to have what we have called *significant* feedback. The voltage gain without feedback is 10 and the resistance ratio R_L/R_S is 6.7. To provide significant feedback, we must make the resistance ratio 3.3 by equation (16.17). Changing R_L does not change the amount of feedback because the voltage gain changes also. Increasing the source resistor does increase the feedback, but the increased dc bias adversely affects the Q-point and a positive gate bias is required to restore the Q-point conditions. Evidently, the 3.3-kΩ resistor is used only for bias.

In a laboratory study of amplifiers the amount of local feedback may be easily estimated by comparing the input circuit voltages and using equation (16.16) as the test for significant feedback. When there are several stages having overall feedback, the amount of feedback has to be determined by comparing the input signal with the gate-to-source signal or its equivalent. The amount of feedback, the ratio of the gain without feedback to the gain with feedback is

$$F_B = \frac{e_I}{e_I - e_S} \tag{16.21}$$

With a large amount of feedback the input and the source voltages tend to become alike, except for the distortion and noise introduced by the amplifier. For this reason the distortion and noise in the gate-to-source voltage, the difference $(e_I - e_S)$, may make it difficult to obtain a precise value of the denominator in equation (16.21). Likewise, the signal observed in the early stages of a feedback amplifier may have a poor wave form. However, the difference (or error) signal often gives a clue to feedback or servo-system difficulties.

16.10 HIGH-GAIN FET AMPLIFIERS

The amplifiers shown in Fig. 16.15 illustrate the advantage of using high figure-of-merit FETs, that is, low V_P devices. A high g_m to drain current ratio makes it possible to use a high-valued drain resistor and to obtain a high voltage gain. In addition, because the dc drain voltage must be at least as high as V_P, a high figure of merit implies that the drain may be operated at low voltages or from a low supply voltage. The second stage using a 2N3086 FET requires a low

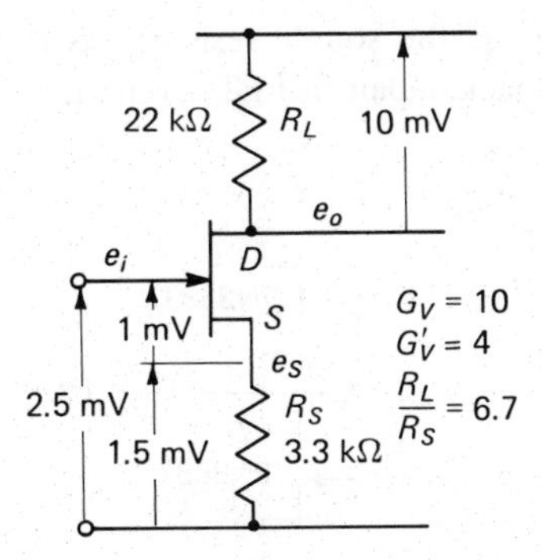

Figure 16.14. Feedback amplifier lacking minimum significant feedback; that is, $e_S = 0.60e_I$ and $G'_V = 0.6\ (R_L/R_S)$.

drain resistor with 6 V on the drain. This stage has a voltage gain of only 2, but with a negative gate bias the drain resistor can be increased until the voltage gain is about 10. The 2N3687 FET stage uses a 10 times higher drain resistor and with zero bias has a voltage gain of 25. A disadvantage of both amplifiers in Fig. 16.15 is that they become inoperative if the drain supply voltage is lowered without lowering the drain resistor. This difficulty can be minimized by using a lower-value drain resistor and accepting a lower gain.

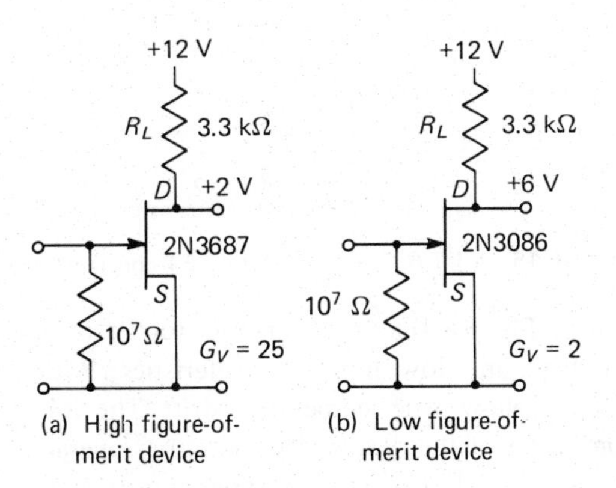

Figure 16.15. JFET *R-C* amplifier with resistor load.

The amplifiers in Fig. 16.16 show that higher voltage gains may be obtained by using a second FET as a constant current load in place of the drain resistor. When a pair is used as in Fig. 16.16, the FETs should be selected so that they are reasonably alike. The small source resistor R_S is used to balance the drain currents and to adjust for a maximum voltage gain. If maximum voltage gain is obtained with the source resistor shorted out, the FETs should be interchanged in order to use the higher I_{DON} device as the amplifier and to permit drain-current balancing. If the source resistor has to be made so high that the gain is reduced by feedback, a pair of FETs even more alike must be selected.

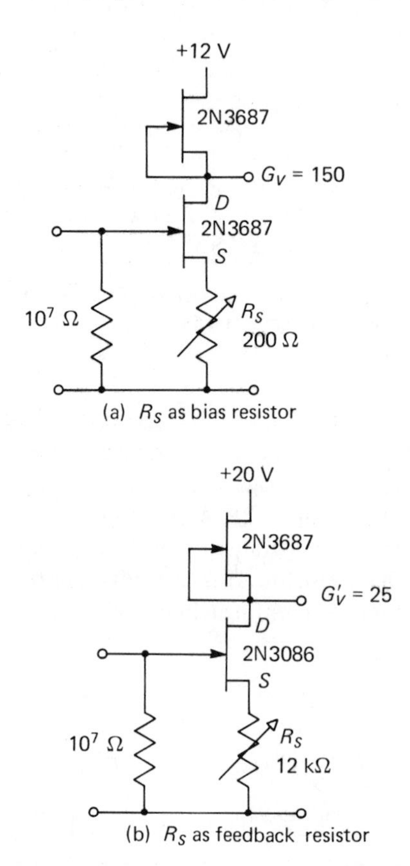

Figure 16.16. JFET *R-C* amplifier with FET constant current load.

The amplifier in Fig. 16.16a is remarkable for its high voltage gain, high temperature stability, and low noise characteristics. With similar FETs the collector *Q*-point will show little temperature drift. The voltage gain is one half the amplification factor of the device. This amplifier operates well with supply voltages of only 4 V, but the voltage gain varies approximately in proportion to the supply voltage. By replacing the lower FET with a high drain current device (Fig. 16.16b), the source feedback resistor may be increased to provide significant feedback. Using a 2N3687 FET in combination with a 2N3086 FET and a 12-kΩ source resistor, we can reduce the voltage gain to 25 from a no-feedback gain of 125. With a 15- to 30-V drain supply—well above the device pinchoff voltage—the gain changes are reduced significantly by the feedback.

16.11 HIGH-GAIN DIRECT-COUPLED FET AMPLIFIER

The amplifier shown in Fig. 16.17 is interesting because of its high voltage gain with feedback, 330, and because the amplifier has over 20-dB feedback. The voltage gain is provided by the input JFET and by the MOS FET, which is operated as a common source stage. The second JFET operates as a constant-current, high-impedance load so that the voltage gain of the input stage is about one half the mu factor of the input JFET. The MOS FET gives an additional voltage gain of about 100. The amplifier is operated open loop with difficulty and requires careful adjustment of the source resistor. With the loop closed, there is adequate feedback and the bias may be easily adjusted to permit satisfactory operation over a wide range of the supply voltage.

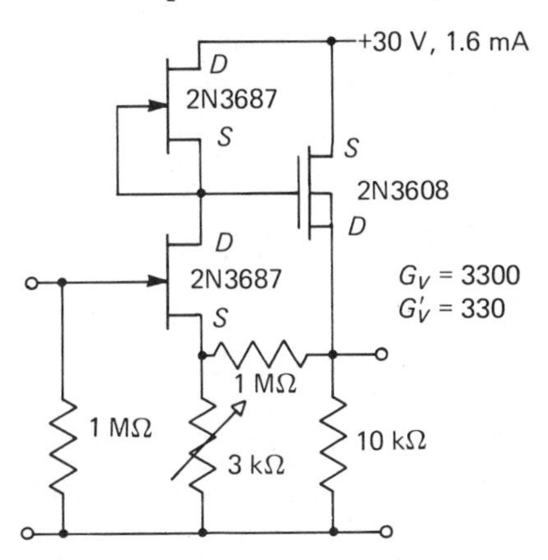

Figure 16.17. High-gain feedback amplifier.

Replacing the MOS FET by an npn emitter follower (2N1711) changes the feedback from negative to positive and reduces the open-loop gain to about 100. The emitter follower cannot have voltage gain, but the positive feedback increases the gain by a factor of 2 or 3 for a closed-loop gain of about 250.

16.12 METER AMPLIFIER

The amplifier shown in Fig. 16.18 illustrates the simplicity with which a pair of FETs may be used to convert a dc voltage, as from a high resistance bridge to a meter reading. The FETs should be selected for similar characteristics, and the meter may be adjusted to read zero or midscale with zero input by shunting one of the drain resistors in order to lower its value. If the FETs have a low-value pinchoff voltage, the battery voltage may be as low as 3 V. Because the amplifier requires only 300 to 400 μA, the meter amplifier operates for long periods on flashlight cells. Once adjusted for balance, the gain and the balance are essentially independent of the supply voltage, and the balance requires only an occasional adjustment. The meter response is about 400-μA/V input to the amplifier (400 μmho).

16.13 FETs IN VIDEO AND RF AMPLIFIERS

Field-effect devices have high voltage-gain-bandwidth products, 100 MHz or more. This means that a broad-band amplifier may be designed with a voltage gain of 10 up to about 10 MHz. Similarly, a tuned RF amplifier may have a voltage gain of 10 with a bandwidth of 10 MHz centered at the resonant frequency. If the voltage gain is reduced to 2 by reducing the circuit impedances, the frequency response may be approximately constant over a frequency band of 50 MHz. This limiting of the bandwidth is caused by the interelectrode and feedback capacities in the FET and in the external circuits. One should note, however, that the gain-bandwidth of a device usually cannot be attained unless that amplifier is suitably designed. For example, we cannot have both a high-impedance level and a high gain-bandwidth.

The JFETs have a drain-to-gate feedback capacitance that is typically 1 to 3 pF, and they may be used in low-frequency radio circuits almost interchangeably with vacuum tubes. However, insulated-gate devices have important advantages in high-frequency applications, and recent improvements make the MOS FETs a better choice for RF circuits. The MOS FETs offer lower noise, lower distortion, and a factor of 10 lower feedback capacitance. For these reasons the MOS FETs seem to be replacing both the bipolar and junction semiconductors in RF applications.

For VHF and UHF applications MOS FETs are currently being used in RF amplifiers, mixers, demodulators, and similar high-frequency circuits. The FETs have the advantage of a square-law transfer characteristic that provides a low value of cross modulation and greatly reduces spurious responses or interference from undesired signals. The triode types of MOS FETs have drain-to-gate feedback capacities that are typically 0.1 to 0.7 pF with transconductance values of 7500 μmho. MOS FETs with dual insulated gates have a feedback capacitance

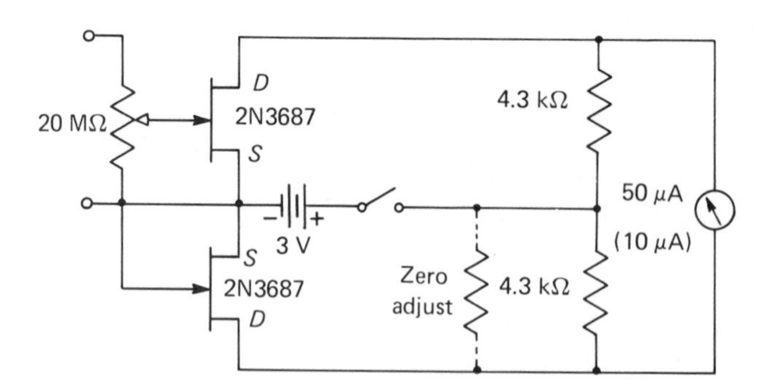

Figure 16.18. Meter amplifier for dc bridge.

as low as 0.1 to 0.2 pF with the transconductance exceeding 10,000 μmho. These devices offer unneutralized power gains of 15 to 20 dB at 200 MHz and 10 to 13 dB at 400 MHz. The dual gate reduces the feedback capacitance, and one gate may be used as the input while the other is used for biasing, AGC control, or for the LO input of a mixer. The low-value leakage current of the insulated gate eliminates input circuit loading caused by high input signals. Thus, with a maximum gain, the insulated gate permits large signal swings without causing either input detuning or loading.

The circuit of a UHF dual-gate MOS amplifier is illustrated in Fig. 16.19. The input circuit is designed and tuned approximately as for a vacuum-tube amplifier, while the output circuit is designed to provide as high a load impedance as the desired stability permits. Bias is provided to the second gate by the voltage divider R_2 and R_3, or the gate may be supplied with an AGC control voltage, or the LO signal for mixing. In a single-gate amplifier the gate may be supplied by returning the ground side of the inductor L_1 to the capacitor C_3. The design of the tuned circuits is essentially the same as outlined in Chapter 12. Circuits with component values and electrode voltages required for a specific FET are usually given in the manufacturer's data sheet.

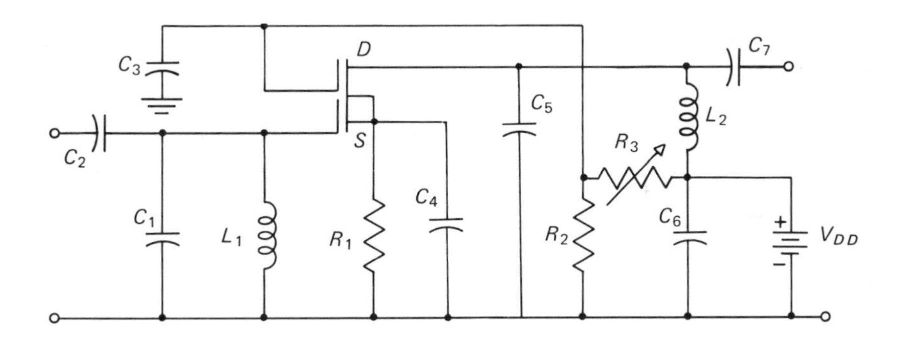

Figure 16.19. Dual-gate MOS FET RF amplifier.

16.14 FET-TRANSISTOR COMPOUNDS

Multistage amplifiers often use FETs and transistors in alternate stages or direct coupled as FET-transistor compounds. The FETs make it possible to use high interstage impedance levels–that is, smaller coupling and filter capacitors–and allow the use of high-impedance potentiometers as controls. Transistors supply high transconductances at low dc collector currents. Thus, amplifiers combining FETs with transistors can be expected to exhibit the advantages of both devices. One of the most useful combinations uses an FET direct coupled to a CE transistor amplifier.

The direct-coupled two-stage amplifier in Fig. 16.20 offers the high input impedance of the FET and a voltage gain of 100 with the Q-point and the overall voltage gain stabilized by significant feedback. These direct-coupled stages are easy to construct and offer high input impedances to above 100 kHz. High figure-of-merit FETs allow the use of high resistance interstage bias resistors and offer high voltage gains. FETs requiring a high drain current have to be loaded by a low-valued bias resistor, and the voltage gain is reduced. Thus, the amplifier will not have significant stabilizing feedback unless, by using a larger valued source resistor, the feedback is increased and the gain is further reduced. The amplifier in Fig. 16.20 is satisfactory for applications at room temperatures, but the high dc gain makes it unsatisfactory for wide temperature range applications.

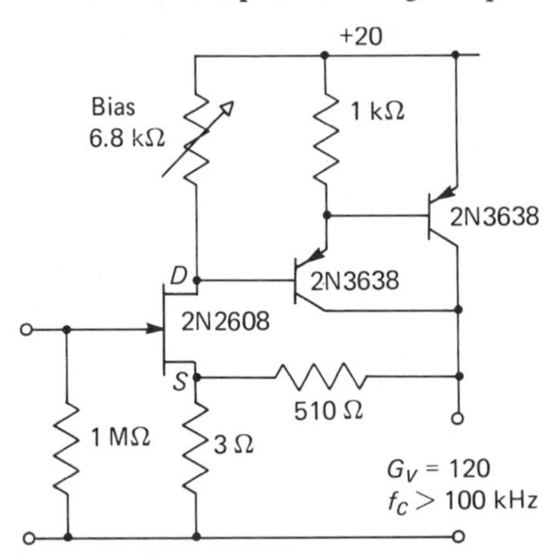

Figure 16.20. High input impedance FET-transistor-pair amplifier.

The amplifier in Fig. 16.21 is designed to be used in ambient temperatures from about −25 to +60°C. The wide temperature range is achieved by eliminating the direct coupling between stages and by introducing temperature compensation in the transistor amplifier to control the Q-point drift. The diodes in series with the transistor base resistor reduce the base bias voltage as the temperature increases and in this way control most of the Q-point drift.

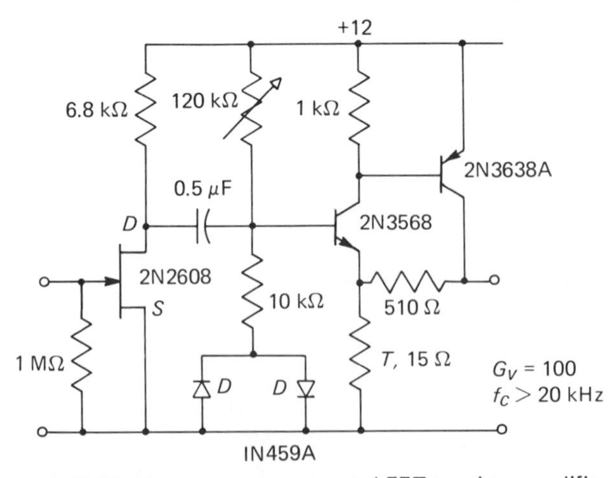

Figure 16.21. Temperature compensated FET-transistor amplifier.

The temperature drift can be further reduced by using a positive temperature coefficient emitter-feedback resistor and by increasing the feedback to reduce the voltage gain. Silicon resistors are available that have a 0.7 per cent/°C positive temperature coefficient. When the voltage across a silicon emitter resistor is 0.15 V, the net voltage change is 1.0 mV per °C. This voltage-temperature change can be adjusted easily by constructing the emitter resistor as a series or parallel combination of a carbon and a silicon resistor.

16.15 INSULATED-GATE FET AMPLIFIERS

The enhancement-mode, insulated-gate FETs makes it possible to construct simple direct-coupled multistage amplifiers. Two such three-stage amplifiers are shown in Figs. 16.22 and 16.23. The amplifier in Fig. 16.22 uses three identical enhancement-mode FETs with a feedback resistor connecting the output stage drain back to the input gate. With like drain resistors the dc feedback holds the

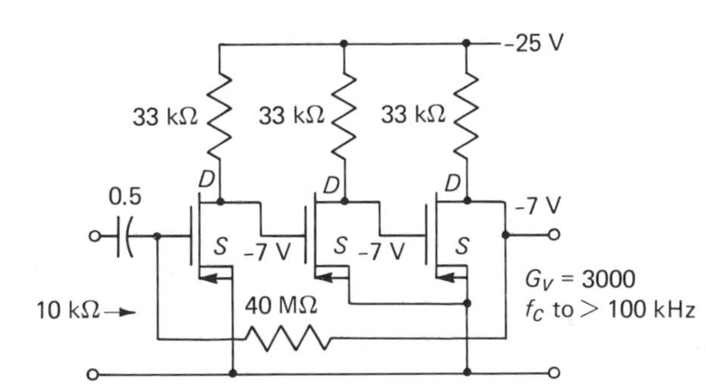

Figure 16.22. High-gain wide-band insulated-gate amplifier; all transistors 2N3608.

Q-points alike and independent of the supply voltage. Because of the feedback the input impedance of the amplifier is equal to the feedback resistor divided by the overall voltage gain; that is,

$$R_I = \frac{R_f}{G_v} \tag{16.22}$$

If the amplifier is driven by a source having an impedance that is less than the input impedance R_I, then the feedback is effectively bypassed and the gain is 3000. With an input capacitor having a reactance small compared with the input impedance, the amplifier has a high ac gain and a frequency response extending to about 100 kHz.

The amplifier in Fig. 16.22 becomes a dc amplifier having a voltage gain of 100 if the input capacitor is replaced by a 40-kΩ resistor R_g. Dc operation requires that the input terminal be connected to a point 7 V negative with respect to ground, and requires a well-regulated dc voltage supply. As a dc amplifier, the voltage gain with feedback equals the ratio of the feedback resistance R_f divided by the series input resistance R_g. The open-loop gain is about 3000 when R_g is zero.

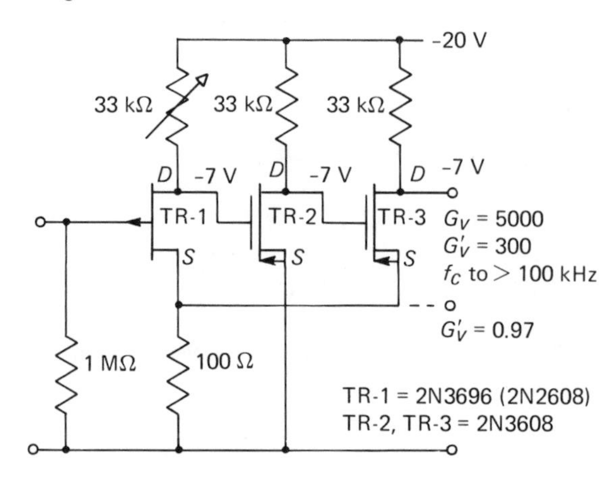

Figure 16.23. High-gain low-noise low-drift dc amplifier.

A similar three-stage amplifier (Fig. 16.23) uses a unipolar or a depletion-mode FET in the input stage in order to permit dc amplification from a grounded source. Feedback is introduced by connecting the first and third FETs to a common source resistor. In this amplifier the input impedance is determined by the gate resistor and the gate-to-drain capacitance. If the signal generator has an impedance of 1 MΩ, the upper cutoff frequency is 20 kHz. The amplifier requires a well-regulated voltage supply unless the dc gain is reduced by increasing the feedback resistor. The first-stage drain resistor serves as a bias adjustment for the second and third stages and will vary with the figure of merit of the FET in the first stage. The better FETs will tolerate a high-valued drain resistor and give a voltage gain of at least 15 in all 3 stages.

16.16 SUMMARY

JFET and MOS FET semiconductors have the advantages of high input impedance, simple biasing circuits, low noise levels, and a relatively low temperature change. Because FETs have gain and impedance characteristics much like the vacuum-tube pentode, most tube circuits may be easily modified to use FETs. The semiconductors have the additional advantage of being available in *n*-channel or *p*-channel types for use with either power-supply polarity.

Because of the high internal output impedance, the voltage gain of a FET stage is approximately $g_m R_L$. However, the mutual conductance g_m varies with the square root of the drain current, and, when biased for large signals, the g_m of the FET is usually one half to two thirds the zero-bias value. Thus, with resistance coupling, the gain of a stage is usually about two thirds the gain calculated with the zero-bias g_m. High voltage gain is obtained in resistance-coupled FET stages by using high figure-of-merit devices and as high a supply voltage as practical. FETs may be used as constant-currrent high-impedance loads in either FET or transistor circuits.

An FET stage has significant feedback only if the ratio of the load resistor divided by the source resistor is at least one third the voltage gain without feedback; and significant feedback actually exists in an operating amplifier if the series-feedback voltage is at least three fourths the input voltage.

MOS FETs have voltage-gain-bandwidth products exceeding 100 MHz with feedback capacities of less than 1 pF and g_m values exceeding 10,000 μmho. Thus, insulated-gate devices are being used in VHF and UHF circuits and have the advantages of low cross modulation, low noise, and simple bias and AGC circuits.

Enhancement-mode MOS FETs may be direct-coupled to produce high-gain dc feedback amplifiers, and FET-transistor compounds have the advantages of high input impedance and a high transconductance with low dc power inputs.

PROBLEMS

16-1. The FET in the amplifier of Fig. P16.1 has a mutual conductance of 800 μmho. (a) If the internal impedance is high enough to be neglected, what is the voltage gain of the amplifier? (b) If the internal impedance is so low that the voltage gain is only two thirds of the gain calculated in (a), what is the magnitude of the internal impedance? In selecting a semiconductor for this amplifier, what value of V_p and of I_{DON} would you recommend?

16-2. A given FET has a figure of merit of 2000 μmho/mA. (a) What is the indicated pinchoff voltage if the zero-bias drain current is 0.8 mA? (b) What is the voltage gain if this FET is operated in the amplifier in Fig. P16.2 by biasing the FET to make the drain voltage 20 V? (c) What is the voltage gain if the FET is used in the same circuit with zero bias?

16-3. Suppose the amplifier shown in Fig. 16.8 is required to have significant feedback after changing the FET but without changing the resistor values and voltages. (a) What is the required g_m of the new device at the indicated Q-point? (b) What is the required figure of merit and I_{DON} of the new device?

16-4. (a) If both FETs in Fig. 16.15 have a gate-to-drain capacitance of 2 pF and a gate-to-source capacitance of 3 pF, what is the input capacitance of each stage? (b) If the driving generator has an internal resistance of 50 kΩ, what is the high-frequency cutoff of each amplifier? (c) How do the voltage-gain-bandwidths compare?

16-5. Repeat Problem 16-4, using Fig. 16.16.

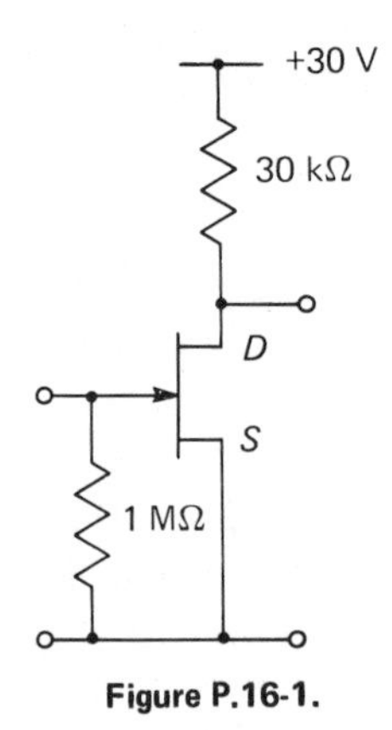

Figure P.16-1.

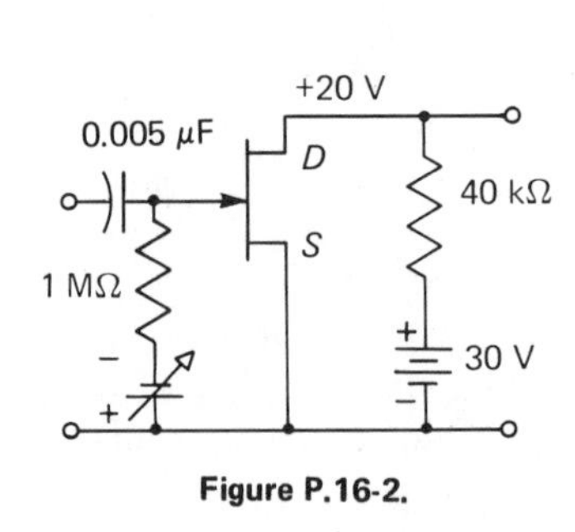

Figure P.16-2.

REFERENCES

16-1. *RCA Transistor Manual*, SC-13, "MOS Field-effect Transistors," pp. 93-109. Harrison, N.J.: Radio Corporation of America, 1967.

16-2. "Understanding and Designing with FETs." Phoenix, Ariz.: Motorola Semiconductor Products, Inc., 1970. (A collection of Application Notes)

DESIGNS

16-1. Construct the stage shown in Fig. 16.15a and determine whether significant feedback can be obtained by inserting a resistor between the source and ground. Determine whether the feedback can be increased by using a positive bias on the gate and increasing the source resistor.

16-2. Obtain an RF type of MOS FET and the manufacturer's data. Design a video amplifier with a 20-MHz bandwidth or a tuned RF amplifier for use at 50 MHz.

APPENDIX

A.1 SEMICONDUCTORS USED IN THIS BOOK

TRANSISTORS

JEDEC No.	Type	TO-	β Min. Max.	V_{CBO} (V)	Power (W)	f_T (MHz)	Use	Source
2N699	Si-*npn*	5	40-120	120	0.6	50	High voltage	F, M
2N1073A	Ge-*pnp*	41	20-60	–80	85	0.45	Audio power	M
2N1177	Ge-*pnp*	45	100-	–30	0.08	140	RF Amplifier	R
2N1304	Ge-*npn*	5	40-200	25	0.15	5	Switch, amp.	R, T
2N1305	Ge-*pnp*	5	40-200	–30	0.15	5	Switch, amp.	R, T
2N1711	Si-*npn*	5	100-300	75	0.8	70	Audio/RF amp.	F, M
2N2870	Ge-*pnp*	3	50-165	–80	30	0.2	Audio power	R
2N3137	Si-*npn*	5	20-120	40	0.6	500	Class-C RF	F, M
2N3215	Ge-*pnp*	37	25-100	–40	10	0.6	Audio Power	D
2N3440	Si-*npn*	5	40-160	300	10	15	High voltage	F, R
2N3569	Si-*npn*	105	100-300	80	0.3	60	Misc., high β	F
2N3638A	Si-*pnp*	105	100-300	–25	0.3	150	Misc., high β	F
2N3739	Si-*npn*	66	40-200	250	20	10	High voltage	M
2N3784	Ge-*pnp*	72	20-200	–30	0.15	700	UHF amp.	M
2N4135	Si-*npn*	72	25-200	30	0.2	425	Low-noise RF	F
2N4258	Si-*pnp*	106	30-120	–12	0.2	700	VHF amplifier	F
2N4355	Si-*pnp*	105	100-400	–60	0.35	100	Low-noise RF	F
2N4889	Si-*pnp*	106	80-300	–150	0.3	40	Low-noise	F
2N5068	Si-*npn*	3	7 @ 5 A	60	87	4	Audio power	M, F
2N5191	Si-*npn*	—	10 @ 4 A	60	40	2	Medium power	M
2N5194	Si-*pnp*	—	10 @ 4 A	–60	40	2	Medium power	M

Note: D-Delco, F-Fairchild, M-Motorola, R-RCA, T-Texas Instruments.

JUNCTION FIELD EFFECT

JEDEC No.	Channel Type	V_P max.	V_{Dmax} (V)	I_{DSS} (mA)	g_m (μmho)	Use	Source
2N2608	*p*	4	–30	0.1-0.5	1100-	Amplifier	Amelco
2N3370	*n*	–3.5	40	0.1-0.6	300-2500	Amplifier	Amelco, Solitron
2N3438	*n*	–2.5	50	0.2-1.0	800-4500	Amplifier	Amelco, Solitron
2N3578	*p*	4.0	–20	0.9-4.5	1200-3500	Amplifier	Siliconix
2N3687	*n*	–1.2	50	0.1-0.5	500-1500	Amplifier	Siliconix
2N4867	*n*	–2.0	40	0.4-1.2	700-2000	Low-noise	Siliconix

INSULATED-GATE FETs (MOS)

JEDEC No.	Type	Basing	V_{Dmax} (V)	I_{DSS} (mA)	g_m (μmho)	Use	Source
2N3608	*p*-enh't	DGBS	–20	1-10	200	Switch/amp.	Philco
2N4351	*n*-enh't	SGDSb	25	1-10	1000	Switch/amp.	Motorola
2N4352	*p*-enh't	SGDSb	–25	1-10	1000	Switch/amp.	Motorola
3N128	*n*-depl.	DSGSb	20	5-30	8000	RF amplifier	RCA
3N140	*n*-depl.	DGGS	20	5-30	12000	Dual gate RF	RCA, Motorola
3N174	*p*-enh't	DGSbS	–30	1-10	400	Amplifier	Texas Instruments

D—Drain, G—Gate, S—Source, Sb—Substrate.

Note: The prices of JFETs and MOS FETs may be unexpectedly high and should be compared before purchase.

DIODES

JEDEC No.	Type	PRV (V)	Current (av. mA)	Use	Manufacturers
1N56	G	30	15	Signal	General Electric
1N60	G	30	5	Signal	General Electric, Sylvania
1N82	S	5	5	Mixer, UHF	General Electric, Sylvania
1N459A	S	175	40	Low leakage	Fairchild, Texas Instruments
1N2326	G	1	100	Compensation	RCA
1N4004	S	400	750	Rectifier	Motorola, Texas Instruments
1N4007	S	1000	750	Rectifier	Motorola, Texas Instruments
1N4728	S	-	-	3.3 V Zener	Motorola
FH1100		5	10	Hot carrier	Fairchild

A.2 SEMICONDUCTOR INTERCHANGEABILITY

Most low-power transistor circuits operate satisfactorily when similar devices with equivalent current gains are substituted, provided the bias is adjusted. Circuits operating at high frequencies or switching at high speeds require transistors that meet the frequency and speed requirements indicated in the data sheets by the cutoff frequency f_B or the switching times. The demands on power transistors are more exacting, and many devices are inferior even though more expensive. The substitution of power transistors is not recommended because the recommended types have been carefully selected. For equipment repairs, a better device should be substituted, if possible, and the bias should always be adjusted.

Field-effect transistors are improving rapidly, and substitutions can be made, providing the g_m and the I_{DSS} ratings are similar. New types are priced reasonably, while the prices of the early types remain high. Amplifier applications require high g_m at low I_{DSS} values and the most inexpensive devices generally prove unsatisfactory.

Silicon diode types are not specified when almost any 400-V, or 200-V, 1-A diode should be satisfactory. For experimenter purposes the voltage rating of diodes should be 3 to 5 times the rms voltage input to the diode. The current rating should be 3 to 5 times the diode dc current. Excellent high-current high-voltage diodes may be obtained at reasonable prices. Diodes produced in small quantities may be surprisingly expensive.

TRANSISTOR-BASING DIAGRAMS

The diagrams are given only as guides because there are many exceptions. All diagrams are viewed towards leads (bottom view).

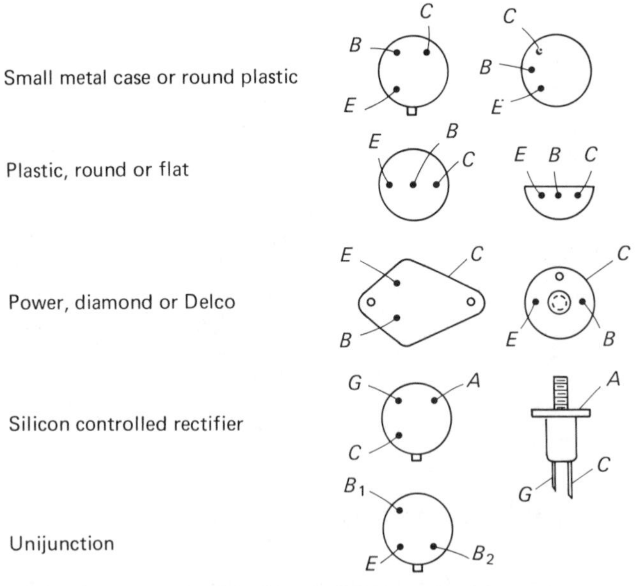

FETs: No standard basing. Identify gate with ohmmeter.

Diodes: The cathode, or plus, is the marked end.

Figure A.2.

A.3 THERMAL RESISTANCES
(Resistance given in °C/W)

Transistor without Heat Sinks	
Epoxy types	200-500
TO-18	200 up
TO-5	150 up
Diamond power	25-50

Insulating Washers	
None	0.2-0.5
Beryllia	0.3-0.5
Aluminum	0.5-1.0
Mica	0.5-1.0

Heat Sinks	
Clip-on for TO-5	25-50
Small-fin types	10-40
Large in still air	1-5

Flat Sheet (One Side)	
3 cm X 3 cm	50
10 cm X 10 cm	8
30 cm X 30 cm	2

MAXIMUM RECOMMENDED TEMPERATURES (°C)

Germanium junction:	100-125
Silicon junction	150-200
J-FETs	150-175
MOS FETs	175-200
Thyristors	100-125
Unijunction	100-125
Silicon diodes	150-200
Zener diodes	150-175

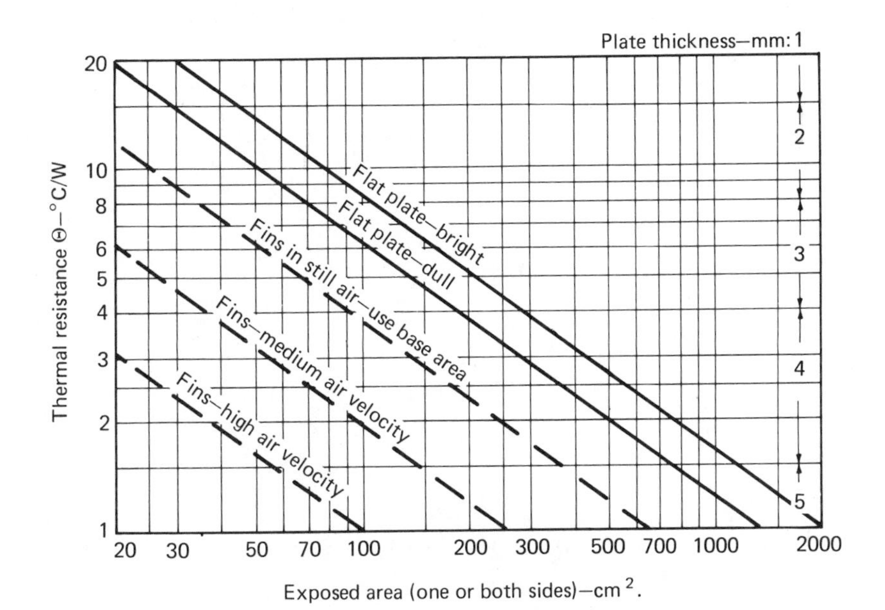

Figure A.3. Thermal resistance of heat sinks

A.4 TUNED IMPEDANCE TRANSFORMERS

Tuned circuits are commonly used to provide an efficient, frequency-selective coupling between unequal impedances. The formulas for the design of impedance transformers tend to be complicated if the equations provide for a wide choice of impedances and Qs. However, if the Q of the circuits is limited and approximations are accepted, the design of impedance-transforming circuits is greatly simplified. The fact that the equations give approximate values for the components is usually unimportant in design because the tuned circuits must be adjusted (that is, tuned) at assembly. The following sections describe the most common types of impedance-transforming networks used in RF circuits. The Qs in the equations give reasonably accurate measures of the frequency bandwidth that is transmitted by the networks. The design of impedance-transforming circuits is often simplified by converting a series resistance and reactance to an equivalent parallel resistance and reactance, and vice versa.

I. EQUIVALENT-SERIES TO PARALLEL IMPEDANCE CONVERSIONS

A series resistance and reactance may be converted to an equivalent parallel resistance and reactance, and vice versa. While each circuit is the equivalent of the other at only one frequency, an equivalent circuit is broadly useful in RF calculations where the frequency band of interest is relatively small compared with the design-center frequency.

If the series resistance and reactance are R_S and X_S, the parallel equivalent resistance R_P and reactance X_P are given by

$$R_p = (1 + Q^2) R_S \quad (1)$$

and

$$X_P = \frac{1 + Q^2}{Q^2} = X_S \quad (2)$$

where

$$Q = \frac{X_S}{R_S} = \frac{R_P}{X_P} \quad (3)$$

For $Q > 3$

$$R_P \cong Q^2 R_S, \qquad X_P \cong X_S \quad (4)$$

For $Q = 1$

$$R_P = 2R_S, \qquad X_P = 2X_S \quad (5)$$

For $Q < 0.3$

$$R_P \cong R_S, \qquad X_P \cong \frac{X_S}{Q^2} \tag{6}$$

Example: Find the parallel equivalent of the series circuit $R_S = 50\ \Omega$ and $X_S = 300\ \Omega$. By equation (3), $Q = 6$. By equation (4), $R_P = 1800\ \Omega$ and $X_P = 300\ \Omega$. The equivalent circuits are

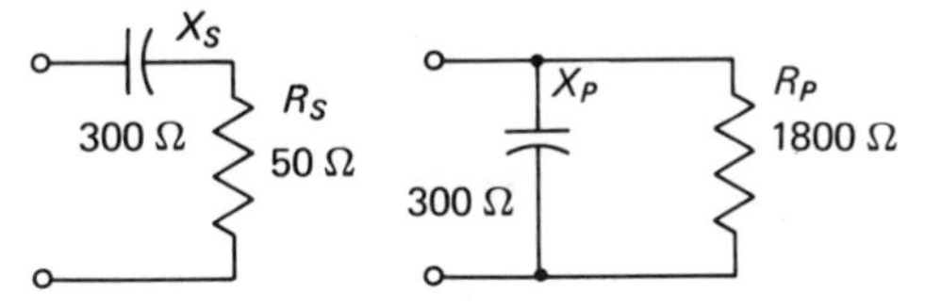

Figure A.4-1. Series and parallel equivalent circuits

II. TUNED-CIRCUIT IMPEDANCES

The resistance R_P measured across a parallel-tuned L-C circuit at resonance, $X_L = X_C$, is approximately Q^2 times the resistance R_S that is in series with either reactance X_L or X_C. Thus, a tuned circuit at resonance may be used to transform a low impedance to a high impedance. Similarly, the resistance R_S measured in a series-resonant circuit is smaller than the resistance R_P that is across either reactance. These characteristics of resonant circuits can be used in the design of tuned circuits for coupling unequal impedances in RF amplifiers.

III. TUNED-CIRCUIT TRANSFORMERS

Assuming $Q > 3$, a tuned L-C circuit transforms a parallel load impedance R_L to a series input impedance R_S that is smaller by the square of the Q-factor. Thus,

$$R_S = \frac{R_L}{Q^2} \tag{1}$$

when

$$X_L = \frac{R_L}{Q} \tag{2}$$

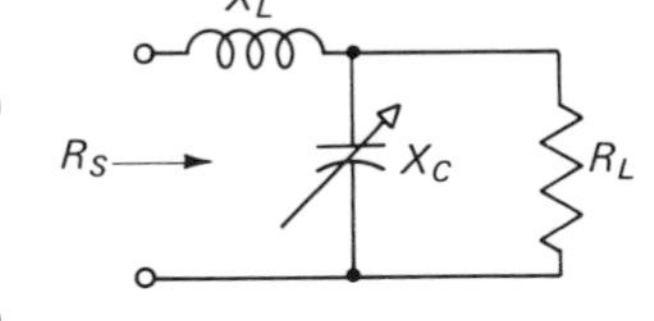

Figure A.4-2. Parallel-to-series transformer

Similarly, a series load impedance R_L is transformed to a shunt input impedance R_P that is larger by the square of the Q-factor. Thus,

$$R_P = Q^2 R_L \tag{3}$$

when

$$X_L = QR_L \tag{4}$$

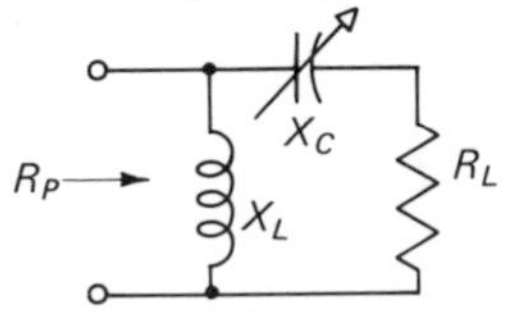

Figure A.4-3. Series to parallel transformer

For efficient power transfer Q should usually be less than 30. With Q restricted to values between 3 and 30, equation (1) may be used for impedance transformations in the range

$$9 < \frac{R_L}{R_S} < 900 \tag{5}$$

and equation (3) permits impedance transformations in the range

$$9 < \frac{R_P}{R_L} < 900 \tag{6}$$

Thus, impedance changes exceeding 9 may be obtained by selecting the circuit Q by equation (1) or (3) and using the inductance value given by equation (2) or (4). Observe that the tuned circuit does not permit an independent choice of the Q and the impedance ratio.

A tuned-circuit impedance transformer is designed as follows:

1. Select a compromise Q by equation (1) or (3).
2. Select X_L by equation (2) or (4).
3. Tune by making $X_C = X_L$.

Example: Design a single-tuned circuit for matching a 50-Ω source to an 1800-Ω load. For $Q > 3$, equation (1) gives $Q = 6$. Equation (2) gives $X_L = 1800/6 = 300$ Ω. The tuned circuit is shown below.

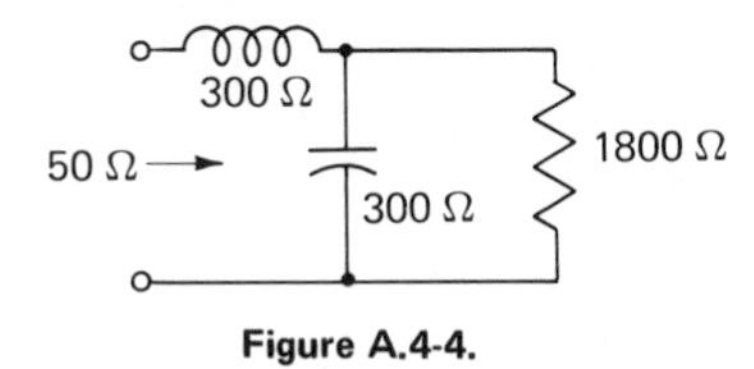

Figure A.4-4.

IV. L-NETWORK IMPEDANCE MATCHING

An L-network has an inductance, with a series R-C circuit on one side and a parallel R-C circuit on the other.

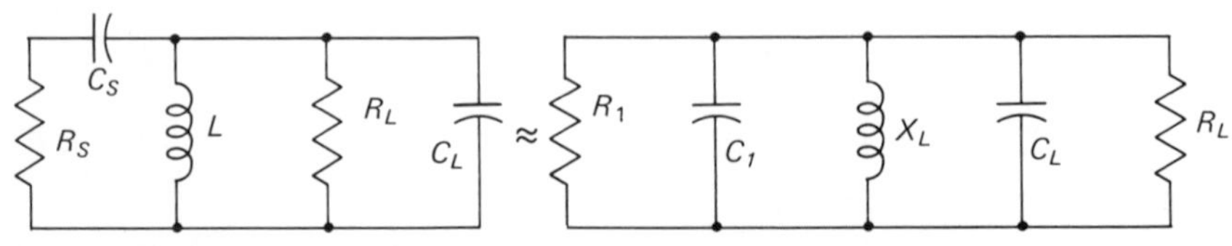

Figure A.4-5. Series R-C and parallel R-C Equivalent with parallel R-C

For an impedance change, the series reactance is selected to make $Q_S^2 R_S = R_L$. Thus,

$$Q_S = \sqrt{\frac{R_L}{R_S}} \tag{1}$$

$$X_S = Q_S R_S \tag{2}$$

The equivalent parallel resistance R_1 matches R_L because $R_1 = Q_S^2 R_S = R_L$. For $R_S < 9R_L$, $C_1 = C_S$, and the reactances of the circuit are resonant when X_L equals the reactance of the parallel capacitors, $C_S + C_L$.

In practical circuits, the L-network is usually constructed with two capacitors C_S and C_P variable. When tuned by C_S and $C_P = C_L$, the resulting mismatch is usually negligible.

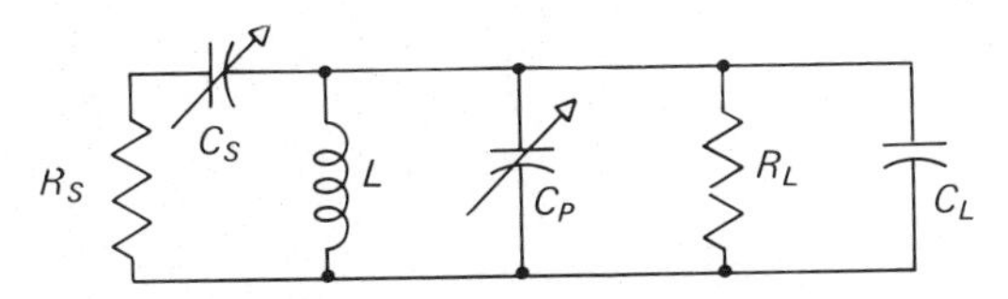

Figure A.4-6. (1) *L* coupling network

If equation (2) requires $Q < 3$, the equations may be used to calculate approximate values of the components, with the circuit adjusted at assembly to give the required impedance match and selectivity.
Example: Given $R_S = 50\ \Omega$ to be matched to $R_L = 1000\ \Omega$. The resonant frequency is 10 MHz, and the load has a parallel input capacitance of 30 pF.

By equation (1), $Q_S = \sqrt{20} = 4.4$. By equation (2), $X_S = 4.4(50) = 220\ \Omega$, and $C_S = 70$ pF. The equivalent circuit has the form

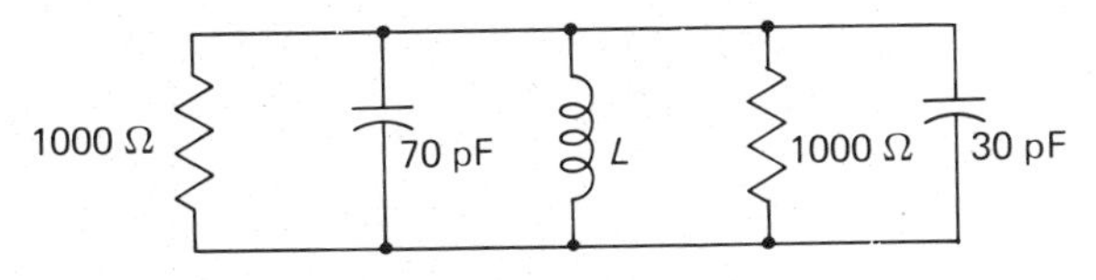

Figure A.4-6. (2) Equivalent circuit

When a 30-pF variable capacitor is connected across the load side of the circuit, the total parallel capacitance is $70 + 30 + 30 = 130$ pF. For resonance at 10 MHz, $L = 2.0\ \mu$H, and the circuit Q is $500/X_L \cong 4$. The required L-network is shown below.

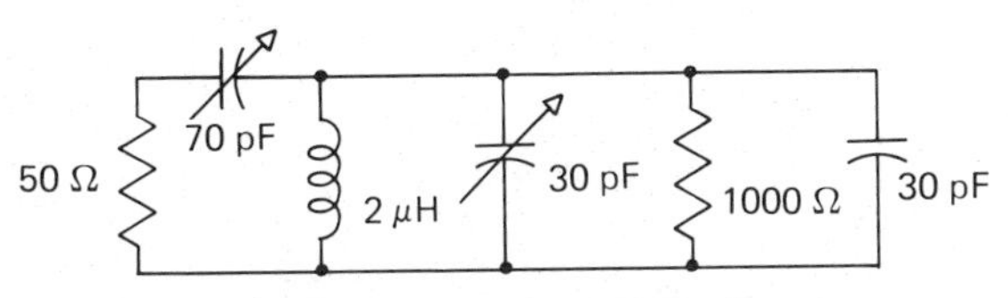

Figure A.4-7. Practical circuit

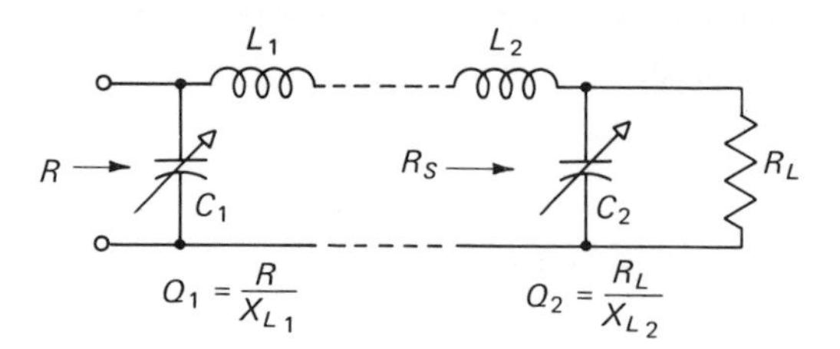

Figure A.4-8. Pi-network as back-to-back resonant circuits

V. PI-NETWORK IMPEDANCE TRANSFORMERS

A pi-network is used for impedance transformers when a low ratio of impedance transformation and a high-Q circuit are required. The pi-network may be designed as a pair of back-to-back tuned-circuit impedance transformers. The output section of the pi makes

$$R_S = \frac{R_L}{Q_2^2} \tag{1}$$

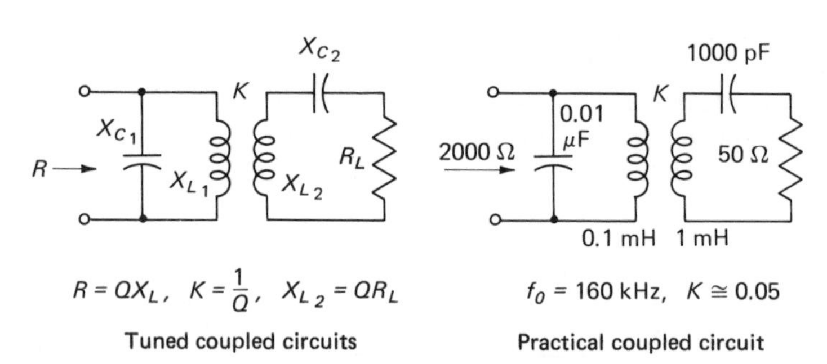

Figure A.4-9.

and the input section makes

$$R = Q_1^2 R_S \tag{2}$$

or

$$R \cong \left(\frac{Q_1}{Q_2}\right)^2 R_L \tag{3}$$

Thus, Q_1 and Q_2 may be selected to obtain the desired bandwidth, with the impedance transformation determined by the ratio Q_1/Q_2 in equation (3). After selecting the desired values of Q_1 and Q_2, the design is the same as outlined for a tuned impedance transformer.

VI. TUNED COUPLED CIRCUITS

Tuned coupled circuits may be designed as a pair of single-tuned circuits. When the circuits are tuned to the same frequency and loosely coupled, the coupling can be adjusted so that the equivalent load resistance is coupled into the input circuit as the required load. When "critically coupled," the response is approximately flat in the band with a 40 per cent greater bandwidth and a higher selectivity than for either circuit alone. The coupling coefficient required for critical coupling varies inversely with the Qs of the tuned circuits. Thus, relatively loose coupling may be used with high-Q circuits. Generally, the tuned circuits are designed with the same Q and resonant frequency.

Example: Design a tuned coupled circuit for coupling a 50-Ω line to a collector that requires a 2000-Ω load. The operating frequency is 160 kHz, and the circuits should have $Q = 20$.

The load circuit is series resonant with a reactance $X_{L2} = 20(50) = 1000\ \Omega$. The collector circuit is designed as a parallel resonant circuit with a reactance $X_{L1} = 2000/20 = 100\ \Omega$. The reactance chart in Appendix A.10 gives the inductance and capacitance values shown in the practical circuit. After tuning the circuits separately, the coupling is adjusted, but not beyond the point of maximum secondary power.

A.5 PHASE-FREQUENCY TABLES

The table gives phase angles in degrees as a function of frequency for gain steps and *R-C* cutoffs. The table is useful for calculating phase-frequency characteristics as with Table IV of Chapter 10. Table III in Chapter 10 is an abbreviated form of the table below.

Phase Angles for Gain Steps and R-C Cutoffs

f/fc	.07	.09	.11	.13	.15	.18	.21	.25	.30	.35	.42	.50	.59	.71	.84	1.0	1.2	1.4	1.7	2.0	2.4	2.8	3.4	4.0	4.8	5.7	6.7	8.0	9.5	11	13
3 DB	1	1	2	2	2	3	3	4	5	5	6	7	8	9	9	10	10	10	9	9	8	7	6	5	5	4	3	3	2	2	2
6 DB	2	3	3	4	4	5	6	7	8	9	11	13	14	16	17	18	19	19	19	18	17	16	14	13	11	9	8	7	6	5	4
9 DB	3	3	4	5	5	6	8	9	11	12	14	17	19	21	24	26	27	28	29	28	27	26	24	21	19	17	14	12	11	9	8
12 DB	3	4	4	5	6	7	9	10	12	14	17	19	22	25	28	31	33	35	36	37	36	35	33	31	28	25	22	19	17	14	12
1-R-C	4	5	6	7	8	10	12	14	17	19	23	27	31	35	40	45	50	55	59	63	67	71	73	76	78	80	82	83	84	85	86
2-R-C	8	10	12	14	17	20	24	28	33	39	46	53	61	71	80	90	100	109	119	127	134	141	147	152	156	160	163	166	168	170	172

Phase Angle and Loss with One R-C Cutoff

f/f_c	0.25	0.35	0.5	0.7	1.0	1.2	1.4	1.7	2.0	2.4	2.8	3.4	4.0	4.8	5.7	6.7	8.0	10.	12.	14.	17.	20.	24.	28.	34.	57.	100
Θ	14	19	27	35	45	50	55	59	63	67	71	73	76	78	80	82	83	84	85	86		87		88		89	
dB	0.3	0.5	1.0	1.8	3.0	3.8	4.8	5.8	7.0	8.2	9.5	11.	12.	14.	15	17	18.	20	22.	23.	25	26.	28	29	31	35	40

Note: For gain steps, the lower corner frequency is at $f/f_c = 1$. The upper corner frequency is where the angle returns to the same value as at the lower corner frequency.

The angles are in degrees lag for an increasing gain and lead for a decreasing gain. By reading the frequency ratios as f_c/f, the table may be used for calculating the phase characteristics of low-frequency gain steps and *R-C* cutoffs.

A.6 INDUCTANCE OF A SINGLE-LAYER SOLENOID

(Microhenries per centimeter diameter)

n \ $\frac{l}{d}$ =	0.25	0.5	0.7	1.0	1.4	2.	4.	8.
10	1.5	1.1	0.84	0.66	0.50	0.38	0.20	0.11
12	2.1	1.5	1.2	0.93	0.71	0.53	0.29	0.15
14	3.0	2.1	1.7	1.3	1.0	0.75	0.40	0.21
17	4.2	3.0	2.4	1.9	1.4	1.1	0.57	0.30
20	6.0	4.2	3.4	2.6	2.0	1.5	0.81	0.42
24	8.5	5.9	4.8	3.7	2.8	2.1	1.1	0.59
28	12.	8.4	6.7	5.3	4.0	3.0	1.6	0.84
34	17.	12.	9.5	7.4	5.7	4.2	2.3	1.2
40	24.	17.	13.	10.	8.0	6.0	3.2	1.7
48	34.	24.	19.	15.	11.	8.5	4.6	2.4
57	48.	34.	27.	21.	16.	12.	6.5	3.4
67	68.	48.	38.	30.	23.	17.	9.1	4.8
80	96.	67.	54.	42.	32.	24.	13.	6.7
100	150.	105.	84.	66.	50.	38.	20.	10.5

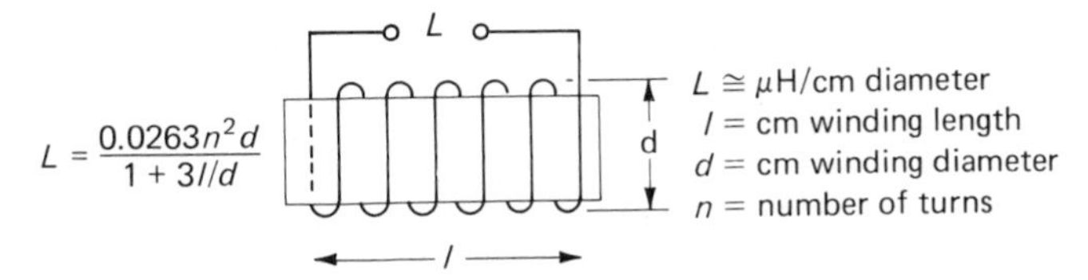

Figure A.6-1.

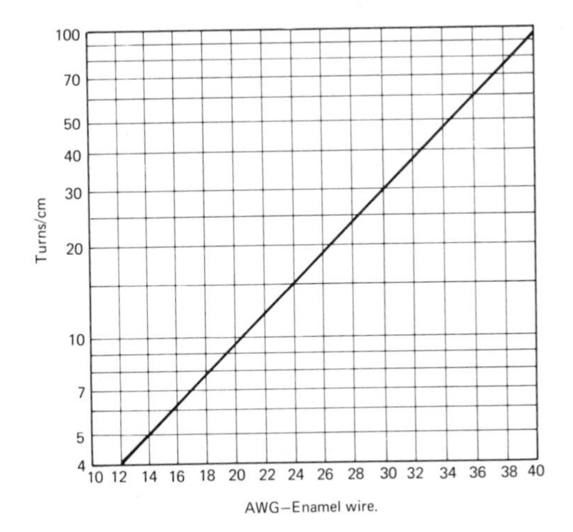

Figure A.6-2. Solenoid turns per centimeter chart.

A.7 FREQUENCY CLASSIFICATIONS

Abbreviation	Frequency	Wavelength	Classification
VLF	-30 kHz	-10 km	Very-low frequencies
LF	30 -300 kHz	10 -1 km	Low frequencies
MF	300 -3000 kHz	1000 -100 m	Medium frequencies
HF	3 -30 MHz	100 -10 m	High frequencies
VHF	30 -300 MHz	10 -1 m	Very-high frequencies
UHF	300 -3000 MHz	100 -10 cm	Ultrahigh frequencies
EHF	3 -30 GHz	10 -1 cm	Superhigh frequencies
EHF	30-300 FHz	10 -1 mm	Extra-high frequencies

AMATEUR RADIO BANDS

80 m	3500 -4000 MHz
40 m	7000 -7300 MHz
20 m	14,000 -14,350 MHz
15 m	21.00 -21.45 MHz
10 m	28.00 -29.7 MHz
6 m	50.0 -54.0 MHz
2 m	144. -148. MHz

TELEVISION BANDS

Channels	Frequencies
2 -4	54 -72 MHz
5 -6	76 -88 MHz
(FM)	(88 -108 MHz)
7 -13	174 -216 MHz
14 -83	470 -890 MHz

MICROWAVE BANDS

P-band	225 -390 MHz	133.3 -76.9 cm
L-band	390 -1550 MHz	76.9 -19.3 cm
S-band	1550 -5200 MHz	19.3 -5.78 cm
C-band	5200 -8500 MHz	5.78 -3.53 cm
X-band	8500 -10,900 MHz	3.53 -2.75 cm
Ku-band	10.9 -17.25 GHz	2.75 -1.74 cm
Ka-band	17.25 -36 GHz	1.74 -0.834 cm
Q-band	36 -46 GHz	8.34 -6.53 mm
V-band	46 -56 GHz	10. -1. mm

SIGNAL VELOCITIES

Free space	300,000,000 m/sec
Twin lead	250,000,000 m/sec
Coaxial cable	200,000,000 m/sec
Toll cable	<50,000,000 m/sec

A.8 DECIBEL FORMULAS AND TABLE

By definition the decibel is a logarithmic measure of power ratios:

$$dB \equiv 10 \log_{10} \frac{P_2}{P_1}$$

Ratios greater than 1 represent a power gain, +(plus) dB. For ratios less than 1, representing a loss, the dB value is found using the reciprocal of the ratio and the dB value is negative, -(minus) dB.

For convenience, the dB is used as a measure of voltage ratios, where

$$dB \equiv 20 \log_{10} \frac{E_2}{E_1}$$

However, voltage ratios expressed in dB units do not represent power ratios unless the voltages are measured across equal resistances.

A Practical Decibel Table

dB	Voltage ratio	Power ratio	dB	Voltage ratio	Power ratio
0	1.00	1.00	14	5.0	25
1	1.12	1.26	16	6.3	40
2	1.26	1.59	17	7.1	50
3	1.41	2.00	20	10.0	100
4	1.58	2.51	25	17.8	316
5	1.78	3.16	30	31.6	10^3
6	2.00	4.00	40	10^2	10^4
8	2.51	6.31	60	10^3	10^6
10	3.16	10.0	80	10^4	10^8
12	4.0	15.9	100	10^5	10^{10}

STANDARD PREFIXES

Multiple	Prefix	Symbol	Multiple	Prefix	Symbol
10^{12}	tera	T	10^{-1}	deci	d
10^9	giga	G	10^{-2}	centi	c
10^6	mega	M	10^{-3}	milli	m
10^3	kilo	k	10^{-6}	micro	μ
10^2	hecto	h	10^{-9}	nano	n
10	deka	dk	10^{-12}	pico	p

A.9 RESISTANCE NOISE CHART

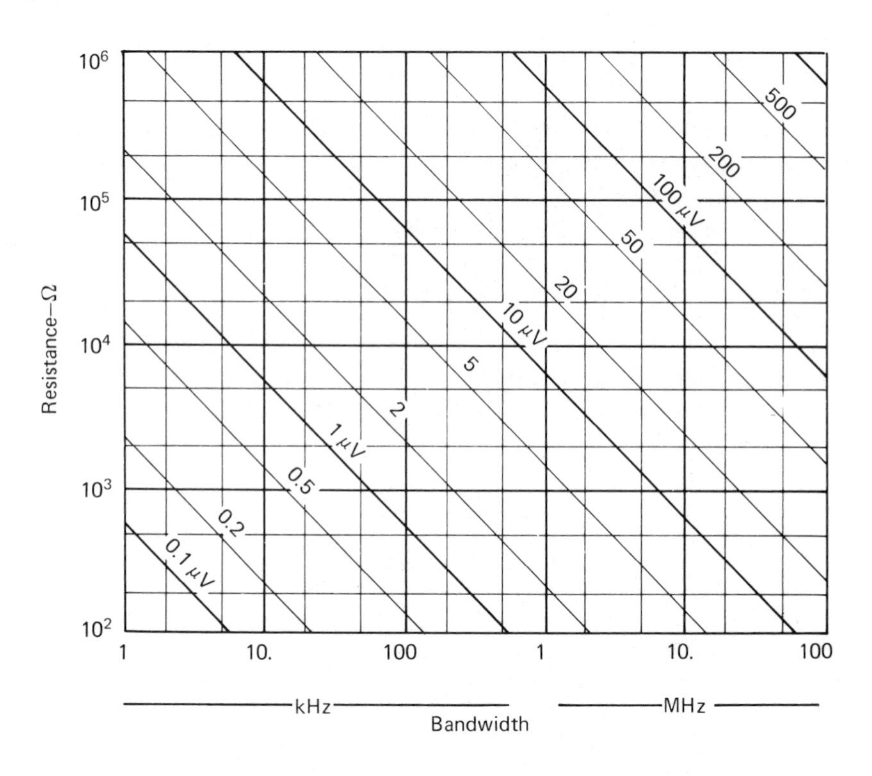

Notes:

1. The resistance noise is in rms microvolts.
2. All intermediate lines are at 2 and 5.
3. For a spot-noise factor of 1, use $BW = 1$ M Hz and read nanovolts instead of microvolts.

Figure A.9

A.10 REACTANCE-FREQUENCY CHARTS

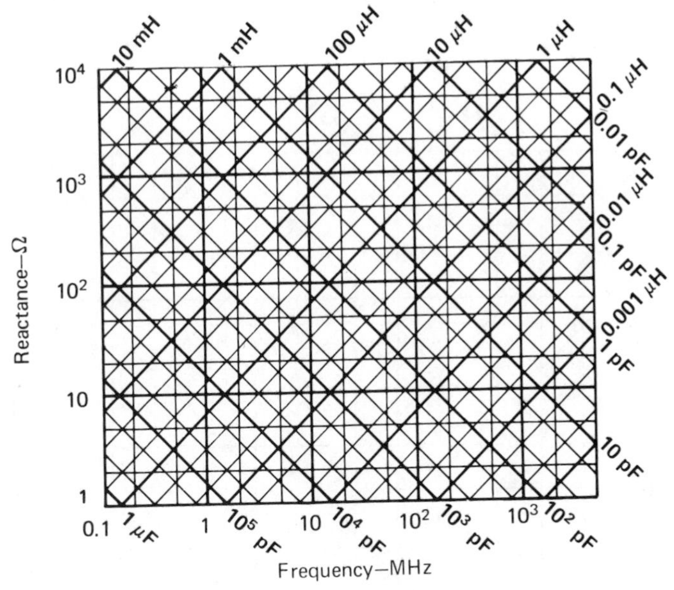

Note: All intermediate lines are at 2 and 5
Reactance chart for frequency above 0.1 M Hz

Figure A. 10-1

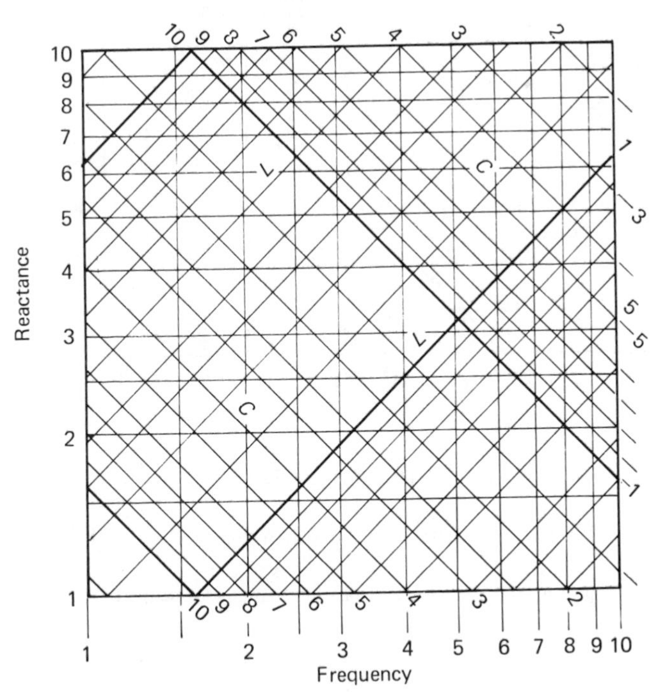

Expanded scale reactance chart

Figure A.10-2

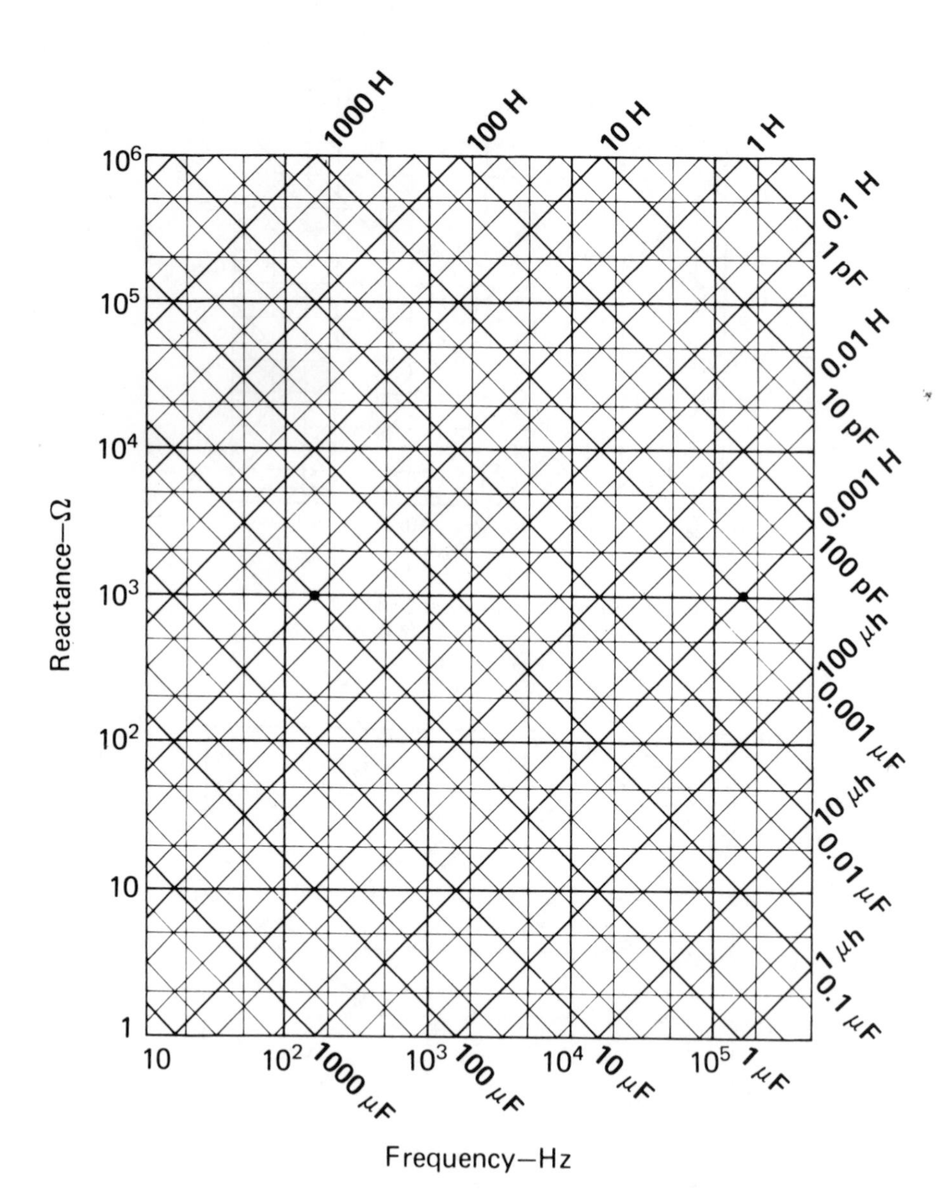

Note: All intermediate lines are at 2 and 5.

Figure A.10-3. Reactance chart for frequencies below 0.5 MHz.

INDEX

G

H

I

L

M

N

O

P

Q

R

S